Springer Proceedings in Mathematics & Statistics

Volume 76

For further volumes:
http://www.springer.com/series/10533

Springer Proceedings in Mathematics & Statistics

This book series features volumes composed of select contributions from workshops and conferences in all areas of current research in mathematics and statistics, including OR and optimization. In addition to an overall evaluation of the interest, scientific quality, and timeliness of each proposal at the hands of the publisher, individual contributions are all refereed to the high quality standards of leading journals in the field. Thus, this series provides the research community with well-edited, authoritative reports on developments in the most exciting areas of mathematical and statistical research today.

Susan M. Cooper • Sean Sather-Wagstaff
Editors

Connections Between Algebra, Combinatorics, and Geometry

Springer

Editors
Susan M. Cooper
Department of Mathematics
Central Michigan University
Mt. Pleasant, MI, USA

Sean Sather-Wagstaff
Department of Mathematics
North Dakota State University
Fargo, ND, USA

ISSN 2194-1009 ISSN 2194-1017 (electronic)
ISBN 978-1-4939-0625-3 ISBN 978-1-4939-0626-0 (eBook)
DOI 10.1007/978-1-4939-0626-0
Springer New York Heidelberg Dordrecht London

Library of Congress Control Number: 2014937796

Mathematics Subject Classification (2010): 14-02, 13A02, 14C20, 13D02, 13D09, 65H10, 68W30, 13D02, 13C05, 05C10, 16E45, 16E05, 13C14, 13A02, 13D02, 13F55

Printed on acid-free paper

Springer is part of Springer Science+Business Media (www.springer.com)

This volume is dedicated to Tony Geramita.
His teaching, mentoring, research, and
friendship have deeply affected us.

Preface

Commutative Algebra is an extremely vibrant field of mathematical research, with connections to many other disciplines. It began with work of David Hilbert in the 1890s and developed with the work of Emmy Noether in the 1920s and Oscar Zariski in the 1930s. Its roots in Algebraic Geometry continue to supply problems and intuition for research on commutative rings. For example, Paul Roberts' proof of the New Intersection Theorem using sophisticated geometric techniques from intersection theory illustrates this beautifully.

Currently, the field of Commutative Algebra is in a period of profound transformation: "Over the last fifteen years commutative algebra has experienced a striking evolution. During that period the outlook of the subject has been altered, new connections to other areas have been established, and powerful techniques have been developed."[1] Today, Commutative Algebraists are involved in intense interaction with researchers in areas as diverse as Algebraic Geometry, Biology, Coding Theory, Combinatorics, Homotopy Theory, K-theory, Representation Theory, and Statistics.

This volume grew from three mathematics research meetings focusing on the interactions between Algebra, Combinatorics, and Geometry:

- Workshop on Connections Between Algebra and Geometry, University of Regina, May 29–June 1, 2012.
- Special Session on Interactions Between Algebraic Geometry and Commutative Algebra, Canadian Mathematical Society Summer Meeting, University of Regina, June 2–3, 2012.
- Conference on Further Connections Between Algebra and Geometry, North Dakota State University, February 2–3, 2013.

[1]By L. Avramov, M. Green, C. Huneke, K. Smith, B. Sturmfels, editors, in the Preface to *Trends in Commutative Algebra*, Proceedings of the Introductory Workshop in Commutative Algebra, MSRI 2002–2003. Cambridge Univ. Press, 2004.

However, this volume is not a conference proceedings. Instead, it is a collection of notes and papers documenting a few manifestations of the connections on which the meetings focused.

The first part of this volume contains lecture notes (including exercises and solutions) from the three courses given at the first meeting listed above. These notes were prepared by the course instructors in collaboration with the problem session assistants and (in one case) one of the workshop participants. These notes provide introductions to three specific areas of interaction: DG homological algebra, fat points, and secant varieties. Each set of notes should be accessible to graduate students and some advanced undergraduate students.

The second part of this volume consists of papers (part survey, part new research) written by some of the speakers from the special session and the conference, and their collaborators. Each paper focuses on interactions between Algebra, Combinatorics, and Geometry. Each one should be accessible to researchers and graduate students with a reasonable background in Commutative Algebra, say, at the level of the texts of Atiyah and MacDonald, Eisenbud, or Matsumura.

Conference Personnel

- Workshop on Connections Between Algebra and Geometry, University of Regina, May 29–June 1, 2012 [2]
  - Organizers:
    - Susan M. Cooper, Central Michigan University
    - Sean Sather-Wagstaff, North Dakota State University
    - Donald Stanley, University of Regina
  - Course instructors:
    - Brian Harbourne, University of Nebraska–Lincoln
    - Enrico Carlini, Politecnico di Torino
    - Sean Sather-Wagstaff, North Dakota State University
  - Problem session assistants:
    - Kristen A. Beck, University of Arizona
    - Susan M. Cooper, Central Michigan University
    - Luke Oeding, University of California, Berkeley
  - Participants:
    - Seyedruhallah Ahmadi, University of Regina
    - Ali Alilooee, Dalhousie University
    - Hannah Altmann, North Dakota State University
    - Marzieh Bayeh, University of Regina
    - Ashwini Bhat, Lakehead University

[2] We include affiliations given at the times of the meetings.

  - Jennifer Biermann, Lakehead University
  - Yuri Delanghe, University of Alberta
  - Annika Denkert, University of Nebraska–Lincoln
  - Aaron Feickert, North Dakota State University
  - Brandon Goodell, North Dakota State University
  - Nathan Grieve, Queen's University
  - Ryan Hines, Central Michigan University
  - Andrew Hoefel, Queen's University
  - Mike Janssen, University of Nebraska–Lincoln
  - Maryam Ehya Jahromi, Dalhousie University
  - Jeremy Lane, University of Regina
  - Melissa Lindsey, Indiana Wesleyan University
  - Sarah Mayes, University of Michigan
  - Augustine O'Keefe, Tulane University
  - Alessandro Oneto, Università di Genova
  - Jared Painter, University of Texas at Arlington
  - Megan Patnott, University of Notre Dame
  - Alexandra Seceleanu, University of Nebraska–Lincoln
  - Stephen Sturgeon, University of Kentucky
  - Jonathan Totushek, North Dakota State University
  - Richard Wicklein, North Dakota State University
  - Mark Wong, Syracuse University
  - Liu Yong, University of Regina

- Special Session on Interactions Between Algebraic Geometry and Commutative Algebra, Canadian Mathematical Society Summer Meeting, University of Regina, June 2–3, 2012
  - Organizers:
    - Susan M. Cooper, Central Michigan University
    - Sean Sather-Wagstaff, North Dakota State University
  - Speakers:
    - Kristen Beck, University of Arizona
    - Jennifer Bierman, Lakehead University
    - Enrico Carlini, Politecnico di Torino
    - Annika Denkert, University of Nebraska–Lincoln
    - Sara Faridi, Dalhousie University
    - Chris Francisco, Oklahoma State University
    - Nathan Grieve, Queen's University
    - Brian Harbourne, University of Nebraska–Lincoln
    - Andrew Hoefel, Queen's University
    - Mike Janssen, University of Nebraska–Lincoln
    - Melissa Lindsey, Indiana Wesleyan University
    - Uwe Nagel, University of Kentucky
    - Luke Oeding, University of California, Berkeley
    - Megan Patnott, University of Notre Dame

  - Chris Peterson, Colorado State University
  - Sandra Spiroff, University of Mississippi
  - Stephen Sturgeon, University of Kentucky

- Conference on Further Connections Between Algebra and Geometry, North Dakota State University, February 2–3, 2013
  - Organizers:
    - Susan M. Cooper, Central Michigan University
    - Sean Sather-Wagstaff, North Dakota State University
  - Speakers:
    - Kristen Beck, University of Arizona
    - Jim Coykendall, North Dakota State University
    - Chris Francisco, Oklahoma State University
    - Tài Huy Hà, Tulane University
    - Brian Harbourne, University of Nebraska–Lincoln
    - Augustine O'Keefe, University of Kentucky
    - Jared Painter, Houston Baptist University
    - Branden Stone, Bard College/Bard Prison Initiative
    - Mark Walker, University of Nebraska–Lincoln
  - Poster presenters:
    - Brandon Goodell, North Dakota State University
    - Sema Gunturkun, University of Kentucky
    - Sara Malec, Georgia State University
    - Saeed Nasseh, North Dakota State University
    - Denise Rangle, University of Texas at Arlington
    - Richard Wicklein, North Dakota State University
  - Participants:
    - Nitin Aggarwal, University of Kansas
    - Saba Al-Kaseasbeh, North Dakota State University
    - Hannah Altmann, North Dakota State University
    - Pye Aung, North Dakota State University
    - Robert Bradford, University of Kansas
    - Cătălin Ciupercă, North Dakota State University
    - Annika Denkert, University of Nebraska–Lincoln
    - Josef Dorfmeister, North Dakota State University
    - Aaron Feickert, North Dakota State University
    - Zachary Flores, University of Kansas
    - Jeffrey Fujioka, Central Michigan University
    - Mike Janssen, University of Nebraska–Lincoln
    - Pin-Hung Kao, Central Michigan University
    - Bethany Kubik, United States Military Academy
    - Kuei-Nuan Lin, University of California, Riverside
    - Haydee Lindo, University of Nebraska–Lincoln
    - Chelsey Morrow, North Dakota State University
    - Megan Patnott, University of Notre Dame

- Jason Pode, Central Michigan University
- William Robinson, University of Kentucky
- William Sanders, University of Kansas
- Piyush Shroff, Texas State University–San Marcos
- Jonathan Totushek, North Dakota State University
- Mark Wong, Syracuse University

Acknowledgments

We gratefully acknowledge the following:

- The Department of Mathematics and Statistics at the University of Regina for its hospitality during the Workshop on Connections Between Algebra and Geometry, in particular, from Allen Herman, Chair.
- Donald Stanley (University of Regina) for co-organizing the Workshop on Connections Between Algebra and Geometry.
- The Canadian Mathematical Society for organizational support during the Special Session on Interactions Between Algebraic Geometry and Commutative Algebra.
- The Mathematics Department at North Dakota State University for its hospitality during the Conference on Further Connections Between Algebra and Geometry.
- Jacalyn Benson (North Dakota State University), Amy Hunter (University of Regina), Teresa Judge (Central Michigan University), and Karen Manders (Pacific Institute for the Mathematical Sciences) for administrative support with the Workshop and the Conference.
- The following funding agencies without which these meetings would not have been possible for us:
  - Central Michigan University, Department of Mathematics
  - National Science Foundation (Award #DMS-1200313)
  - National Security Agency
  - North Dakota State University, Department of Mathematics
  - North Dakota State University, College of Science and Mathematics
  - Pacific Institute for the Mathematical Sciences
  - University of Regina, Faculty of Science
  - University of Regina, Department of Mathematics and Statistics.

Mount Pleasant, MI
Fargo, ND
November 2013

Susan M. Cooper
Sean Sather-Wagstaff

Contents

Contributors

Kristen A. Beck Department of Mathematics, The University of Arizona, Tucson, AZ, USA

Enrico Carlini School of Mathematical Sciences, Monash University, Melbourne, VIC, Australia

Susan M. Cooper Department of Mathematics, Central Michigan University, Mount Pleasant, MI, USA

Sara Faridi School of Mathematics and Statistics, Dalhousie University, Halifax, NS, Canada

Christopher A. Francisco Department of Mathematics, Oklahoma State University, Stillwater, OK, USA

Nathan Grieve Department of Mathematics and Statistics, McGill University, Montreal, QC, Canada

Zachary A. Griffin Harold Vance Department of Petroleum Engineering, Texas A&M University, College Station, TX, USA

Huy Tài Hà Department of Mathematics, Tulane University, New Orleans, LA, USA

Brian Harbourne Department of Mathematics, University of Nebraska, Lincoln, NE, USA

Jonathan D. Hauenstein Department of Mathematics, North Carolina State University, Raleigh, NC, USA

Jeffrey Mermin Department of Mathematics, Oklahoma State University, Stillwater, OK, USA

Luke Oeding Department of Mathematics and Statistics, Auburn University, Auburn, AL, USA

Jared Painter Houston Baptist University, Fondren Rd., Houston, TX, USA

Chris Peterson Department of Mathematics, Colorado State University, Fort Collins, CO, USA

Sean Sather-Wagstaff Department of Mathematics, North Dakota State University, Fargo, ND, USA

Jay Schweig Department of Mathematics, Oklahoma State University, Stillwater, OK, USA

Andrew J. Sommese Department of Applied and Computational Mathematics and Statistics, University of Notre Dame, Notre Dame, IN, USA

Branden Stone Mathematics Program, Bard College, Annandale-on-the-Hudson, NY, USA

Part I
Lecture Notes

This part of the volume contains lecture notes from three short-courses given in Summer 2012 at the Workshop on Connections Between Algebra and Geometry held at the University of Regina:

- Differential Graded Commutative Algebra
- Secant Varieties
- Fat Points and Symbolic Powers

A Somewhat Gentle Introduction to Differential Graded Commutative Algebra

Kristen A. Beck and Sean Sather-Wagstaff

Dedicated with much respect to Tony Geramita

Abstract Differential graded (DG) commutative algebra provides powerful techniques for proving theorems about modules over commutative rings. These notes are a somewhat colloquial introduction to these techniques. In order to provide some motivation for commutative algebraists who are wondering about the benefits of learning and using these techniques, we present them in the context of a recent result of Nasseh and Sather-Wagstaff. These notes were used for the course "Differential Graded Commutative Algebra" that was part of the Workshop on Connections Between Algebra and Geometry at the University of Regina, May 29–June 1, 2012.

1 Introduction

Convention 1.1. The term "ring" is short for "commutative noetherian ring with identity," and "module" is short for "unital module." Let R be a ring.

These are notes for the course "Differential Graded Commutative Algebra" that was part of the Workshop on Connections Between Algebra and Geometry held at the University of Regina, May 29–June 1, 2012. They represent our attempt

K.A. Beck
Department of Mathematics, The University of Arizona, 617 N. Santa Rita Ave., P.O. Box 210089, Tucson, AZ 85721, USA
e-mail: kbeck@math.arizona.edu

S. Sather-Wagstaff (✉)
Department of Mathematics, NDSU Dept # 2750, P.O. Box 6050, Fargo, ND 58108-6050, USA
e-mail: sean.sather-wagstaff@ndsu.edu

S.M. Cooper and S. Sather-Wagstaff (eds.), *Connections Between Algebra, Combinatorics, and Geometry*, Springer Proceedings in Mathematics & Statistics 76,
DOI 10.1007/978-1-4939-0626-0_1,

to provide a small amount of (1) motivation for commutative algebraists who are wondering about the benefits of learning and using Differential Graded (DG) techniques, and (2) actual DG techniques.

1.1 DG Algebra

DG commutative algebra provides a useful framework for proving theorems about rings and modules, the statements of which have no reference to the DG universe. For instance, a standard theorem says the following:

Theorem 1.2 ([20, Corollary 1]). *Let $(R, \mathfrak{m}) \to (S, \mathfrak{n})$ be a flat local ring homomorphism, that is, a ring homomorphism making S into a flat R-module such that $\mathfrak{m}S \subseteq \mathfrak{n}$. Then S is Gorenstein if and only if R and $S/\mathfrak{m}S$ are Gorenstein. Moreover, there is an equality of Bass series $I_S(t) = I_R(t)I_{S/\mathfrak{m}S}(t)$.*

(See Definition 9.2 for the term "Bass series.") Of course, the flat hypothesis is very important here. On the other hand, the use of DG algebras allows for a slight (or vast, depending on your perspective) improvement of this:

Theorem 1.3 ([7, Theorem A]). *Let $(R, \mathfrak{m}) \to (S, \mathfrak{n})$ be a local ring homomorphism of finite flat dimension, that is, a local ring homomorphism such that S has a bounded resolution by flat R-module. Then there is a formal Laurent series $I_\varphi(t)$ with nonnegative integer coefficients such that $I_S(t) = I_R(t)I_\varphi(t)$. In particular, if S is Gorenstein, then so is R.*

In this result, the series $I_\varphi(t)$ is the *Bass series* of φ. It is the Bass series of the "homotopy closed fibre" of φ (instead of the usual closed fibre $S/\mathfrak{m}S$ of φ that is used in Theorem 1.2) which is the commutative DG algebra $S \otimes^{\mathbf{L}}_R R/\mathfrak{m}$. In particular, when S is flat over R, this is the usual closed fibre $S/\mathfrak{m}S \cong S \otimes_R R/\mathfrak{m}$, so one recovers Theorem 1.2 as a corollary of Theorem 1.3.

Furthermore, DG algebra comes equipped with constructions that can be used to replace your given ring with one that is nicer in some sense. To see how this works, consider the following strategy for using completions.

To prove a theorem about a given local ring R, first show that the assumptions ascend to the completion $\hat{R}$, prove the result for the complete ring $\hat{R}$, and show how the conclusion for $\hat{R}$ implies the desired conclusion for R. This technique is useful since frequently $\hat{R}$ is nicer than R. For instance, $\hat{R}$ is a homomorphic image of a power series ring over a field or a complete discrete valuation ring, so it is universally catenary (whatever that means) and it has a dualizing complex (whatever that is), while the original ring R may not have either of these properties.

When R is Cohen–Macaulay and local, a similar strategy sometimes allows one to mod out by a maximal R-regular sequence $\mathbf{x}$ to assume that R is artinian. The regular sequence assumption is like the flat condition for $\hat{R}$ in that it (sometimes) allows for the transfer of hypotheses and conclusions between R and the quotient $\overline{R} := R/(\mathbf{x})$. The artinian hypothesis is particularly nice, for instance, when R contains a field because then $\overline{R}$ is a finite dimensional algebra over a field.

The DG universe contains a construction $\tilde{R}$ that is similar to $\overline{R}$, with an advantage and a disadvantage. The advantage is that it is more flexible than $\overline{R}$ because it does not require the ring to be Cohen–Macaulay, and it produces a finite dimensional algebra over a field regardless of whether or not R contains a field. The disadvantage is that $\tilde{R}$ is a DG algebra instead of just an algebra, so it is graded commutative (almost, but not quite, commutative) and there is a bit more data to track when working with $\tilde{R}$. However, the advantages outweigh the disadvantages in that $\tilde{R}$ allows us to prove results for arbitrary local rings that can only be proved (as we understand things today) in special cases using $\overline{R}$. One such result is the following:

Theorem 1.4 ([32, Theorem A]). *A local ring has only finitely many semidualizing modules up to isomorphism.*

Even if you don't know what a semidualizing module is, you can see the point. Without DG techniques, we only know how to prove this result for Cohen–Macaulay rings that contain a field; see Theorem 2.13. With DG techniques, you get the unqualified result, which answers a question of Vasconcelos [41].

1.2 What These Notes Are

Essentially, these notes contain a sketch of the proof of Theorem 1.4; see 5.32, 7.38, and 8.17 below. Along the way, we provide a big-picture view of some of the tools and techniques in DG algebra (and other areas) needed to get a basic understanding of this proof. Also, since our motivation comes from the study of semidualizing modules, we provide a bit of motivation for the study of those gadgets in Sect. 9. In particular, we do not assume that the reader is familiar with the semidualizing world.

Since these notes are based on a course, they contain many exercises; sketches of solutions are contained in Sect. 10. They also contain a number of examples and facts that are presented without proof. A diligent reader may also wish to consider many of these as exercises.

1.3 What These Notes Are Not

These notes do not contain a great number of details about the tools and techniques in DG algebra. There are already excellent sources available for this, particularly, the seminal works [4, 6, 10]. The interested reader is encouraged to dig into these sources for proofs and constructions not given here. Our goal is to give some idea of what the tools look like and how they get used to solve problems. (To help readers in their digging, we provide many references for properties that we use.)

1.4 Notation

When it is convenient, we use notation from [11, 31]. Here we specify our conventions for some notions that have several notations:

$\mathrm{pd}_R(M)$: projective dimension of an R-module M
$\mathrm{id}_R(M)$: injective dimension of an R-module M
$\mathrm{len}_R(M)$: length of an R-module M
S_n: the symmetric group on $\{1,\dots,n\}$.
$\mathrm{sgn}(\iota)$: the signum of an element $\iota \in S_n$.

2 Semidualizing Modules

This section contains background material on semidualizing modules. It also contains a special case of Theorem 1.4; see Theorem 2.13. Further survey material can be found in [37, 39] and Sect. 9.

Definition 2.1. A finitely generated R-module C is *semidualizing* if the natural homothety map $\chi_C^R\colon R \to \mathrm{Hom}_R(C,C)$ given by $r \mapsto [c \mapsto rc]$ is an isomorphism and $\mathrm{Ext}^i_R(C,C) = 0$ for all $i \geqslant 1$. A *dualizing* R-module is a semidualizing R-module such that $\mathrm{id}_R(C) < \infty$. The set of isomorphism classes of semidualizing R-modules is denoted $\mathfrak{S}_0(R)$.

Remark 2.2. The symbol $\mathfrak{S}$ is an S, as in `\mathfrak{S}`.

Example 2.3. The free R-module R^1 is semidualizing.

Fact 2.4. The ring R has a dualizing module if and only if it is Cohen–Macaulay and a homomorphic image of a Gorenstein ring; when these conditions are satisfied, a dualizing R-module is the same as a "canonical" R-module. See Foxby [19, Theorem 4.1], Reiten [34, (3) Theorem], and Sharp [40, 2.1 Theorem (i)].

Remark 2.5. To the best of our knowledge, semidualizing modules were first introduced by Foxby [19]. They have been rediscovered independently by several authors who seem to all use different terminology for them. A few examples of this, presented chronologically, are:

Author(s)	Terminology	Context
Foxby [19]	"PG-module of rank 1"	Commutative algebra
Vasconcelos [41]	"Spherical module"	Commutative algebra
Golod [23]	"Suitable module"[a]	Commutative algebra
Wakamatsu [43]	"Generalized tilting module"	Representation theory
Christensen [14]	"Semidualizing module"	Commutative algebra
Mantese and Reiten [30]	"Wakamatsu tilting module"	Representation theory

[a]Apparently, another translation of the Russian term Golod used is "comfortable" module

The following facts are quite useful in practice.

Fact 2.6. Assume that $(R, \mathfrak{m})$ is local, and let C be a semidualizing R-module. If R is Gorenstein, then $C \cong R$. The converse holds if C is a dualizing R-module. See [14, (8.6) Corollary].

If $\mathrm{pd}_R(C) < \infty$, then $C \cong R$ by Sather-Wagstaff [37, Fact 1.14]. Here is a sketch of the proof. The isomorphism $\mathrm{Hom}_R(C, C) \cong R$ implies that $\mathrm{Supp}_R(C) = \mathrm{Spec}(R)$ and $\mathrm{Ass}_R(C) = \mathrm{Ass}_R(R)$. In particular, an element $x \in \mathfrak{m}$ is C-regular if and only if it is R-regular. If x is R-regular, it follows that C/xC is semidualizing over R/xR. By induction on $\mathrm{depth}(R)$, we conclude that $\mathrm{depth}_R(C) = \mathrm{depth}(R)$. The Auslander–Buchsbaum formula implies that C is projective, so it is free since R is local. Finally, the isomorphism $\mathrm{Hom}_R(C, C) \cong R$ implies that C is free of rank 1, that is, $C \cong R$.

Fact 2.7. Let $\varphi\colon R \to S$ be a ring homomorphism of finite flat dimension. (For example, this is so if φ is flat or surjective with kernel generated by an R-regular sequence.) If C is a semidualizing R-module, then $S \otimes_R C$ is a semidualizing S-module. The converse holds when φ is faithfully flat or local. The functor $S \otimes_R -$ induces a well-defined function $\mathfrak{S}_0(R) \to \mathfrak{S}_0(S)$ which is injective when φ is local. See [21, Theorems 4.5 and 4.9].

Exercise 2.8. Verify the conclusions of Fact 2.7 when φ is flat. That is, let $\varphi\colon R \to S$ be a flat ring homomorphism. Prove that if C is a semidualizing R-module, then the base-changed module $S \otimes_R C$ is a semidualizing S-module. Prove that the converse holds when φ is faithfully flat, e.g., when φ is local.

The next lemma is for use in Theorem 2.13, which is a special case of Theorem 1.4. See Remark 2.10 and Question 2.11 for further perspective.

Lemma 2.9. *Assume that R is local and artinian. Then there is an integer ρ depending only on R such that $\mathrm{len}_R(C) \leqslant \rho$ for every semidualizing R-module C.*

Proof. Let k denote the residue field of R. We show that the integer $\rho = \mathrm{len}_R(R)\mu_R^0$ satisfies the conclusion where $\mu_R^0 = \mathrm{rank}_k(\mathrm{Hom}_R(k, R))$. (This is the 0th Bass number of R; see Definition 9.2.) Let C be a semidualizing R-module. Set $\beta = \mathrm{rank}_k(k \otimes_R C)$ and $\mu = \mathrm{rank}_k(\mathrm{Hom}_R(k, C))$. Since R is artinian and C is finitely generated, it follows that $\mu \geqslant 1$. Also, the fact that R is local implies that there is an R-module epimorphism $R^\beta \to C$, so we have $\mathrm{len}_R(C) \leqslant \mathrm{len}_R(R)\beta$. Thus, it remains to show that $\beta \leqslant \mu_R^0$.

The next sequence of isomorphisms uses adjointness and tensor cancellation:

$$
\begin{aligned}
k^{\mu_R^0} &\cong \mathrm{Hom}_R(k, R)\\
&\cong \mathrm{Hom}_R(k, \mathrm{Hom}_R(C, C))\\
&\cong \mathrm{Hom}_R(C \otimes_R k, C)\\
&\cong \mathrm{Hom}_R(k \otimes_k (C \otimes_R k), C)
\end{aligned}
$$

$$\cong \operatorname{Hom}_k(C \otimes_R k, \operatorname{Hom}_R(k, C))$$
$$\cong \operatorname{Hom}_k(k^{\beta}, k^{\mu})$$
$$\cong k^{\beta\mu}.$$

Since $\mu \geqslant 1$, it follows that $\beta \leqslant \beta\mu = \mu_R^0$, as desired. □

Remark 2.10. Assume that R is local and Cohen–Macaulay. If D is dualizing for R, then there is an equality $e_R(D) = e(R)$ of Hilbert–Samuel multiplicities with respect to the maximal ideal of R. See, e.g., [11, Proposition 3.2.12.e.i and Corollary 4.7.8]. It is unknown whether the same equality holds for an arbitrary *semi*dualizing R-module. Using the "additivity formula" for multiplicities, this boils down to the following. Some progress is contained in [17].

Question 2.11. Assume that R is local and artinian. For every semidualizing R-module C, must one have $\operatorname{len}_R(C) = \operatorname{len}(R)$?

While we are in the mood for questions, here is a big one. In every explicit calculation of $\mathfrak{S}_0(R)$, the answer is "yes"; see [35, 38].

Question 2.12. Assume that R is local. Must $|\mathfrak{S}_0(R)|$ be 2^n for some $n \in \mathbb{N}$?

Next, we sketch the proof of Theorem 1.4 when R is Cohen–Macaulay and contains a field. This sketch serves to guide the proof of the result in general.

Theorem 2.13 ([15, Theorem 1]). *Assume that $(R, \mathfrak{m}, k)$ is Cohen–Macaulay local and contains a field. Then $|\mathfrak{S}_0(R)| < \infty$.*

Proof. Case 1: R is artinian, and k is algebraically closed. In this case, Cohen's structure theorem implies that R is a finite dimensional k-algebra. Since k is algebraically closed, a result of Happel [25, proof of first proposition in Sect. 3] says that for each $n \in \mathbb{N}$ the following set is finite.

$$T_n = \{\text{isomorphism classes of } R\text{-modules } N \mid \operatorname{Ext}_R^1(N, N) = 0 \text{ and } \operatorname{len}_R(N) = n\}$$

Lemma 2.9 implies that there is a $\rho \in \mathbb{N}$ such that $\mathfrak{S}_0(R)$ is contained in the finite set $\bigcup_{n=1}^{\rho} T_n$, so $\mathfrak{S}_0(R)$ is finite.

Case 2: k is algebraically closed. Let $\mathbf{x} = x_1, \ldots, x_n \in \mathfrak{m}$ be a maximal R-regular sequence. Since R is Cohen–Macaulay, the quotient $R' = R/(\mathbf{x})$ is artinian. Also, R' has the same residue field as R, so Case 1 implies that $\mathfrak{S}_0(R')$ is finite. Since R is local, Fact 2.7 provides an injection $\mathfrak{S}_0(R) \hookrightarrow \mathfrak{S}_0(R')$, so $\mathfrak{S}_0(R)$ is finite as well.

Case 3: the general case. A result of Grothendieck [24, Théorèm 19.8.2(ii)] provides a flat local ring homomorphism $R \to \overline{R}$ such that $\overline{R}/\mathfrak{m}\overline{R}$ is algebraically closed. In particular, since R and $\overline{R}/\mathfrak{m}\overline{R}$ are Cohen–Macaulay, it follows that $\overline{R}$ is

Cohen–Macaulay. The fact that R contains a field implies that $\overline{R}$ also contains a field. Hence, Case 2 shows that $\mathfrak{S}_0(\overline{R})$ is finite. Since R is local, Fact 2.7 provides an injective function $\mathfrak{S}_0(R) \hookrightarrow \mathfrak{S}_0(R')$, so $\mathfrak{S}_0(R)$ is finite as well. □

Remark 2.14. Happel's result uses some deep ideas from algebraic geometry and representation theory. The essential point comes from a theorem of Voigt [42] (see also Gabriel [22, 1.2 Corollary]). We'll need a souped-up version of this result for the full proof of Theorem 1.4. This is the point of Sect. 8.

Remark 2.15. The proof of Theorem 2.13 uses the extra assumptions (extra compared to Theorem 1.4) in crucial places. The Cohen–Macaulay assumption is used in the reduction to the artinian case. And the fact that R contains a field is used in order to invoke Happel's result. In order to remove these assumptions for the proof of Theorem 1.4, we find an algebra U that is finite dimensional over an algebraically closed field such that there is an injective function $\mathfrak{S}_0(R) \hookrightarrow \mathfrak{S}(U)$. The trick is that U is a DG algebra, and $\mathfrak{S}(U)$ is a set of equivalence classes of semidualizing DG U-modules. So, we need to understand the following:

(a) What are DG algebras, and how is U constructed?
(b) What are semidualizing DG modules, and how is the map $\mathfrak{S}_0(R) \hookrightarrow \mathfrak{S}(U)$ constructed?
(c) Why is $\mathfrak{S}(U)$ finite?

This is the point of the rest of the notes. See Sects. 5, 7, and 8.

3 Hom Complexes

This section has several purposes. First, we set some notation and terminology. Second, we make sure that the reader is familiar with some notions that we need later in the notes. One of the main points of this section is Fact 3.18.

3.1 Complexes

The following gadgets form the foundation for homological algebra, and we shall use them extensively.

Definition 3.1. An *R-complex*[1] is a sequence of R-module homomorphisms

$$X = \cdots \xrightarrow{\partial^X_{i+1}} X_i \xrightarrow{\partial^X_i} X_{i-1} \xrightarrow{\partial^X_{i-1}} \cdots$$

[1]Readers more comfortable with notations like $X_\bullet$ or X_* for complexes should feel free to decorate their complexes as they see fit.

such that $\partial_i^X \partial_{i+1}^X = 0$ for all i. For each $x \in X_i$, the *degree* of x is $|x| := i$. The *ith homology module* of X is $\mathrm{H}_i(X) := \mathrm{Ker}(\partial_i^X)/\mathrm{Im}(\partial_{i+1}^X)$. A *cycle* in X_i is an element of $\mathrm{Ker}(\partial_i^X)$.

We use the following notation for augmented resolutions in several places below.

Example 3.2. Let M be an R-module. We consider M as an R-complex "concentrated in degree 0":

$$M = \quad 0 \to M \to 0.$$

Given an augmented projective resolution

$$P^+ = \cdots \xrightarrow{\partial_2^P} P_1 \xrightarrow{\partial_1^P} P_0 \xrightarrow{\tau} M \to 0$$

the truncated resolution

$$P = \cdots \xrightarrow{\partial_2^P} P_1 \xrightarrow{\partial_1^P} P_0 \to 0$$

is an R-complex such that $\mathrm{H}_0(P) \cong M$ and $\mathrm{H}_i(P) = 0$ for all $i \neq 0$. Similarly, given an augmented injective resolution

$$^+I = \quad 0 \to M \xrightarrow{\epsilon} I_0 \xrightarrow{\partial_0^I} I_{-1} \xrightarrow{\partial_{-1}^I} \cdots$$

the truncated resolution

$$I = \quad 0 \to I_0 \xrightarrow{\partial_0^I} I_{-1} \xrightarrow{\partial_{-1}^I} \cdots$$

is an R-complex such that $\mathrm{H}_0(I) \cong M$ and $\mathrm{H}_i(I) = 0$ for all $i \neq 0$.

3.2 The Hom Complex

The next constructions are used extensively in these notes. For instance, the chain maps are the morphisms in the category of R-complexes.

Definition 3.3. Let X and Y be R-complexes. The *Hom complex* $\mathrm{Hom}_R(X,Y)$ is defined as follows. For each integer n, set $\mathrm{Hom}_R(X,Y)_n := \prod_{p\in\mathbb{Z}} \mathrm{Hom}_R(X_p, Y_{p+n})$ and $\partial_n^{\mathrm{Hom}_R(X,Y)}(\{f_p\}) := \{\partial_{p+n}^Y f_p - (-1)^n f_{p-1}\partial_p^X\}$. A *chain map* $X \to Y$ is a cycle in $\mathrm{Hom}_R(X,Y)_0$, i.e., an element of $\mathrm{Ker}(\partial_0^{\mathrm{Hom}_R(X,Y)})$. An element in $\mathrm{Hom}_R(X,Y)_0$ is *null-homotopic* if it is in $\mathrm{Im}(\partial_1^{\mathrm{Hom}_R(X,Y)})$. An *isomorphism* $X \xrightarrow{\cong} Y$ is a chain map $X \to Y$ with a two-sided inverse. We sometimes write f in place of $\{f_p\}$.

Exercise 3.4. Let X and Y be R-complexes.

(a) Prove that $\operatorname{Hom}_R(X, Y)$ is an R-complex.
(b) Prove that a chain map $X \to Y$ is a sequence of R-module homomorphisms $\{f_p : X_p \to Y_p\}$ making the following diagram commute:

$$\begin{array}{ccccccc}
\cdots & \xrightarrow{\partial^X_{i+1}} & X_i & \xrightarrow{\partial^X_i} & X_{i-1} & \xrightarrow{\partial^X_{i-1}} & \cdots \\
 & & \downarrow{\scriptstyle f_i} & & \downarrow{\scriptstyle f_{i-1}} & & \\
\cdots & \xrightarrow{\partial^Y_{i+1}} & Y_i & \xrightarrow{\partial^Y_i} & Y_{i-1} & \xrightarrow{\partial^Y_{i-1}} & \cdots .
\end{array}$$

(c) Prove that if $\{f_p\} \in \operatorname{Hom}_R(X, Y)_0$ is null-homotopic, then it is a chain map.
(d) Prove that a sequence $\{f_p\} \in \operatorname{Hom}_R(X, Y)_0$ is null-homotopic if and only if there is a sequence $\{s_p : X_p \to Y_{p+1}\}$ of R-module homomorphisms such that $f_p = \partial^Y_{p+1} s_p + s_{p-1} \partial^X_p$ for all $p \in \mathbb{Z}$.

The following exercises contain useful properties of these constructions.

Exercise 3.5 ("Hom Cancellation"). Let X be an R-complex. Prove that the map $\tau : \operatorname{Hom}_R(R, X) \to X$ given by $\tau_n(\{f_p\}) = f_n(1)$ is an isomorphism of R-complexes.

Exercise 3.6. Let X be an R-complex, and let M be an R-module.

(a) Prove that $\operatorname{Hom}_R(M, X)$ is isomorphic to the following complex:

$$\cdots \xrightarrow{(\partial^X_{n+1})_*} (X_n)_* \xrightarrow{(\partial^X_n)_*} (X_{n-1})_* \xrightarrow{(\partial^X_{n-1})_*} \cdots$$

where $(-)_* = \operatorname{Hom}_R(M, -)$ and $(\partial^X_n)_*(f) = \partial^X_n f$.
(b) Prove that $\operatorname{Hom}_R(X, M)$ is isomorphic to the following complex:

$$\cdots \xrightarrow{(\partial^X_n)^*} X^*_n \xrightarrow{(\partial^X_{n+1})^*} X^*_{n+1} \xrightarrow{(\partial^X_{n+2})^*} \cdots .$$

where $(-)^* = \operatorname{Hom}_R(-, M)$ and $(\partial^X_n)^*(f) = f \partial^X_n$. [Hint: Mind the signs.]

Exercise 3.7. Let $f : X \to Y$ be a chain map.

(a) Prove that for each $i \in \mathbb{Z}$, the chain map f induces a well-defined R-module homomorphism $\mathrm{H}_i(f) : \mathrm{H}_i(X) \to \mathrm{H}_i(Y)$ given by $\mathrm{H}_i(f)(\overline{x}) := \overline{f_i(x)}$.
(b) Prove that if f is null-homotopic, then $\mathrm{H}_i(f) = 0$ for all $i \in \mathbb{Z}$.

The following concept is central for homological algebra; see, e.g., Remark 3.11.

Definition 3.8. A chain map $f : X \to Y$ is a *quasiisomorphism* if for all $i \in \mathbb{Z}$ the induced map $\mathrm{H}_i(f) : \mathrm{H}_i(X) \to \mathrm{H}_i(Y)$ is an isomorphism. We use the symbol $\simeq$ to identify quasiisomorphisms.

Exercise 3.9. Prove that each isomorphism of R-complexes is a quasiisomorphism.

Exercise 3.10. Let M be an R-module with augmented projective resolution P^+ and augmented injective resolution ${}^+I$; see the notation from Example 3.2. Prove that τ and ϵ induce quasiisomorphisms $P \xrightarrow{\simeq} M \xrightarrow{\simeq} I$.

Remark 3.11. Let M and N be R-modules. The fact that $\mathrm{Ext}^i_R(M, N)$ can be computed using a projective resolution P of M or an injective resolution I of N is called the "balance" property for Ext. It can be proved by showing that there are quasiisomorphisms $\mathrm{Hom}_R(P, N) \xrightarrow{\simeq} \mathrm{Hom}_R(P, I) \xleftarrow{\simeq} \mathrm{Hom}_R(M, I)$. See Fact 3.15.

3.3 Hom and Chain Maps (Functoriality)

Given that the chain maps are the morphisms in the category of R-complexes, the next construction and the subsequent exercise indicate that $\mathrm{Hom}_R(Z, -)$ and $\mathrm{Hom}_R(-, Z)$ are functors.

Definition 3.12. Given a chain map $f: X \to Y$ and an R-complex Z, we define $\mathrm{Hom}_R(Z, f): \mathrm{Hom}_R(Z, X) \to \mathrm{Hom}_R(Z, Y)$ as follows: each $\{g_p\} \in \mathrm{Hom}_R(Z, X)_n$ is mapped to $\{f_{p+n} g_p\} \in \mathrm{Hom}_R(Z, Y)_n$. Similarly, define $\mathrm{Hom}_R(f, Z): \mathrm{Hom}_R(Y, Z) \to \mathrm{Hom}_R(X, Z)$ by the formula $\{g_p\} \mapsto \{g_p f_p\}$.

Remark 3.13. We do not use a sign-change in this definition because $|f| = 0$.

Exercise 3.14. Given a chain map $f: X \to Y$ and an R-complex Z, prove that $\mathrm{Hom}_R(Z, f)$ and $\mathrm{Hom}_R(f, Z)$ are chain maps.

Fact 3.15. Let $f: X \xrightarrow{\simeq} Y$ be a quasiisomorphism, and let Z be an R-complex. In general, the chain map $\mathrm{Hom}_R(Z, f): \mathrm{Hom}_R(Z, X) \to \mathrm{Hom}_R(Z, Y)$ is not a quasiisomorphism. However, if Z is a complex of projective R-modules such that $Z_i = 0$ for $i \ll 0$, then $Z \otimes_R f$ is a quasiisomorphism. Similarly, $\mathrm{Hom}_R(f, Z): \mathrm{Hom}_R(Y, Z) \to \mathrm{Hom}_R(X, Z)$ is not a quasiisomorphism. However, if Z is a complex of injective R-modules such that $Z_i = 0$ for $i \gg 0$, then $\mathrm{Hom}_R(f, Z)$ is a quasiisomorphism.

3.4 Homotheties and Semidualizing Modules

We next explain how the Hom complex relates to the semidualizing modules from Sect. 2.

Exercise 3.16. Let X be an R-complex, and let $r \in R$. For each $p \in \mathbb{Z}$, let $\mu^{X,r}_p: X_p \to X_p$ be given by $x \mapsto rx$. (Such a map is a "homothety." When it is convenient, we denote this map as $X \xrightarrow{r} X$.)

Prove that $\mu^{X,r} := \{\mu_p^{X,r}\} \in \operatorname{Hom}_R(X,X)_0$ is a chain map. Prove that for all $i \in \mathbb{Z}$ the induced map $\operatorname{H}_i(\mu^{X,r})\colon \operatorname{H}_i(X) \to \operatorname{H}_i(X)$ is multiplication by r.

Exercise 3.17. Let X be an R-complex. We use the notation from Exercise 3.16. Define $\chi_0^X\colon R \to \operatorname{Hom}_R(X,X)$ by the formula $\chi_0^X(r) := \mu^{X,r} \in \operatorname{Hom}_R(X,X)_0$. Prove that this determines a chain map $\chi^X\colon R \to \operatorname{Hom}_R(X,X)$. The chain map χ^X is the "homothety morphism" for X.

Fact 3.18. Let M be a finitely generated R-module. We use the notation from Exercise 3.17. The following conditions are equivalent:

(i) M is a semidualizing R-module.
(ii) For each projective resolution P of M, the chain map $\chi^P\colon R \to \operatorname{Hom}_R(P,P)$ is a quasiisomorphism.
(iii) For some projective resolution P of M, the chain map $\chi^P\colon R \to \operatorname{Hom}_R(P,P)$ is a quasiisomorphism.
(iv) For each injective resolution I of M, the chain map $\chi^I\colon R \to \operatorname{Hom}_R(I,I)$ is a quasiisomorphism.
(v) For some injective resolution I of M, the chain map $\chi^I\colon R \to \operatorname{Hom}_R(I,I)$ is a quasiisomorphism.

In some sense, the point is that the homologies of the complexes $\operatorname{Hom}_R(P,P)$ and $\operatorname{Hom}_R(I,I)$ are exactly the modules $\operatorname{Ext}_R^i(M,M)$ by Fact 3.15.

4 Tensor Products and the Koszul Complex

Tensor products for complexes are as fundamental for complexes as they are for modules. In this section, we use them to construct the Koszul complex; see Definition 4.10. In Sect. 7, we use them for base change; see, e.g., Exercise 7.10.

4.1 *Tensor Product of Complexes*

As with the Hom complex, the sign convention in the next construction guarantees that it is an R-complex; see Exercise 4.2 and Remark 4.3. Note that Remark 4.4 describes a notational simplification.

Definition 4.1. Fix R-complexes X and Y. The *tensor product complex $X \otimes_R Y$* is defined as follows. For each integer n, set $(X \otimes_R Y)_n := \bigoplus_{p\in\mathbb{Z}} X_p \otimes_R Y_{n-p}$ and let $\partial_n^{X\otimes_R Y}$ be given on generators by the formula $\partial_n^{X\otimes_R Y}(\ldots,0,x_p \otimes y_{n-p},0,\ldots) := (\ldots,0,\partial_p^X(x_p)\otimes y_{n-p},(-1)^p x_p \otimes \partial_{n-p}^Y(y_{n-p}),0,\ldots)$.

Exercise 4.2. Let X, Y, and Z be R-complexes.

(a) Prove that $X \otimes_R Y$ is an R-complex.
(b) Prove that there is a "tensor cancellation" isomorphism $R \otimes_R X \cong X$.

(c) Prove that there is a "commutativity" isomorphism $X \otimes_R Y \cong Y \otimes_R X$. (Hint: Mind the signs. This isomorphism is given by $x \otimes y \mapsto (-1)^{|x||y|} y \otimes x$.)
(d) Verify the "associativity" isomorphism $X \otimes_R (Y \otimes_R Z) \cong (X \otimes_R Y) \otimes_R Z$.

Remark 4.3. There is a rule of thumb for sign conventions like the one in the hint for Exercise 4.2: whenever two factors u and v are commuted in an expression, you multiply by $(-1)^{|u||v|}$. This can already be seen in $\partial^{\mathrm{Hom}_R(X,Y)}$ and $\partial^{X \otimes_R Y}$. This graded commutativity is one of the keys to DG algebra. See Sect. 5.

Remark 4.4. After working with the tensor product of complexes for a few moments, one realizes that the sequence notation $(\ldots, 0, x_p \otimes y_{n-p}, 0, \ldots)$ is unnecessarily cumbersome. We use the sequence notation in a few of the solutions in Sect. 10, but not for many of them. Similarly, from now on, instead of writing $(\ldots, 0, x_p \otimes y_{n-p}, 0, \ldots)$, we write the simpler $x_p \otimes y_{n-p}$. As we note in 10.11, one needs to be somewhat careful with this notation, as elements $u \otimes v$ and $x \otimes y$ only live in the same summand when $|u| = |x|$ and $|v| = |y|$.

Fact 4.5. Given R-complexes $X^1, \ldots, X^n$ an induction argument using the associativity isomorphism from Exercise 4.2 shows that the n-fold tensor product $X^1 \otimes_R \cdots \otimes_R X^n$ is well defined (up to isomorphism).

4.2 *Tensor Products and Chain Maps (Functoriality)*

As for Hom, the next items indicate that $Z \otimes_R -$ and $- \otimes_R Z$ are functors.

Definition 4.6. Consider a chain map $f: X \to Y$ and an R-complex Z. We define $Z \otimes_R f: Z \otimes_R X \to Z \otimes_R Y$ by the formula $z \otimes y \mapsto z \otimes f(y)$. Similarly, define the map $f \otimes_R Z: X \otimes_R Z \to Y \otimes_R Z$ by the formula $x \otimes z \mapsto f(x) \otimes z$.

Remark 4.7. We do not use a sign-change in this definition because $|f| = 0$.

Exercise 4.8. Consider a chain map $f: X \to Y$ and an R-complex Z. Then the maps $Z \otimes_R f: Z \otimes_R X \to Z \otimes_R Y$ and $f \otimes_R Z: X \otimes_R Z \to Y \otimes_R Z$ are chain maps.

Fact 4.9. Let $f: X \xrightarrow{\simeq} Y$ be a quasiisomorphism, and let Z be an R-complex. In general, the chain map $Z \otimes_R f: Z \otimes_R X \to Z \otimes_R Y$ is not a quasiisomorphism. However, if Z is a complex of projective R-modules such that $Z_i = 0$ for $i \ll 0$, then $Z \otimes_R f$ is a quasiisomorphism.

4.3 *The Koszul Complex*

Here begins our discussion of the prototypical DG algebra.

Definition 4.10. Let $\mathbf{x} = x_1, \ldots, x_n \in R$. For $i = 1, \ldots, n$ set

$$K^R(x_i) = \quad 0 \to R \xrightarrow{x_i} R \to 0.$$

Using Remark 4.5, we set

$$K^R(\mathbf{x}) = K^R(x_1, \ldots, x_n) = K^R(x_1) \otimes_R \cdots \otimes_R K^R(x_n).$$

Exercise 4.11. Let $x, y, z \in R$. Write out explicit formulas, using matrices for the differentials, for $K^R(x, y)$ and $K^R(x, y, z)$.

Exercise 4.12. Let $\mathbf{x} = x_1, \ldots, x_n \in R$. Prove that $K^R(\mathbf{x})_i \cong R^{\binom{n}{i}}$ for all $i \in \mathbb{Z}$. (Here we use the convention $\binom{n}{i} = 0$ for all $i < 0$ and $i > n$.)

Exercise 4.13. Let $\mathbf{x} = x_1, \ldots, x_n \in R$. Let $\sigma \in S_n$, and set $\mathbf{x}' = x_{\sigma(1)}, \ldots, x_{\sigma(n)}$. Prove that $K^R(\mathbf{x}) \cong K^R(\mathbf{x}')$.

Given a generating sequence $\mathbf{x}$ for the maximal ideal of a local ring R, one concludes from the next lemma that each homology module $\mathrm{H}_i(K^R(\mathbf{x}))$ has finite length. This is crucial for the proof of Theorem 1.4.

Lemma 4.14. *If* $\mathbf{x} = x_1, \ldots, x_n \in R$ *and* $\mathfrak{a} = (\mathbf{x})R$, *then* $\mathfrak{a}\,\mathrm{H}_i(K^R(\mathbf{x})) = 0$ *for all* $i \in \mathbb{Z}$.

Proof (Sketch of Proof). It suffices to show that for $j = 1, \ldots, n$ and for all $i \in \mathbb{Z}$ we have $x_j\,\mathrm{H}_i(K^R(\mathbf{x})) = 0$. By symmetry (Exercise 4.13) it suffices to show that $x_1\,\mathrm{H}_i(K^R(\mathbf{x})) = 0$ for all $i \in \mathbb{Z}$. The following diagram shows that the chain map $K^R(x_1) \xrightarrow{x_1} K^R(x_1)$ is null-homotopic.

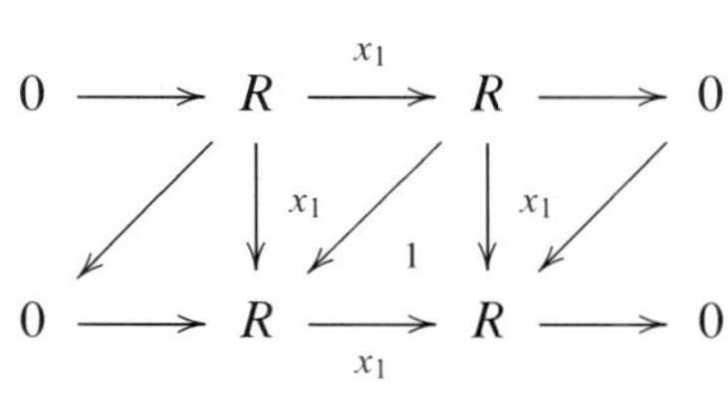

It is routine to show that this implies that the induced map $K^R(\mathbf{x}) \xrightarrow{x_1} K^R(\mathbf{x})$ is null-homotopic. The desired conclusion now follows from Exercise 3.7. □

The next construction allows us to push our complexes around.

Definition 4.15. Let X be an R-complex, and let $n \in \mathbb{Z}$. The *nth suspension* (or *shift*) of X is the complex $\Sigma^n X$ such that $(\Sigma^n X)_i := X_{i-n}$ and $\partial_i^{\Sigma^n X} = (-1)^n \partial_{i-n}^X$. We set $\Sigma X := \Sigma^1 X$.

The next fact is in general quite useful, though we do not exploit it here.

Fact 4.16. Let $\mathbf{x} = x_1, \ldots, x_n \in R$. The Koszul complex $K^R(\mathbf{x})_i$ is "self-dual," that is, that there is an isomorphism of R-complexes $\mathrm{Hom}_R(K^R(\mathbf{x}), R) \cong \Sigma^n K^R(\mathbf{x})$. (This fact is not trivial.)

Exercise 4.17. Verify the isomorphism from Fact 4.16 for $n = 1, 2, 3$.

The following result gives the first indication of the utility of the Koszul complex. We use it explicitly in the proof of Theorem 1.4.

Lemma 4.18. *Let* $\mathbf{x} = x_1, \ldots, x_n \in R$. *If* $\mathbf{x}$ *is* R*-regular, then* $K^R(\mathbf{x})$ *is a free resolution of* $R/(\mathbf{x})$ *over* R.

Proof. Argue by induction on n.

Base case: $n = 1$. Assume that x_1 is R-regular. Since $K^R(x_1)$ has the form $0 \to R \xrightarrow{x_1} R \to 0$, the fact that x_1 is R-regular implies that $\mathrm{H}_i(K^R(x_1)) = 0$ for all $i \neq 0$. As each module in $K^R(x_1)$ is free, it follows that $K^R(x_1)$ is a free resolution of $R/(x_1)$.

Inductive step: Assume that $n \geqslant 2$ and that the result holds for regular sequences of length $n-1$. Assume that $\mathbf{x}$ is R-regular (of length n). Thus, the sequence $\mathbf{x}' = x_1, \ldots, x_{n-1}$ is R-regular, and x_n is $R/(\mathbf{x}')$-regular. The first condition implies that $K' := K^R(\mathbf{x}')$ is a free resolution of $R/(\mathbf{x}')$ over R. By definition, we have $K := K^R(\mathbf{x}) = K' \otimes_R K^R(x_n)$. Further, by definition of $K' \otimes_R K^R(x_n)$, we have

$$K \cong \cdots \xrightarrow{\begin{pmatrix}\partial^{K'}_{i+1} & (-1)^i x_n \\ 0 & \partial^{K'}_{i}\end{pmatrix}} \begin{matrix}K'_i \\ \bigoplus \\ K'_{i-1}\end{matrix} \xrightarrow{\begin{pmatrix}\partial^{K'}_{i} & (-1)^{i-1} x_n \\ 0 & \partial^{K'}_{i-1}\end{pmatrix}} \begin{matrix}K'_{i-1} \\ \bigoplus \\ K'_{i-2}\end{matrix} \xrightarrow{\begin{pmatrix}\partial^{K'}_{i-1} & (-1)^{i-2} x_n \\ 0 & \partial^{K'}_{i-2}\end{pmatrix}} \cdots.$$

(Note that the term $K'_i \bigoplus K'_{i-1}$ is shorthand for $(K'_i \otimes_R R1) \bigoplus (K'_{i-1} \otimes_R Re)$.) Using this, there is a short exact sequence of R-complexes and chain maps[2]

$$0 \longrightarrow K' \longrightarrow K \longrightarrow K'' \longrightarrow 0$$

$$\begin{array}{ccccccccc}
 & & \vdots & & \vdots & & \vdots & & \\
 & & \downarrow{\scriptstyle \partial^{K'}_{i+1}} & & \downarrow{\scriptstyle \partial^{K}_{i+1}} & & \downarrow{\scriptstyle \partial^{K'}_{i}} & & \\
0 & \longrightarrow & K'_i & \longrightarrow & K_i & \longrightarrow & K'_{i-1} & \longrightarrow & 0 \\
 & & \downarrow{\scriptstyle \partial^{K'}_{i}} & & \downarrow{\scriptstyle \partial^{K}_{i}} & & \downarrow{\scriptstyle \partial^{K'}_{i-1}} & & \\
0 & \longrightarrow & K'_{i-1} & \longrightarrow & K_{i-1} & \longrightarrow & K'_{i-2} & \longrightarrow & 0 \\
 & & \downarrow{\scriptstyle \partial^{K'}_{i-1}} & & \downarrow{\scriptstyle \partial^{K}_{i-1}} & & \downarrow{\scriptstyle \partial^{K'}_{i-2}} & & \\
 & & \vdots & & \vdots & & \vdots & &
\end{array}$$

[2]Readers who know the mapping cone description of K should not be surprised by this argument.

where K'' is obtained by shifting K'.[3] Furthermore, it can be shown that the long exact sequence in homology has the form

$$\cdots \mathrm{H}_i(K') \xrightarrow{(-1)^i x_n} \mathrm{H}_i(K') \to \mathrm{H}_i(K) \to \mathrm{H}_i(K') \xrightarrow{(-1)^{i-1} x_n} \mathrm{H}_i(K') \to \cdots .$$

Since K' is a free resolution of $R/(\mathbf{x}')$ over R, we have $\mathrm{H}_i(K') = 0$ for all $i \neq 0$, and $\mathrm{H}_0(K') \cong R/(\mathbf{x}')$. As x_n is $R/(\mathbf{x}')$-regular, an analysis of the long exact sequence shows that $\mathrm{H}_i(K) = 0$ for all $i \neq 0$, and $\mathrm{H}_0(K) \cong R/(\mathbf{x})$. It follows that K is a free resolution of $R/(\mathbf{x})$, as desired. □

4.4 *Alternate Description of the Koszul Complex*

The following description of $K^R(\mathbf{x})$ says that $K^R(\mathbf{x})$ is given by the "exterior algebra" on R^n; see Fact 4.22.

Definition 4.19. Let $\mathbf{x} = x_1, \ldots, x_n \in R$. Fix a basis $e_1, \ldots, e_n \in R^n$. For $i > 1$, set $\bigwedge^i R^n := R^{\binom{n}{i}}$ with basis given by the set of formal symbols $e_{j_1} \wedge \cdots \wedge e_{j_i}$ such that $1 \leqslant j_1 < \cdots < j_i \leqslant n$. This extends to all $i \in \mathbb{Z}$ as follows: $\bigwedge^1 R^n = R^n$ with basis $e_1, \ldots, e_n$ and $\bigwedge^0 R^n = R^1$ with basis 1; for $i < 0$, set $\bigwedge^i R^n = R^{\binom{n}{i}} = 0$.

Define $\tilde{K}^R(\mathbf{x})$ as follows. For all $i \in \mathbb{Z}$ set $\tilde{K}^R(\mathbf{x})_i = \bigwedge^i R^n$, and let $\partial_i^{\tilde{K}^R(\mathbf{x})}$ be given on basis vectors by the formula

$$\partial_i^{\tilde{K}^R(\mathbf{x})}(e_{j_1} \wedge \cdots \wedge e_{j_i}) = \sum_{p=1}^{i} (-1)^{p-1} x_{j_p} e_{j_1} \wedge \cdots \wedge \widehat{e_{j_p}} \wedge \cdots \wedge e_{j_i}$$

where the notation $\widehat{e_{j_p}}$ indicates that e_{j_p} has been removed from the list. In the case $i = 1$, the formula reads as $\partial_1^{\tilde{K}^R(\mathbf{x})}(e_j) = x_j$.

Remark 4.20. Our definition of $\bigwedge^i R^n$ is ad hoc. A better way to think about it (in some respects) is in terms of a universal mapping property for alternating multilinear maps. A basis-free construction can be given in terms of a certain quotient of the i-fold tensor product $R^n \otimes_R \cdots \otimes_R R^n$.

Exercise 4.21. Let $\mathbf{x} = x_1, \ldots, x_n \in R$. Write out explicit formulas, using matrices for the differentials, for $\tilde{K}^R(\mathbf{x})$ in the cases $n = 1, 2, 3$.

Fact 4.22. Let $\mathbf{x} = x_1, \ldots, x_n \in R$. There is an isomorphism of R-complexes $K^R(\mathbf{x}) \cong \tilde{K}^R(\mathbf{x})$. (This fact is not trivial. For perspective on this, compare the

[3]Note that K'' is technically not equal to the complex $\Sigma K'$ from Definition 4.15, since there is no sign on the differential. On the other hand the complexes K'' and $\Sigma K'$ are isomorphic.

solutions to Exercises 4.11 and 4.21 in 10.13 and 10.17.) In light of this fact, we do not distinguish between $K^R(\mathbf{x})$ and $\tilde{K}^R(\mathbf{x})$ for the remainder of these notes.

Remark 4.23. A third description of $K^R(\mathbf{x})$ involves the mapping cone. Even though it is extremely useful, we do not discuss it in detail here.

4.5 Algebra Structure on the Koszul Complex

In our estimation, the Koszul complex is one of the most important constructions in commutative algebra. When the sequence $\mathbf{x}$ is R-regular, it is an R-free resolution of $R/(\mathbf{x})$, by Lemma 4.18. In general, it detects depth and has all scads of other magical properties. For us, one of its most important features is its algebra structure, which we describe next.

Definition 4.24. Let $n \in \mathbb{N}$ and let $e_1, \ldots, e_n \in R^n$ be a basis. In $\bigwedge^2 R^n$, define

$$e_{j_2} \wedge e_{j_1} := \begin{cases} -e_{j_1} \wedge e_{j_2} & \text{whenever } 1 \leqslant j_1 < j_2 \leqslant n \\ 0 & \text{whenever } 1 \leqslant j_1 = j_2 \leqslant n. \end{cases}$$

Extending this bilinearly, we define $\alpha \wedge \beta$ for all $\alpha, \beta \in \bigwedge^1 R^n = R^n$: write $\alpha = \sum_p \alpha_p e_p$ and $\beta = \sum_q \beta_q e_q$, and define

$$\alpha \wedge \beta = \left(\sum_p \alpha_p e_p\right) \wedge \left(\sum_q \beta_q e_q\right) = \sum_{p,q} \alpha_p \beta_q e_p \wedge e_q = \sum_{p<q} (\alpha_p \beta_q - \alpha_q \beta_p) e_p \wedge e_q.$$

This extends to a multiplication $\bigwedge^1 R^n \times \bigwedge^t R^n \to \bigwedge^{1+t} R^n$ using the following formula, assuming that $1 \leqslant i \leqslant n$ and $1 \leqslant j_1 < \cdots < j_t \leqslant n$:

$$e_i \wedge (e_{j_1} \wedge \cdots \wedge e_{j_t}) := \begin{cases} 0 & \text{if } i = j_p \text{ for some } p \\ e_i \wedge e_{j_1} \wedge \cdots \wedge e_{j_t} & \text{if } i < j_1 \\ (-1)^p e_{j_1} \wedge \cdots \wedge e_{j_p} \wedge e_i \wedge \cdots \wedge e_{j_t} & \text{if } j_p < i < j_{p+1} \\ (-1)^t e_{j_1} \wedge \cdots \wedge e_{j_t} \wedge e_i & \text{if } j_t < i. \end{cases}$$

This extends (by induction on s) to a multiplication $\bigwedge^s R^n \times \bigwedge^t R^n \to \bigwedge^{s+t} R^n$ using the following formula when $i_1 < \ldots < i_s$ and $1 \leqslant j_1 < \cdots < j_t \leqslant n$:

$$(e_{i_1} \wedge \cdots \wedge e_{i_s}) \wedge (e_{j_1} \wedge \cdots \wedge e_{j_t}) := e_{i_1} \wedge [(e_{i_2} \wedge \cdots \wedge e_{i_s}) \wedge (e_{j_1} \wedge \cdots \wedge e_{j_t})].$$

This multiplication is denoted as $(\alpha, \beta) \mapsto \alpha \wedge \beta$. When $s = 0$, since $\bigwedge^0 R^n = R$, the usual scalar multiplication $R \times \bigwedge^t R^n \to \bigwedge^t R^n$ describes the rule for multiplication $\bigwedge^0 R^n \times \bigwedge^t R^n \to \bigwedge^t R^n$, and similarly when $t = 0$. This further extends to a well-defined multiplication on $\bigwedge R^n := \bigoplus_i \bigwedge^i R^n$.

Remark 4.25. According to Definition 4.24, for $0 \in \bigwedge^s R^n$ and $\beta \in \bigwedge^t R^n$, we have $0 \wedge \beta = 0 = \beta \wedge 0$.

Example 4.26. We compute a few products in $\bigwedge R^4$:

$$\begin{aligned}
(e_1 \wedge e_2) \wedge (e_3 \wedge e_4) &= e_1 \wedge e_2 \wedge e_3 \wedge e_4 \\
(e_1 \wedge e_2) \wedge (e_2 \wedge e_3) &= 0 \\
(e_1 \wedge e_3) \wedge (e_2 \wedge e_4) &= e_1 \wedge [e_3 \wedge (e_2 \wedge e_4)] \\
&= e_1 \wedge [-e_2 \wedge e_3 \wedge e_4] \\
&= -e_1 \wedge e_2 \wedge e_3 \wedge e_4.
\end{aligned}$$

Exercise 4.27. Write out multiplication tables (for basis vectors only) for $\bigwedge R^n$ with $n = 1, 2, 3$.

Definition 4.24 suggests the next notation, which facilitates many computations. The subsequent exercises exemplify this, culminating in the important Exercise 4.33

Definition 4.28. Let $n \in \mathbb{N}$, let $j_1, \dots, j_t \in \{1, \dots, n\}$, and let $e_1, \dots, e_n \in R^n$ be a basis. Since multiplication of basis vectors in $\bigwedge R^n$ is defined inductively, the following element (also defined inductively) is well defined.

$$e_{j_1} \wedge \cdots \wedge e_{j_t} := \begin{cases} 0 & \text{if } j_p = j_q \text{ for some } p \neq q \\ e_{j_1} \wedge (e_{j_2} \wedge \cdots \wedge e_{j_t}) & \text{if } j_p \neq j_q \text{ for all } p \neq q \end{cases}$$

Exercise 4.29. Let $n \in \mathbb{N}$, and let $e_1, \dots, e_n \in R^n$ be a basis. Let $j_1, \dots, j_t \in \{1, \dots, n\}$ such that $j_p \neq j_q$ for all $p \neq q$, and let $\iota \in S_n$ such that ι fixes all elements of $\{1, \dots, n\} \smallsetminus \{j_1, \dots, j_t\}$. Prove that

$$e_{\iota(j_1)} \wedge \cdots \wedge e_{\iota(j_t)} = \operatorname{sgn}(\iota) e_{j_1} \wedge \cdots \wedge e_{j_t}$$

Exercise 4.30. Let $n \in \mathbb{N}$, and let $e_1, \dots, e_n \in R^n$ be a basis. Prove that for all $i_1, \dots, i_s, j_1, \dots, j_t \in \{1, \dots, n\}$ we have

$$(e_{i_1} \wedge \cdots \wedge e_{i_s}) \wedge (e_{j_1} \wedge \cdots \wedge e_{j_t}) = e_{i_1} \wedge \cdots \wedge e_{i_s} \wedge e_{j_1} \wedge \cdots \wedge e_{j_t}$$

and that this is 0 if there is a repetition in the list $i_1, \dots, i_s, j_1, \dots, j_t$. (The points here are the order on the subscripts matter in Definition 4.24 and do not matter in Definition 4.28, so one needs to make sure that the signs that occur from Definition 4.24 agree with those from Definition 4.28.)

Exercise 4.31. Let $n \in \mathbb{N}$ and let $e_1, \ldots, e_n \in R^n$ be a basis. Prove that the multiplication from Definition 4.24 makes $\bigwedge R^n$ into a graded commutative R-algebra:

(a) multiplication in $\bigwedge R^n$ is associative, distributive, and unital;
(b) for elements $\alpha \in \bigwedge^s R^n$ and $\beta \in \bigwedge^t R^n$, we have $\alpha \wedge \beta = (-1)^{st} \beta \wedge \alpha$;
(c) for $\alpha \in \bigwedge^s R^n$, if s is odd, then $\alpha \wedge \alpha = 0$; and
(d) the composition $R \xrightarrow{\cong} \bigwedge^0 R^n \xrightarrow{\subseteq} \bigwedge R^n$ is a ring homomorphism, the image of which is contained in the center of $\bigwedge R^n$.

Hint: The distributive law holds essentially by definition. For the other properties in (a) and (b), prove the desired formula for basis vectors, then verify it for general elements using distributivity.

Exercise 4.32. Let $n \in \mathbb{N}$ and let $e_1, \ldots, e_n \in R^n$ be a basis. Let $\mathbf{x} = x_1, \ldots, x_n \in R$, and let $j_1, \ldots, j_t \in \{1, \ldots, n\}$. Prove that the element $e_{j_1} \wedge \cdots \wedge e_{j_t}$ from Definition 4.28 satisfies the following formula:

$$\partial_i^{\tilde{K}^R(\mathbf{x})}(e_{j_1} \wedge \cdots \wedge e_{j_t}) = \sum_{s=1}^{t} (-1)^{s+1} x_{j_s} e_{j_1} \wedge \cdots \wedge \widehat{e_{j_s}} \wedge \cdots \wedge e_{j_t}.$$

(Note that if $j_1 < \cdots < j_t$, then this is the definition of $\partial_i^{\tilde{K}^R(\mathbf{x})}(e_{j_1} \wedge \cdots \wedge e_{j_t})$. However, we are not assuming that $j_1 < \cdots < j_t$.)

Exercise 4.33. Let $n \in \mathbb{N}$ and let $e_1, \ldots, e_n \in R^n$ be a basis. Let $\mathbf{x} = x_1, \ldots, x_n \in R$. Prove that the multiplication from Definition 4.24 satisfies the "Leibniz rule": for elements $\alpha \in \bigwedge^s R^n$ and $\beta \in \bigwedge^t R^n$, we have

$$\partial_{s+t}^{\tilde{K}^R(\mathbf{x})}(\alpha \wedge \beta) = \partial_s^{\tilde{K}^R(\mathbf{x})}(\alpha) \wedge \beta + (-1)^s \alpha \wedge \partial_t^{\tilde{K}^R(\mathbf{x})}(\beta).$$

Hint: Prove the formula for basis vectors and verify it for general elements using distributivity and linearity.

5 DG Algebras and DG Modules I

This section introduces the main tools for the proof of Theorem 1.4. This proof is outlined in 5.32. From the point of view of this proof, the first important example of a DG algebra is the Koszul complex; see Example 5.3. However, the proof showcases another important example, namely the DG algebra resolution of Definition 5.29.

5.1 DG Algebras

The first change of perspective required for the proof of Theorem 1.4 is the change from rings to DG algebras.

Definition 5.1. A *commutative differential graded algebra over* R (*DG* R*-algebra* for short) is an R-complex A equipped with a binary operation $A \times A \to A$, written as $(a, b) \mapsto ab$ and called the *product* on A, satisfying the following properties[4]:

- associative: for all $a, b, c \in A$ we have $(ab)c = a(bc)$;
- distributive: for all $a, b, c \in A$ such that $|a| = |b|$ we have $(a+b)c = ac + bc$ and $c(a+b) = ca + cb$;
- unital: there is an element $1_A \in A_0$ such that for all $a \in A$ we have $1_A a = a$;
- graded commutative: for all $a, b \in A$ we have $ba = (-1)^{|a||b|}ab \in A_{|a|+|b|}$, and $a^2 = 0$ when $|a|$ is odd;
- positively graded: $A_i = 0$ for $i < 0$; and
- Leibniz Rule: for all $a, b \in A$ we have

$$\partial^A_{|a|+|b|}(ab) = \partial^A_{|a|}(a)b + (-1)^{|a|} a \partial^A_{|b|}(b).$$

Given a DG R-algebra A, the *underlying algebra* is the graded commutative R-algebra $A^\natural = \bigoplus_{i=0}^{\infty} A_i$. When R is a field and $\operatorname{rank}_R(\bigoplus_{i\geqslant 0} A_i) < \infty$, we say that A is *finite-dimensional* over R.

It should be helpful for the reader to keep the next two examples in mind for the remainder of these notes.

Example 5.2. The ring R, considered as a complex concentrated in degree 0, is a DG R-algebra such that $R^\natural = R$.

Example 5.3. Given a sequence $\mathbf{x} = x_1, \cdots, x_n \in R$, the Koszul complex $K = K^R(\mathbf{x})$ is a DG R-algebra such that $K^\natural = \bigwedge R^n$; see Exercises 4.31 and 4.33. In particular, if $n = 1$, then $K^\natural \cong R[X]/(X^2)$ where $|X| = 1$.

The following exercise is a routine interpretation of the Leibniz Rule. It is also an important foreshadowing of the final part of the proof of Theorem 1.4 which is given in Sect. 8. See also Exercise 5.15.

Exercise 5.4. Prove that if A is a DG R-algebra, then there is a well-defined chain map $\mu^A \colon A \otimes_R A \to A$ given by $\mu^A(a \otimes b) = ab$, and that A_0 is an R-algebra. (Interested readers may wish to formulate and prove the converse to this statement also.)

The next notion will allow us to transfer information from one DG algebra to another as in the arguments for $R \to \hat{R}$ and $R \to \overline{R}$ described in Sect. 1.

---

[4]We assume that readers at this level are familiar with associative laws and the like. However, given that the DG universe is riddled with sign conventions, we explicitly state these laws for the sake of clarity.

Definition 5.5. A *morphism* of DG R-algebras is a chain map $f: A \to B$ between DG R-algebras respecting products and multiplicative identities: $f(aa') = f(a)f(a')$ and $f(1_A) = 1_B$. A morphism of DG R-algebras that is also a quasiisomorphism is a *quasiisomorphism of DG R-algebras*.

Part (a) of the next exercise contains the first morphism of DG algebras that we use in the proof of Theorem 1.4.

Exercise 5.6. Let A be a DG R-algebra.

(a) Prove that the map $R \to A$ given by $r \mapsto r \cdot 1_A$ is a morphism of DG R-algebras. As a special case, given a sequence $\mathbf{x} = x_1, \cdots, x_n \in R$, the natural map $R \to K = K^R(\mathbf{x})$ given by $r \mapsto r \cdot 1_K$ is a morphism of DG R-algebras.
(b) Prove that the natural inclusion map $A_0 \to A$ is a morphism of DG R-algebras.
(c) Prove that the natural map $K \to R/(\mathbf{x})$ is a morphism of DG R-algebras that is a quasiisomorphism if $\mathbf{x}$ is R-regular; see Exercise 3.10 and Lemma 4.18.

Part (b) of the next exercise is needed for the subsequent definition.

Exercise 5.7. Let A be a DG R-algebra.

(a) Prove that the condition $A_{-1} = 0$ implies that A_0 surjects onto $\mathrm{H}_0(A)$ and that $\mathrm{H}_0(A)$ is an A_0-algebra.
(b) Prove that A_i is an A_0-module, and $\mathrm{H}_i(A)$ is an $\mathrm{H}_0(A)$-module for each i.

Remark 5.8. Let A be a DG R-algebra. Then the subset $Z(A) := \bigoplus_{i=0}^{\infty} \mathrm{Ker}(\partial_i^A)$ is a graded R-algebra. Also, the submodule $B(A) := \bigoplus_{i=0}^{\infty} \mathrm{Im}(\partial_{i+1}^A)$ is a graded ideal of $Z(A)$, so the quotient $\mathrm{H}(A) = B(A)/Z(A) = \bigoplus_{i=0}^{\infty} \mathrm{H}_i(A)$ is a graded R-algebra.

Definition 5.9. Let A be a DG R-algebra. We say that A is *noetherian* if $\mathrm{H}_0(A)$ is noetherian and $\mathrm{H}_i(A)$ is finitely generated over $\mathrm{H}_0(A)$ for all $i \geqslant 0$.

Exercise 5.10. Given a sequence $\mathbf{x} = x_1, \cdots, x_n \in R$, prove that the Koszul complex $K^R(\mathbf{x})$ is a noetherian DG R-algebra. Moreover, prove that any DG R-algebra A such that each A_i is finitely generated over R is noetherian.

5.2 DG Modules

In the passage from rings to DG algebras, modules and complexes change to DG modules, which we describe next.

Definition 5.11. Let A be a DG R-algebra, and let i be an integer. A *differential graded module over A* (*DG A-module* for short) is an R-complex M equipped with a binary operation $A \times M \to M$, written as $(a, m) \mapsto am$ and called the *scalar multiplication* of A on M, satisfying the following properties:

- associative: for all $a, b \in A$ and $m \in M$ we have $(ab)m = a(bm)$;
- distributive: for all $a, b \in A$ and $m, n \in M$ such that $|a| = |b|$ and $|m| = |n|$, we have $(a + b)m = am + bm$ and $a(m + n) = am + an$;

- unital: for all $m \in M$ we have $1_A m = m$;
- graded: for all $a \in A$ and $m \in M$ we have $am \in M_{|a|+|m|}$;
- Leibniz Rule: for all $a \in A$ and $m \in M$ we have

$$\partial^A_{|a|+|m|}(am) = \partial^A_{|a|}(a)m + (-1)^{|a|}a\partial^M_{|m|}(m).$$

The *underlying* $A^\natural$*-module* associated with M is the $A^\natural$-module $M^\natural = \bigoplus_{j=-\infty}^{\infty} M_j$.

The ith *suspension* of a DG A-module M is the DG A-module $\Sigma^i M$ defined by $(\Sigma^i M)_n := M_{n-i}$ and $\partial_n^{\Sigma^i M} := (-1)^i \partial^M_{n-i}$. The scalar multiplication on $\Sigma^i M$ is defined by the formula $a * m := (-1)^{i|a|}am$. The notation ΣM is short for $\Sigma^1 M$.

The next exercise contains examples that should be helpful to keep in mind.

Exercise 5.12. (a) Prove that DG R-module is just an R-complex.
(b) Given a DG R-algebra A, prove that the complex A is a DG A-module where the scalar multiplication is just the internal multiplication on A.
(c) Given a morphism $A \to B$ of DG R-algebras, prove that every DG B-module is a DG A-module by restriction of scalars. As a special case, given a sequence $\mathbf{x} = x_1, \cdots, x_n \in R$, every $R/(\mathbf{x})$-complex is a DG $K^R(\mathbf{x})$-module; see Exercise 5.6.

The operation $X \mapsto A \otimes_R X$ described next is "base change," which is crucial for our passage between DG algebras in the proof of Theorem 1.4.

Exercise 5.13. Let $\mathbf{x} = x_1, \cdots, x_n \in R$, and set $K = K^R(\mathbf{x})$. Given an R-module M, prove that the complex $K \otimes_R M$ is a DG K-module via the multiplication $a(b\otimes m) := (ab)\otimes m$. More generally, given an R-complex X and a DG R-algebra A, prove that the complex $A \otimes_R X$ is a DG A-module via the multiplication $a(b \otimes x) := (ab) \otimes x$.

Exercise 5.14. Let A be a DG R-algebra, let M be a DG A-module, and let $i \in \mathbb{Z}$. Prove that $\Sigma^i M$ is a DG A-module.

The next exercise further foreshadows important aspects of Sect. 8.

Exercise 5.15. Let A be a DG R-algebra, and let M be a DG A-module. Prove that there is a well-defined chain map $\mu^M : A \otimes_R M \to M$ given by $\mu^M(a \otimes m) = am$.

We consider the following example throughout these notes. It is simple but demonstrates our constructions. And even it has some nontrivial surprises.

Example 5.16. We consider the trivial Koszul complex $U = K^R(0)$:

$$U = \quad 0 \to Re \xrightarrow{0} R1 \to 0.$$

The notation indicates that we are using the basis $e \in U_1$ and $1 = 1_U \in U_0$.

Exercise 5.13 shows that R is a DG U-module. Another example is the following, again with specified basis in each degree:

$$G = \cdots \xrightarrow{1} Re_3 \xrightarrow{0} R1_2 \xrightarrow{1} Re_1 \xrightarrow{0} R1_0 \to 0.$$

The notation for the bases is chosen to help remember the DG U-module structure:

$$1 \cdot 1_{2n} = 1_{2n} \qquad 1 \cdot e_{2n+1} = e_{2n+1}$$

$$e \cdot 1_{2n} = e_{2n+1} \qquad e \cdot e_{2n+1} = 0.$$

One checks directly that G satisfies the axioms to be a DG U-module. It is worth noting that $\mathrm{H}_0(G) \cong R$ and $\mathrm{H}_i(G) = 0$ for all $i \neq 0$.[5]

We continue with Example 5.16, but working over a field F instead of R.

Example 5.17. We consider the trivial Koszul complex $U = K^F(0)$:

$$U = \quad 0 \to Fe \xrightarrow{0} F1 \to 0.$$

Consider the graded vector space $W = \bigoplus_{i \in \mathbb{Z}} W_i$, where $W_0 = F\eta_0 \cong F$ with basis η_0 and $W_i = 0$ for $i \neq 0$:

$$W = \quad 0 \bigoplus F\nu_0 \bigoplus 0.$$

We are interested in identifying all the possible DG U-module structures on W, that is, all possible differentials

$$0 \to F\nu_0 \to 0$$

and rules for scalar multiplication making this into a DG U-module. See Sect. 8 for more about this.

The given vector space W has exactly one DG U-module structure. To see this, first note that we have no choice for the differential since it maps $W_i \to W_{i-1}$ and at least one of these modules is 0; hence, $\partial_i = 0$ for all i. Also, we have no choice for the scalar multiplication: multiplication by $1 = 1_U$ must be the identity, and multiplication by e maps $W_i \to W_{i+1}$ and at least one of these modules is 0. (This example is trivial, but it will be helpful later.)

Similarly, we consider the graded vector space

$$W' = \quad 0 \bigoplus F\eta_1 \bigoplus F\eta_0 \bigoplus 0.$$

[5]For perspective, G is modeled on the free resolution $\cdots \xrightarrow{e} R[e]/(e^2) \xrightarrow{e} R[e]/(e^2) \to 0$ of R over the ring $R[e]/(e^2)$.

This vector space allows for one possibly nontrivial differential

$$\partial_1' \in \operatorname{Hom}_F(F\eta_1, F\eta_0) \cong F.$$

So, in order to make W' into an R-complex, we need to choose an element $x_1 \in F$:

$$(W', x_1) = \quad 0 \to F\eta_1 \xrightarrow{x_1} F\eta_0 \to 0.$$

To be explicit, this means that $\partial_1'(\eta_1) = x_1\eta_0$, and hence $\partial_1'(r\eta_1) = x_1 r\eta_0$ for all $r \in F$. Since W' is concentrated in degrees 0 and 1, this is an R-complex.

For the scalar multiplication of U on the complex (W', x_1), again multiplication by 1 must be the identity, but multiplication by e has one nontrivial option

$$\mu_0' \in \operatorname{Hom}_F(F\eta_0, F\eta_1) \cong F$$

which we identify with an element $x_0 \in F$. To be explicit, this means that $e\eta_0 = x_0\eta_1$, and hence $er\eta_0 = x_0 r\eta_1$ for all $r \in F$.

For the Leibniz Rule to be satisfied, we must have

$$\partial_{i+1}'(e \cdot \eta_i) = \partial_1^U(e) \cdot \eta_i + (-1)^{|e|} e \cdot \partial_i'(\eta_i)$$

for $i = 0, 1$. We begin with $i = 0$:

$$\begin{aligned}
\partial_1'(e \cdot \eta_0) &= \partial_1^U(e) \cdot \eta_0 + (-1)^{|e|} e \cdot \partial_0'(\eta_0) \\
\partial_1'(x_0\eta_1) &= 0 \cdot \eta_0 - e \cdot 0 \\
x_0\partial_1'(\eta_1) &= 0 \\
x_0x_1\eta_0 &= 0
\end{aligned}$$

so we have $x_0x_1 = 0$, that is, either $x_0 = 0$ or $x_1 = 0$. For $i = 1$, we have

$$\begin{aligned}
\partial_2'(e \cdot \eta_1) &= \partial_1^U(e) \cdot \eta_1 + (-1)^{|e|} e \cdot \partial_1'(\eta_1) \\
0 &= 0 \cdot \eta_1 - e \cdot (x_1\eta_0) \\
0 &= -x_1 e \cdot \eta_0 \\
0 &= -x_1x_0\eta_1
\end{aligned}$$

so we again conclude that $x_0 = 0$ or $x_1 = 0$. One can check the axioms from Definition 5.11 to see that either of these choices gives rise to a DG U-module structure on W'. In other words, the DG U-module structures on W' are parameterized by the following algebraic subset of $F^2 = \mathbb{A}_F^2$

$$\{(x_0, x_1) \in \mathbb{A}_F^2 \mid x_0x_1 = 0\} = V(x_0) \cup V(x_1)$$

which is the union of the two coordinate axes in $\mathbb{A}^2_F$. This is one of the fundamental points of Sect. 8, that DG module structures on a fixed finite-dimensional graded vector space are parametrized by algebraic varieties.

Homologically finite DG modules, defined next, take the place of finitely generated modules in our passage to the DG universe.

Definition 5.18. Let A be a DG R-algebra. A DG A-module M is *bounded below* if $M_n = 0$ for all $n \ll 0$; and it is *homologically finite* if each $\mathrm{H}_0(A)$-module $\mathrm{H}_n(M)$ is finitely generated and $\mathrm{H}_n(M) = 0$ for $|n| \gg 0$.

Example 5.19. In Exercise 5.13, the DG K-module $R/(\mathbf{x})$ is bounded below and homologically finite. In Example 5.16, the DG U-modules R and G are bounded below and homologically finite. In Example 5.17, the DG U-module structures on W and W' are bounded below and homologically finite.

5.3 Morphisms of DG Modules

In the passage from modules and complexes to DG modules, homomorphisms and chain maps are replaced with morphisms.

Definition 5.20. A *morphism* of DG A-modules is a chain map $f: M \to N$ between DG A-modules that respects scalar multiplication: $f(am) = af(m)$. Isomorphisms in the category of DG A-modules are identified by the symbol $\cong$. A *quasiisomorphism* of DG A-modules is a morphism $M \to N$ such that each induced map $\mathrm{H}_i(M) \to \mathrm{H}_i(N)$ is an isomorphism, i.e., a morphism of DG A-modules that is a quasiisomorphism of R-complexes; these are identified by the symbol $\simeq$.

Remark 5.21. A morphism of DG R-modules is simply a chain map, and a quasiisomorphism of DG R-modules is simply a quasiisomorphism in the sense of Definition 3.8. Given a DG R-algebra A, a morphism of DG A-modules is an isomorphism if and only if it is injective and surjective.

The next exercise indicates that base change is a functor.

Exercise 5.22. Let $\mathbf{x} = x_1, \cdots, x_n \in R$, and set $K = K^R(\mathbf{x})$. (See Exercise 5.13.)

(a) Given an R-linear map $f: M \to N$, prove that the chain map $K \otimes_R f: K \otimes_R M \to K \otimes_R N$ is a morphism of DG K-modules. More generally, given a chain map of R-complexes $g: X \to Y$ and a DG R-algebra A, prove that the chain map $A \otimes_R g: A \otimes_R X \to A \otimes_R Y$ is a morphism of DG A-modules.

(b) Give an example showing that if g is a quasiisomorphism, then $A \otimes_R g$ need not be a quasiisomorphism. (Note that if A_i is R-projective for each i (e.g., if $A = K$), then g being a quasiisomorphism implies that $A \otimes_R g$ is a quasiisomorphism by Fact 4.9.)

(c) Prove that the natural map $K \to R/(\mathbf{x})$ is a morphism of DG K-modules. More generally, prove that every morphism $A \to B$ of DG R-algebras is a morphism of DG A-modules, where B is a DG A-module via restriction of scalars.

Next, we use our running example to provide some morphisms of DG modules.

Example 5.23. We continue with the notation of Example 5.16.

Let $f: G \to \Sigma R$ be a morphism of DG U-modules:

$$\begin{array}{ccccccccccc}
G = & \cdots & \xrightarrow{1} & Re_3 & \xrightarrow{0} & R1_2 & \xrightarrow{1} & Re_1 & \xrightarrow{0} & R1_0 & \longrightarrow 0 \\
{\scriptstyle f}\downarrow & & & & & \downarrow & & \downarrow{\scriptstyle f_1} & & \downarrow & \\
\Sigma R & & & & & 0 & \longrightarrow & R & \longrightarrow & 0. &
\end{array}$$

Commutativity of the first square shows that $f = 0$. One can also see this from the following computation:

$$f_1(e_1) = f_1(e \cdot 1_0) = ef_0(1_0) = 0.$$

That is, the only morphism of DG U-modules $G \to \Sigma R$ is the zero-morphism. The same conclusion holds for any morphism $G \to \Sigma^{2n+1} R$ with $n \in \mathbb{Z}$.

On the other hand, for each $n \in \mathbb{N}$, every element $r \in R$ determines a morphism $g^{r,n}: G \to \Sigma^{2n} R$, via multiplication by r. For instance in the case $n = 1$:

$$\begin{array}{ccccccccccc}
G = & \cdots & \xrightarrow{1} & Re_3 & \xrightarrow{0} & R1_2 & \xrightarrow{1} & Re_1 & \xrightarrow{0} & R1_0 & \longrightarrow 0 \\
{\scriptstyle g^{r,1}}\downarrow & & & \downarrow & & {\scriptstyle g_2^{r,1}}\downarrow{\scriptstyle =r\cdot} & & \downarrow & & & \\
\Sigma^2 R = & & & 0 & \longrightarrow & R & \longrightarrow & 0. & & &
\end{array}$$

Each square commutes, and the linearity condition is from the next computations:

$$\begin{aligned}
g_2^{r,1}(1 \cdot 1_2) &= g_2^{r,1}(1_2) = r = 1 \cdot r = 1 \cdot g_2^{r,1}(1_2) \\
g_2^{r,1}(e \cdot e_1) &= g_2^{r,1}(0) = 0 = e \cdot 0 = e \cdot g_1^{r,1}(e_1) \\
g_3^{r,1}(e \cdot 1_2) &= g_3^{r,1}(e_3) = 0 = e \cdot r = e \cdot g_2^{r,1}(1_2).
\end{aligned}$$

Further, the isomorphism $\operatorname{Hom}_R(R, R) \cong R$ shows that each morphism $G \to \Sigma^{2n} R$ is of the form $g^{r,n}$. Also, one checks readily that the map $G \xrightarrow{g^{u,0}} R$ is a quasiisomorphism for each unit u of R.

5.4 Truncations of DG Modules

The next operation allows us to swap a given DG module with a "shorter" one; see Exercise 5.27(b).

Definition 5.24. Let A be a DG R-algebra, and let M be a DG A-module. The *supremum* of M is

$$\sup(M) := \sup\{i \in \mathbb{Z} \mid \mathrm{H}_i(M) \neq 0\}.$$

Given an integer n, the *nth soft left truncation of M* is the complex

$$\tau(M)_{(\leqslant n)} := \quad 0 \to M_n / \operatorname{Im}(\partial^M_{n+1}) \to M_{n-1} \to M_{n-2} \to \cdots$$

with differential induced by ∂^M.

Example 5.25. We continue with the notation of Example 5.16. For each $n \geqslant 1$, set $i = \lfloor n/2 \rfloor$ and prove that

$$\tau(G)_{(\leqslant n)} = \quad 0 \to R1_{2i} \xrightarrow{1} Re_{2i-1} \xrightarrow{0} \cdots \xrightarrow{1} Re_3 \xrightarrow{0} R1_2 \xrightarrow{1} Re_1 \xrightarrow{0} R1_0 \to 0.$$

Remark 5.26. Let P be a projective resolution of an R-module M, with P^+ denoting the augmented resolution as in Example 3.2. Then one has $\tau(P)_{(\leqslant 0)} \cong M$. In particular, P is a projective resolution of $\tau(P)_{(\leqslant 0)}$.

Exercise 5.27. Let A be a DG R-algebra, let M be a DG A-module, and let $n \in \mathbb{Z}$.

(a) Prove that the truncation $\tau(M)_{(\leqslant n)}$ is a DG A-module with the induced scalar multiplication, and the natural chain map $M \to \tau(M)_{(\leqslant n)}$ is a morphism of DG A-modules.
(b) Prove that the morphism from part (a) is a quasiisomorphism if and only if $n \geqslant \sup(M)$.

5.5 DG Algebra Resolutions

The following fact provides the final construction needed to give an initial sketch of the proof of Theorem 1.4.

Fact 5.28. Let $Q \to R$ be a ring epimorphism. Then there is a quasiisomorphism $A \xrightarrow{\simeq} R$ of DG Q-algebras such that each A_i is finitely generated and projective over R and $A_i = 0$ for $i > \mathrm{pd}_Q(R)$. See, e.g., [4, Proposition 2.2.8].

Definition 5.29. In Fact 5.28, the quasiisomorphism $A \xrightarrow{\simeq} R$ is a *DG algebra resolution* of R over Q.

Remark 5.30. When $\mathbf{y} \in Q$ is a Q-regular sequence, the Koszul complex $K^Q(\mathbf{y})$ is a DG algebra resolution of $Q/(\mathbf{y})$ over Q by Lemma 4.18 and Example 5.3. Section 6 contains other classical examples.

Example 5.31. Let Q be a ring, and consider an ideal $I \subsetneq Q$. Assume that the quotient $R := Q/I$ has $\mathrm{pd}_Q(R) \leqslant 1$. Then every projective resolution of R over Q of the form $A = (0 \to A_1 \to Q \to 0)$ has the structure of a DG algebra resolution.

We conclude this section with the beginning of the proof of Theorem 1.4. The rest of the proof is contained in 7.38 and 8.17.

5.32 (First Part of the Proof of Theorem 1.4). There is a flat local ring homomorphism $R \to R'$ such that R' is complete with algebraically closed residue field, as in the proof of Theorem 2.13. Since there is a 1-1 function $\mathfrak{S}_0(R) \hookrightarrow \mathfrak{S}_0(R')$ by Fact 2.7, we can replace R with R' and assume without loss of generality that R is complete with algebraically closed residue field.

Since R is complete and local, Cohen's structure theorem provides a ring epimorphism $\tau\colon (Q, \mathfrak{n}, k) \to (R, \mathfrak{m}, k)$ where Q is a complete regular local ring such that $\mathfrak{m}$ and $\mathfrak{n}$ have the same minimal number of generators. Let $\mathbf{y} = y_1, \ldots, y_n \in \mathfrak{n}$ be a minimal generating sequence for $\mathfrak{n}$, and set $\mathbf{x} = x_1, \ldots, x_n \in \mathfrak{m}$ where $x_i := \tau(y_i)$. It follows that we have $K^R(\mathbf{x}) \cong K^Q(\mathbf{y}) \otimes_Q R$. Since Q is regular and $\mathbf{y}$ is a minimal generating sequence for $\mathfrak{n}$, the Koszul complex $K^Q(\mathbf{y})$ is a minimal Q-free resolution of k by Lemma 4.18.

Fact 5.28 provides a DG algebra resolution $A \xrightarrow{\simeq} R$ of R over Q. Note that $\mathrm{pd}_Q(R) < \infty$ since Q is regular. We consider the following diagram of morphisms of DG Q-algebras:

$$R \to K^R(\mathbf{x}) \cong K^Q(\mathbf{y}) \otimes_Q R \xleftarrow{\simeq} K^Q(\mathbf{y}) \otimes_Q A \xrightarrow{\simeq} k \otimes_Q A =: U. \tag{1}$$

The first map is from Exercise 5.6. The isomorphism is from the previous paragraph. The first quasiisomorphism comes from an application of $K^Q(\mathbf{y}) \otimes_Q -$ to the quasiisomorphism $R \xleftarrow{\simeq} A$, using Fact 4.9. The second quasiisomorphism comes from an application of $- \otimes_Q A$ to the quasiisomorphism $K^Q(\mathbf{y}) \xrightarrow{\simeq} k$. Note that $k \otimes_Q A$ is a finite dimensional DG k-algebra because of the assumptions on A.

We show in 7.38 below how this provides a diagram

$$\mathfrak{S}_0(R) \hookrightarrow \mathfrak{S}(R) \xrightarrow{\equiv} \mathfrak{S}(K^R(\mathbf{x})) \xleftarrow{\equiv} \mathfrak{S}(K^Q(\mathbf{y}) \otimes_Q A) \xrightarrow{\equiv} \mathfrak{S}(U) \tag{2}$$

where $\equiv$ identifies bijections of sets. We then show in 8.17 that $\mathfrak{S}(U)$ is finite, and it follows that $\mathfrak{S}_0(R)$ is finite, as desired.

6 Examples of Algebra Resolutions

Remark 5.30 and Example 5.31 provide constructions of DG algebra resolutions, in particular, for rings of projective dimension at most 1. The point of this section is to extend this to rings of projective dimension 2 and rings of projective dimension 3 determined by Gorenstein ideals.

Definition 6.1. Let I be an ideal of a local ring $(R, \mathfrak{m})$. The *grade* of I in R, denoted $\operatorname{grade}_R(I)$, is defined to be the length of the longest regular sequence of R contained in I. Equivalently, we have

$$\operatorname{grade}_R(I) := \min\left\{i \mid \operatorname{Ext}^i_R(R/I, R) \neq 0\right\}$$

and it follows that $\operatorname{grade}_R(I) \leqslant \operatorname{pd}_R(R/I)$. We say that I is *perfect* if $\operatorname{grade}_R(I) = \operatorname{pd}_R(R/I) < \infty$. In this case, $\operatorname{Ext}^i_R(R/I, R)$ is nonvanishing precisely when $i = \operatorname{pd}_R(R/I)$. If, in addition, this single nonvanishing cohomology module is isomorphic to R/I, then I is said to be *Gorenstein*.

Notation 6.2. Let A be a matrix[6] over R and $J, K \subset \mathbb{N}$. The submatrix of A obtained by deleting columns indexed by J and rows indexed by K is denoted A^J_K. We abbreviate $A^{\emptyset}_{\{i\}}$ as A_i, and so on. Let $I_n(A)$ be the ideal of R generated by the "$n \times n$ minors" of A, that is, the determinants of the $n \times n$ matrices of the form A^J_K.

6.1 Resolutions of Length Two

The following result, known as the Hilbert–Burch Theorem, provides a characterization of perfect ideals of grade two. It was first proven by Hilbert in 1890 in the case that R is a polynomial ring [27]; the more general statement was proven by Burch in 1968 [13, Theorem 5].

Theorem 6.3 ([13, 27]). *Let I be an ideal of the local ring $(R, \mathfrak{m})$.*

(a) *If* $\operatorname{pd}_R(R/I) = 2$, *then*

(1) *there is a non-zerodivisor $a \in R$ such that R/I has a projective resolution of the form $0 \to R^n \xrightarrow{A} R^{n+1} \xrightarrow{B} R \to 0$ where B is the $1 \times (n+1)$ matrix with ith column given by $(-1)^{i-1} a \det(A_i)$,*

(2) *one has $I = aI_n(A)$, and*

(3) *the ideal $I_n(A)$ is perfect of grade 2.*

(b) *Conversely, if A is an $(n+1) \times n$ matrix over R such that* $\operatorname{grade}(I_n(A)) \geqslant 2$, *then $R/I_n(A)$ has a projective resolution of the form $0 \to R^n \xrightarrow{A} R^{n+1} \xrightarrow{B} R \to 0$ where B is the $1 \times (n+1)$ matrix with ith column given by $(-1)^{i-1} \det(A_i)$.*

Exercise 6.4. Use Theorem 6.3 to build a grade two perfect ideal I in $R = k[\![x, y]\!]$.

Herzog [26] showed that Hilbert–Burch resolutions can be endowed with DG algebra structures.

---

[6]While much of the work in this section can be done basis-free, the formulations are somewhat more transparent when bases are specified and matrices are used to represent homomorphisms.

Theorem 6.5 ([26]). *Given an* $(n+1)\times n$ *matrix* A *over* R*, let* B *be the* $1\times(n+1)$ *matrix with* i*th column given by* $(-1)^{i-1}a\det(A_i)$ *for some non-zerodivisor* $a\in R$*. Then the* R*-complex*

$$0\to\bigoplus_{\ell=1}^{n}Rf_\ell\xrightarrow{A}\bigoplus_{\ell=1}^{n+1}Re_\ell\xrightarrow{B}R1\to 0$$

has the structure of a DG R*-algebra with the following multiplication relations:*

(1) $e_i^2=0=f_jf_k$ *and* $e_if_j=0=f_je_i$ *for all* i,j,k*, and*

(2) $e_ie_j=-e_je_i=a\sum_{k=1}^{n}(-1)^{i+j+k}\det(A_{i,j}^{k})f_k$ *for all* $1\leqslant i<j\leqslant n+1$.[7]

Exercise 6.6. Verify the Leibniz rule for the product defined in Theorem 6.5.

Exercise 6.7. Using the ideal I from Exercise 6.4, build a (minimal) free resolution for R/I, then specify the relations giving this resolution a DG R-algebra structure.

6.2 Resolutions of Length Three

We turn our attention to resolutions of length three, first recalling needed machinery.

Definition 6.8. A square matrix A over R is *alternating* if it is skew-symmetric and has all 0's on its diagonal. Let A be an $n\times n$ alternating matrix over R. If n is even, then there is an element $\mathrm{Pf}(A)\in R$ such that $\mathrm{Pf}(A)^2=\det(A)$. If n is odd, then $\det(A)=0$, so we set $\mathrm{Pf}(A)=0$. The element $\mathrm{Pf}(A)$ is called the *Pfaffian* of A. (See, e.g., [11, Sect. 3.4] for more details.) We denote by $\mathrm{Pf}_{n-1}(A)$ the ideal of R generated by the submaximal Pfaffians of A, that is,

$$\mathrm{Pf}_{n-1}(A):=\big(\mathrm{Pf}(A_i^i)\mid 1\leqslant i\leqslant n\big)R.$$

Example 6.9. Let $x,y,z\in R$. For the matrix $A=\begin{bmatrix}0&x\\-x&0\end{bmatrix}$, we have $\det(A)=x^2$, so $\mathrm{Pf}(A)=x$, and $\mathrm{Pf}_1(A)=0$.

For the matrix $B=\begin{bmatrix}0&x&y\\-x&0&z\\-y&-z&0\end{bmatrix}$ we have $\det(B)=0=\mathrm{Pf}(B)$ and

$$\mathrm{Pf}_2(B)=\left(\mathrm{Pf}\left(\begin{bmatrix}0&x\\-x&0\end{bmatrix}\right),\mathrm{Pf}\left(\begin{bmatrix}0&y\\-y&0\end{bmatrix}\right),\mathrm{Pf}\left(\begin{bmatrix}0&z\\-z&0\end{bmatrix}\right)\right)R=(x,y,z)R.$$

Buchsbaum and Eisenbud [12] study the structure of resolutions of length three. Specifically, they characterize such resolutions and exhibit their DG structure.

[7]Note that the sign in this expression differs from the one found in [4, Example 2.1.2].

Theorem 6.10 ([12, Theorem 2.1]). *Let I be an ideal of the local ring $(R,\mathfrak{m})$.*

(a) *If I is Gorenstein and $\operatorname{pd}_R(R/I) = 3$, then there is an odd integer $n \geqslant 3$ and an $n \times n$ matrix A over $\mathfrak{m}$ such that $I = \operatorname{Pf}_{n-1}(A)$ and R/I has a minimal free resolution of the form $0 \to R \xrightarrow{B^T} R^n \xrightarrow{A} R^n \xrightarrow{B} R \to 0$ where B is the $1 \times n$ matrix with ith column given by $(-1)^i \operatorname{Pf}(A_i)$.*

(b) *Conversely, if A is an $n \times n$ alternating matrix over $\mathfrak{m}$ such that $\operatorname{rank}(A) = n-1$, then n is odd and $\operatorname{grade}(\operatorname{Pf}_{n-1}(A)) \leqslant 3$; if $\operatorname{grade}(\operatorname{Pf}_{n-1}(A)) = 3$, then $R/\operatorname{Pf}_{n-1}(A)$ has a minimal free resolution of the form*

$$0 \to R \xrightarrow{B^T} R^n \xrightarrow{A} R^n \xrightarrow{B} R \to 0$$

where B is the $1 \times n$ matrix with ith column given by $(-1)^{i-1} \operatorname{Pf}(A_i)$.

It follows from Theorem 6.10 that the minimal number of generators of a grade-3 Gorenstein ideal must be odd.

Example 6.11. Let $R = k[\![x,y,z]\!]$ and consider the alternating matrix $B = \begin{bmatrix} 0 & x & y \\ -x & 0 & z \\ -y & -z & 0 \end{bmatrix}$ from Example 6.9 with

$$I = \operatorname{Pf}_2(B) = \left(\operatorname{Pf}\left(\begin{bmatrix} 0 & x \\ -x & 0 \end{bmatrix}\right), \operatorname{Pf}\left(\begin{bmatrix} 0 & y \\ -y & 0 \end{bmatrix}\right), \operatorname{Pf}\left(\begin{bmatrix} 0 & z \\ -z & 0 \end{bmatrix}\right)\right) R = (x,y,z)R.$$

Theorem 6.10 implies that a minimal free resolution of R/I over R is of the form

$$0 \to Rg \xrightarrow{\begin{bmatrix} z \\ -y \\ x \end{bmatrix}} Rf_1 \oplus Rf_2 \oplus Rf_3 \xrightarrow{\begin{bmatrix} 0 & x & y \\ -x & 0 & z \\ -y & -z & 0 \end{bmatrix}} Re_1 \oplus Re_2 \oplus Re_3 \xrightarrow{[z\ -y\ x]} R1 \to 0.$$

(Compare this to the Koszul complex $K^R(x,-y,z)$.)

The next two examples of Buchsbaum and Eisenbud [12] illustrate that there exists, for any odd $n \geqslant 3$, a grade-3 Gorenstein ideal which is n-generated.

Example 6.12 ([12, Proposition 6.2]). Let $R = k[\![x,y,z]\!]$. For $n \geqslant 3$ odd, define M_n to be the $n \times n$ alternating matrix whose entries *above* the diagonal are

$$(M_n)_{ij} := \begin{cases} x & \text{if } i \text{ is odd and } j = i+1 \\ y & \text{if } i \text{ is even and } j = i+1 \\ z & \text{if } j = n-i+1 \\ 0 & \text{otherwise.} \end{cases}$$

For instance, we have

$$M_3 = \begin{bmatrix} 0 & x & z \\ -x & 0 & y \\ -z & -y & 0 \end{bmatrix} \qquad \mathrm{Pf}_2(M_3) = (x, y, z)R$$

$$M_5 = \begin{bmatrix} 0 & x & 0 & 0 & z \\ -x & 0 & y & z & 0 \\ 0 & -y & 0 & x & 0 \\ 0 & -z & -x & 0 & y \\ -z & 0 & 0 & -y & 0 \end{bmatrix} \qquad \mathrm{Pf}_4(M_5) = (y^2, xz, xy + z^2, yz, x^2)R.$$

Then $\mathrm{Pf}_{n-1}(M_n)$ is an n-generated Gorenstein ideal of grade 3 in R.

Example 6.13 ([12, Proposition 6.1]). Let M_n be a generic $(2n + 1) \times (2n + 1)$ alternating matrix over the ring $Q_n := R\,[\![x_{i,j} \mid 1 \leqslant i < j \leqslant 2n + 1]\!]$. That is:

$$M_n := \begin{bmatrix} 0 & x_{1,2} & x_{1,3} & \cdots & x_{1,2n+1} \\ -x_{1,2} & 0 & x_{2,3} & \cdots & x_{2,2n+1} \\ -x_{1,3} & -x_{2,3} & 0 & & \\ \vdots & \vdots & & \ddots & \\ -x_{1,2n+1} & -x_{2,2n+1} & & & 0 \end{bmatrix}$$

Then, for every $n \geqslant 1$, the ideal $\mathrm{Pf}_{2n}(M_n)$ is Gorenstein of grade 3 in Q_n.

The next result provides an explicit description of the DG algebra structure on the resolution from Theorem 6.10. Note that our result is the more elemental version from [4, Example 2.1.3]. It is worth noting that Buchsbaum and Eisenbud show that every free resolution over R of the form $0 \to F_3 \to F_2 \to F_1 \to R \to 0$ has the structure of a DG R-algebra.

Theorem 6.14 ([12, Theorem 4.1]). *Let A be an $n \times n$ alternating matrix over R, and let B be the $1 \times n$ matrix whose ith column is given by $(-1)^{i-1}\,\mathrm{Pf}(A_i^i)$. Then the graded R-complex*

$$0 \to Rg \xrightarrow{B^{\mathrm{T}}} \bigoplus_{\ell=1}^{n} Rf_\ell \xrightarrow{A} \bigoplus_{\ell=1}^{n} Re_\ell \xrightarrow{B} R1 \to 0$$

admits the structure of a DG R-algebra, with the following products:

(1) $e_i^2 = 0$ *and* $e_i f_j = f_j e_i = \delta_{ij} g$ *for all* $1 \leqslant i, j \leqslant n$, *and*

(2) $e_i e_j = -e_j e_i = \sum_{k=1}^{n} (-1)^{i+j+k} \rho_{ijk}\,\mathrm{Pf}(A_{ijk}^{ijk}) f_k$ *for all* $1 \leqslant i \neq j \leqslant n$, *where* $\rho_{ijk} = -1$ *whenever* $i < k < j$, *and* $\rho_{ijk} = 1$ *otherwise.*

Exercise 6.15. Let $R = k[x, y, z]$ and consider the graded R-complex given by

$$0 \to Rg \xrightarrow{\begin{bmatrix} z \\ -y \\ x \end{bmatrix}} Rf_1 \oplus Rf_2 \oplus Rf_3 \xrightarrow{\begin{bmatrix} 0 & x & y \\ -x & 0 & z \\ -y & -z & 0 \end{bmatrix}} Re_1 \oplus Re_2 \oplus Re_3 \xrightarrow{[z\ -y\ x]} R1 \to 0$$

Using Theorem 6.14, write the product relations that give this complex the structure of a DG R-algebra.

6.3 Longer Resolutions

In general, resolutions of length greater than 3 are not guaranteed to possess a DG algebra structure, as the next example of Avramov shows.

Example 6.16 ([4, Theorem 2.3.1]). Consider the local ring $R = k[\![w, x, y, z]\!]$. Then the minimal free resolutions over R of the quotients $R/(w^2, wx, xy, yz, z^2)R$ and $R/(w^2, wx, xy, yz, z^2, wy^6, x^7, x^6z, y^7)R$ do not admit DG R-algebra structures.

On the other hand, resolutions of ideals I with $\mathrm{pd}_R(R/I) \geqslant 4$ that are sufficiently nice do admit DG algebra structures. For instance, Kustin and Miller prove the following in [28, Theorem] and [29, 4.3 Theorem].

Example 6.17 ([28, 29]). Let I be a Gorenstein ideal of a local ring R. If $\mathrm{pd}_R(R/I) = 4$, then the minimal R-free resolution of R/I has the structure of a DG R-algebra.

7 DG Algebras and DG Modules II

In this section, we describe the notions needed to define semidualizing DG modules and to explain some of their base-change properties. This includes a discussion of two types of Ext for DG modules. The section concludes with another piece of the proof of Theorem 1.4; see 8.17.

Convention 7.1. Throughout this section, A is a DG R-algebra, and L, M, and N are DG A-modules.

7.1 Hom for DG Modules

The semidualizing property for R-modules is defined in part by a Hom condition, so we begin our treatment of the DG-version with Hom.

Definition 7.2. Given an integer i, a *DG A-module homomorphism of degree n* is an element $f \in \mathrm{Hom}_R(M, N)_n$ such that $f_{i+j}(am) = (-1)^{ni} a f_j(m)$ for all $a \in A_i$ and $m \in M_j$. The graded submodule of $\mathrm{Hom}_R(M, N)$ consisting of all DG A-module homomorphisms $M \to N$ is denoted $\mathrm{Hom}_A(M, N)$.

Part (b) of the next exercise gives another hint of the semidualizing property for DG modules.

Exercise 7.3. (a) Prove that $\mathrm{Hom}_A(M, N)$ is a DG A-module via the action

$$(af)_j(m) := a(f_j(m)) = (-1)^{|a||f|} f_{j+|a|}(am)$$

and using the differential from $\mathrm{Hom}_R(M, N)$.

(b) Prove that for each $a \in A$ the multiplication map $\mu^{M,a}\colon M \to M$ given by $m \mapsto am$ is a homomorphism of degree $|a|$.

(c) Prove that $f \in \mathrm{Hom}_A(M, N)_0$ is a morphism if and only if it is a cycle, that is, if and only if $\partial_0^{\mathrm{Hom}_A(M,N)}(f) = 0$.

Example 7.4. We continue with the notation of Example 5.16. From computations like those in Example 5.23, it follows that $\mathrm{Hom}_U(G, R)$ has the form

$$\mathrm{Hom}_U(G, R) = \quad 0 \to R \to 0 \to R \to 0 \to \cdots$$

where the copies of R are in even nonpositive degrees. Multiplication by e is 0 on $\mathrm{Hom}_U(G, R)$, by degree considerations, and multiplication by 1 is the identity.

Next, we give an indication of the functoriality of $\mathrm{Hom}_A(N, -)$ and $\mathrm{Hom}_A(-, N)$.

Definition 7.5. Given a morphism $f\colon L \to M$ of DG A-modules, we define the map $\mathrm{Hom}_A(N, f)\colon \mathrm{Hom}_A(N, L) \to \mathrm{Hom}_A(N, M)$ as follows: each sequence $\{g_p\} \in \mathrm{Hom}_A(N, L)_n$ is mapped to $\{f_{p+n} g_p\} \in \mathrm{Hom}_A(N, M)_n$. Similarly, define the map $\mathrm{Hom}_A(f, N)\colon \mathrm{Hom}_A(M, N) \to \mathrm{Hom}_A(L, N)$ by the formula $\{g_p\} \mapsto \{g_p f_p\}$.

Remark 7.6. We do not use a sign-change in this definition because $|f| = 0$.

Exercise 7.7. Given a morphism $f\colon L \to M$ of DG A-modules, prove that the maps $\mathrm{Hom}_A(N, f)$ and $\mathrm{Hom}_A(f, N)$ are well-defined morphisms of DG A-modules.

7.2 *Tensor Product for DG Modules*

As with modules and complexes, we use the tensor product to base change DG modules along a morphism of DG algebras.

Definition 7.8. The *tensor product* $M \otimes_A N$ is the quotient $(M \otimes_R N)/U$ where U is generated over R by the elements of the form $(am) \otimes n - (-1)^{|a||m|} m \otimes (an)$. Given an element $m \otimes n \in M \otimes_R N$, we denote the image in $M \otimes_A N$ as $m \otimes n$.

Exercise 7.9. Prove that the tensor product $M \otimes_A N$ is a DG A-module via the scalar multiplication

$$a(m \otimes n) := (am) \otimes n = (-1)^{|a||m|} m \otimes (an).$$

The next exercises describe base change and some canonical isomorphisms for DG modules.

Exercise 7.10. Let $A \to B$ be a morphism of DG R-algebras. Prove that $B \otimes_A M$ has the structure of a DG B-module by the action $b(b' \otimes m) := (bb') \otimes m$. Prove that this structure is compatible with the DG A-module structure on $B \otimes_A M$ via restriction of scalars.

Exercise 7.11. Verify the following isomorphisms of DG A-modules:

$\operatorname{Hom}_A(A, L) \cong L$	Hom cancellation
$A \otimes_A L \cong L$	tensor cancellation
$L \otimes_A M \cong M \otimes_A L$	tensor commutativity

In particular, there are DG A-module isomorphisms $\operatorname{Hom}_A(A, A) \cong A \cong A \otimes_A A$.

Fact 7.12. There is a natural "Hom tensor adjointness" DG A-module isomorphism $\operatorname{Hom}_A(L \otimes_A M, N) \cong \operatorname{Hom}_A(M, \operatorname{Hom}_A(L, N))$.

Next, we give an indication of the functoriality of $N \otimes_A -$ and $- \otimes_A N$.

Definition 7.13. Given a morphism $f: L \to M$ of DG A-modules, we define the map $N \otimes_A f: N \otimes_A L \to N \otimes_A M$ by the formula $z \otimes y \mapsto z \otimes f(y)$. Define the map $f \otimes_A N: L \otimes_A N \to M \otimes_A N$ by the formula $x \otimes z \mapsto f(x) \otimes z$.

Remark 7.14. We do not use a sign-change in this definition because $|f| = 0$.

Exercise 7.15. Given a morphism $f: L \to M$ of DG A-modules, prove that the maps $N \otimes_A f$ and $f \otimes_R N$ are well-defined morphisms of DG A-modules.

7.3 Semi-free Resolutions

Given the fact that the semidualizing property includes an Ext-condition, it should come as no surprise that we need a version of free resolutions in the DG setting.

Definition 7.16. A subset E of L is called a *semibasis* if it is a basis of the underlying $A^\natural$-module $L^\natural$. If L is bounded below, then L is called *semi-free* if it has a semibasis.[8] A *semi-free resolution* of a DG A-module M is a quasiisomorphism $F \xrightarrow{\simeq} M$ of DG A-modules such that F is semi-free.

The next exercises and example give some semi-free examples to keep in mind.

Exercise 7.17. Prove that a semi-free DG R-module is simply a bounded below complex of free R-modules. Prove that each free resolution F of an R-module M gives rise to a semi-free resolution $F \xrightarrow{\simeq} M$; see Exercise 3.10.

Exercise 7.18. Prove that M is exact (as an R-complex) if and only if $0 \xrightarrow{\simeq} M$ is a semi-free resolution. Prove that the DG A-module $\Sigma^n A$ is semi-free for each $n \in \mathbb{Z}$, as is $\bigoplus_{n \geqslant n_0} \Sigma^n A^{\beta_n}$ for all $n_0 \in \mathbb{Z}$ and $\beta_n \in \mathbb{N}$.

Exercise 7.19. Let $\mathbf{x} = x_1, \cdots, x_n \in R$, and set $K = K^R(\mathbf{x})$.

(a) Given a bounded below complex F of free R-modules, prove that the complex $K \otimes_R F$ is a semi-free DG K-module.
(b) If $F \xrightarrow{\simeq} M$ is a free resolution of an R-module M, prove that $K \otimes_R F \xrightarrow{\simeq} K \otimes_R M$ is a semi-free resolution of the DG K-module $K \otimes_R M$. More generally, if $F \xrightarrow{\simeq} M$ is a semi-free resolution of a DG R-module M, prove that $K \otimes_R F \xrightarrow{\simeq} K \otimes_R M$ is a semi-free resolution of the DG K-module $K \otimes_R M$. See Fact 4.9.

Example 7.20. In the notation of Example 5.16, the natural map $g^{1,0}\colon G \to R$ is a semi-free resolution of R over U; see Example 5.23. The following display indicates why G is semi-free over U, that is, why $G^\natural$ is free over $U^\natural$:

$$
\begin{aligned}
U &= \quad 0 \to Re \xrightarrow{0} R1 \to 0 \\
U^\natural &= \quad Re \bigoplus R1 \\
G &= \cdots \xrightarrow{1} Re_3 \xrightarrow{0} R1_2 \xrightarrow{1} Re_1 \xrightarrow{0} R1_0 \to 0 \\
G^\natural &= \cdots (Re_3 \bigoplus R1_2) \bigoplus (Re_1 \bigoplus R1_0).
\end{aligned}
$$

The next item compares to Remark 5.26.

Remark 7.21. If L is semi-free, then the natural map $L \to \tau(L)_{(\leqslant n)}$ is a semi-free resolution for each $n \geqslant \sup(L)$.

The next facts contain important existence results for semi-free resolutions. Notice that the second paragraph applies when A is a Koszul complex over R or is finite dimensional over a field, by Exercise 5.10.

[8] As is noted in [10], when L is not bounded below, the definition of "semi-free" is more technical. However, our results do not require this level of generality, so we focus only on this case.

Fact 7.22. The DG A-module M has a semi-free resolution if and only if $\mathrm{H}_i(M) = 0$ for $i \ll 0$, by [10, Theorem 2.7.4.2].

Assume that A is noetherian, and let j be an integer. Assume that each module $\mathrm{H}_i(M)$ is finitely generated over $\mathrm{H}_0(A)$ and that $\mathrm{H}_i(M) = 0$ for $i < j$. Then M has a semi-free resolution $F \xrightarrow{\simeq} M$ such that $F^\natural \cong \bigoplus_{i=j}^\infty \Sigma^i (A^\natural)^{\beta_i}$ for some integers β_i, and so $F_i = 0$ for all $i < j$; see [1, Proposition 1]. In particular, homologically finite DG A-modules admit "degree-wise finite, bounded below" semi-free resolutions.

Fact 7.23. Assume that L and M are semi-free. If there is a quasiisomorphism $L \xrightarrow{\simeq} M$, then there is also a quasiisomorphism $M \xrightarrow{\simeq} L$ by [4, Proposition 1.3.1].

The previous fact explains why the next relations are symmetric. The fact that they are reflexive and transitive are straightforward to verify.

Definition 7.24. Two semi-free DG A-modules L and M are *quasiisomorphic* if there is a quasiisomorphism $L \xrightarrow{\simeq} M$; this equivalence relation is denoted by the symbol $\simeq$. Two semi-free DG A-modules L and M are *shift-quasiisomorphic* if there is an integer m such that $L \simeq \Sigma^m M$; this equivalence relation is denoted by $\sim$.

7.4 Semidualizing DG Modules

For Theorem 1.4, we use a version of Christensen and Sather-Wagstaff's notion of semidualizing DG modules from [16], defined next.

Definition 7.25. The *homothety morphism* $\chi^A_M : A \to \mathrm{Hom}_A(M, M)$ is given by the formula $(\chi^A_M)_{|a|}(a) := \mu^{M,a}$, i.e., $(\chi^A_M)_{|a|}(a)_{|m|}(m) = am$.

Assume that A is noetherian. Then M is a *semidualizing* DG A-module if M is homologically finite and semi-free such that $\chi^A_M : A \to \mathrm{Hom}_A(M, M)$ is a quasiisomorphism. Let $\mathfrak{S}(A)$ denote the set of shift-quasiisomorphism classes of semidualizing DG A-modules, that is, the set of equivalence classes of semidualizing DG A-modules under the relation $\sim$ from Definition 7.24.

Exercise 7.26. Prove that the homothety morphism $\chi^A_M : A \to \mathrm{Hom}_A(M, M)$ is a well-defined morphism of DG A-modules.

The following fact explains part of diagram (2).

Fact 7.27. Let M be an R-module with projective resolution P. Then Fact 3.18 shows that M is a semidualizing R-module if and only if P is a semidualizing DG R-module. It follows that we have an injection $\mathfrak{S}_0(R) \hookrightarrow \mathfrak{S}(R)$.

The next example justifies our focus on shift-quasiisomorphism classes of semidualizing DG A-modules.

Example 7.28. Let B and C be semi-free DG A-modules such that $B \sim C$. Then B is semidualizing over A if and only if C is semidualizing over A. The point is the following. The condition $B \simeq \Sigma^i C$ tells us that B is homologically finite if and only if $\Sigma^i C$ is homologically finite, that is, if and only if C is homologically finite. Fact 7.23 provides a quasiisomorphism $B \xrightarrow[\simeq]{f} \Sigma^i C$. Thus, there is a commutative diagram of morphisms of DG A-modules:

$$\begin{array}{ccccc}
\mathrm{Hom}_A(C,C) & \xleftarrow{\chi^A_C} & A & \xrightarrow{\chi^A_B} & \mathrm{Hom}_A(B,B) \\
& \searrow^{\cong} & \downarrow{\scriptstyle \chi^A_{\Sigma^i C}} & & {\scriptstyle \simeq}\downarrow{\scriptstyle \mathrm{Hom}_A(B,f)} \\
& & \mathrm{Hom}_A(\Sigma^i C, \Sigma^i C) & \xrightarrow[\simeq]{\mathrm{Hom}_A(f,\Sigma^i C)} & \mathrm{Hom}_A(B, \Sigma^i C).
\end{array}$$

The unspecified isomorphism follows from a bookkeeping exercise.[9] The morphisms $\mathrm{Hom}_A(f, C)$ and $\mathrm{Hom}_A(B, f)$ are quasiisomorphisms by [4, Propositions 1.3.2 and 1.3.3] because B and $\Sigma^i C$ are semi-free and f is a quasiisomorphism. It follows that χ^A_B is a quasiisomorphism if and only if $\chi^A_{\Sigma^i C}$ is a quasiisomorphism if and only if χ^A_C is a quasiisomorphism.

The following facts explain other parts of (2).

Fact 7.29. Assume that $(R, \mathfrak{m})$ is local. Fix a list of elements $\mathbf{x} \in \mathfrak{m}$ and set $K = K^R(\mathbf{x})$. Base change $K \otimes_R -$ induces an injective map $\mathfrak{S}(R) \hookrightarrow \mathfrak{S}(K)$ by [16, A.3. Lemma]; if R is complete, then this map is bijective by [33, Corollary 3.10].

Fact 7.30. Let $\varphi \colon A \xrightarrow{\simeq} B$ be a quasiisomorphism of noetherian DG R-algebras. Base change $B \otimes_A -$ induces a bijection from $\mathfrak{S}(A)$ to $\mathfrak{S}(B)$ by [32, Lemma 2.22(c)].

7.5 *Ext for DG Modules*

One subtlety in the proof of Fact 7.29 is found in the behavior of Ext for DG modules, which we describe next.

Definition 7.31. Given a semi-free resolution $F \xrightarrow{\simeq} M$, for each integer i we set $\mathrm{Ext}^i_A(M, N) := \mathrm{H}_{-i}(\mathrm{Hom}_A(F, N))$.[10]

The next two items are included in our continued spirit of providing perspective.

[9] The interested reader may wish to show how this isomorphism is defined and to check the commutativity of the diagram. If this applies to you, make sure to mind the signs.

[10] One can also define $\mathrm{Tor}^R_i(M, N) := \mathrm{H}_i(F \otimes_A N)$, but we do not need this here.

Exercise 7.32. Given R-modules M and N, prove that the module $\mathrm{Ext}^i_R(M,N)$ defined in 7.31 is the usual $\mathrm{Ext}^i_R(M,N)$; see Exercise 7.17.

Example 7.33. In the notation of Examples 5.16, 7.4 and 7.20 imply

$$\mathrm{Ext}^i_U(R,R) = \mathrm{H}_{-i}(\mathrm{Hom}_U(G,R)) = \begin{cases} R & \text{if } i \geqslant 0 \text{ is even} \\ 0 & \text{otherwise.} \end{cases}$$

Contrast this with the equality $\mathrm{Ext}^i_{U^\natural}(R,R) = R$ for all $i \geqslant 0$. This shows that U is fundamentally different from $U^\natural \cong R[X]/(X^2)$, even though U is obtained using a trivial differential on $R[X]/(X^2)$ with the natural grading.

The next result compares with the fact that $\mathrm{Ext}^i_R(M,N)$ is independent of the choice of free resolution when M and N are modules.

Fact 7.34. For each index i, the module $\mathrm{Ext}^i_A(M,N)$ is independent of the choice of semi-free resolution of M by [4, Proposition 1.3.3].

Remark 7.35. An important fact about $\mathrm{Ext}^1_R(M,N)$ for R-modules M and N is the following: the elements of $\mathrm{Ext}^1_R(M,N)$ are in bijection with the equivalence classes of short exact sequences (i.e., "extensions") of the form $0 \to N \to X \to M \to 0$. For DG modules over a DG R-algebra A, things are a bit more subtle.

Given DG A-modules M and N, one defines the notion of a short exact sequence of the form $0 \to N \to X \to M \to 0$ in the naive way: the arrows are morphisms of DG A-modules such that for each $i \in \mathbb{Z}$ the sequence $0 \to N_i \to X_i \to M_i \to 0$ is exact. One defines an equivalence relation on the set of short exact sequences of this form (i.e., "extensions") in the natural way: two extensions $0 \to N \xrightarrow{f} X \xrightarrow{g} M \to 0$ and $0 \to N \xrightarrow{f'} X' \xrightarrow{g'} M \to 0$ are equivalent if there is a commutative diagram

$$\begin{array}{ccccccccc} 0 & \longrightarrow & N & \xrightarrow{f} & X & \xrightarrow{g} & M & \longrightarrow & 0 \\ & & \Big\downarrow{=} & & \Big\downarrow{h} & & \Big\downarrow{=} & & \\ 0 & \longrightarrow & N & \xrightarrow{f'} & X' & \xrightarrow{g'} & M & \longrightarrow & 0 \end{array}$$

of morphisms of DG A-modules. Let $\mathrm{YExt}^1_A(M,N)$ denote the set of equivalence classes of such extensions. (The "Y" is for "Yoneda.") As with R-modules, one can define an abelian group structure on $\mathrm{YExt}^1_A(M,N)$. However, in general one has $\mathrm{YExt}^1_A(M,N) \not\cong \mathrm{Ext}^1_A(M,N)$, even when $A = R$, as the next example shows.

Example 7.36. Let $R = k[\![X]\!]$, and consider the following exact sequence of DG R-modules, i.e., exact sequence of R-complexes:

$$0 \longrightarrow \underline{R} \longrightarrow \underline{R} \longrightarrow \underline{k} \longrightarrow 0$$

$$\begin{array}{ccccccccc}
 & & 0 & & 0 & & 0 & & \\
 & & \downarrow & & \downarrow & & \downarrow & & \\
0 & \longrightarrow & R & \xrightarrow{X} & R & \longrightarrow & k & \longrightarrow & 0 \\
 & & {\scriptstyle 1}\downarrow & & {\scriptstyle 1}\downarrow & & {\scriptstyle 1}\downarrow & & \\
0 & \longrightarrow & R & \xrightarrow{X} & R & \longrightarrow & k & \longrightarrow & 0 \\
 & & \downarrow & & \downarrow & & \downarrow & & \\
 & & 0 & & 0 & & 0. & &
\end{array}$$

This sequence does not split over R (it is not even degree-wise split) so it gives a nontrivial class in $\mathrm{YExt}^1_R(\underline{k}, \underline{R})$, and we conclude that $\mathrm{YExt}^1_R(\underline{k}, \underline{R}) \neq 0$. On the other hand, $\underline{k}$ is homologically trivial, so we have $\mathrm{Ext}^1_R(\underline{k}, \underline{R}) = 0$ since $0 \xrightarrow{\simeq} \underline{k}$ is a semi-free resolution.

For our proof of Theorem 1.4, the following connection between Ext and YExt is quite important; see [32, Corollary 3.8 and Proposition 3.12].

Fact 7.37. If L is semi-free, then we have $\mathrm{YExt}^1_A(L, M) \cong \mathrm{Ext}^1_A(L, M)$; if furthermore $\mathrm{Ext}^1_R(L, L) = 0$, then for each $n \geqslant \sup(L)$, one has

$$\mathrm{YExt}^1_A(L, L) = 0 = \mathrm{YExt}^1_A(\tau(L)_{(\leqslant n)}, \tau(L)_{(\leqslant n)}).$$

We conclude this section with the second part of the proof of Theorem 1.4. The rest of the proof is contained in 8.17.

7.38 (Second Part of the Proof of Theorem 1.4). We continue with the notation established in 5.32. The properties of diagram (2) follow from diagram (1) because of Facts 7.29 and 7.30. Thus, it remains to show that $\mathfrak{S}(k \otimes_Q A)$ is finite. This is shown in 8.17.

8 A Version of Happel's Result for DG Modules

This section contains the final steps of the proof of Theorem 1.4; see 8.17. The idea, from [2,22,25,42] is a bold one: use algebraic geometry to study all possible module structures on a fixed set. A simple case of this is in Example 5.17. We begin with some notation for use throughout this section.

Notation 8.1. Let F be an algebraically closed field, and let

$$U := (0 \to U_q \xrightarrow{\partial^U_q} U_{q-1} \xrightarrow{\partial^U_{q-1}} \cdots \xrightarrow{\partial^U_1} U_0 \to 0)$$

be a finite-dimensional DG F-algebra. Let $\dim_F(U_i) = n_i$ for $i = 0, \ldots, q$. Let

$$W := \bigoplus_{i=0}^{s} W_i$$

be a graded F-vector space with $r_i := \dim_F(W_i)$ for $i = 0, \ldots, s$.

A DG U-module structure on W consists of two pieces of data. First, we need a differential ∂. Second, once the differential ∂ has been chosen, we need a scalar multiplication μ. Let $\mathrm{Mod}^U(W)$ denote the set of all ordered pairs (∂, μ) making W into a DG U-module. Let $\mathrm{End}_F(W)_0$ denote the set of F-linear endomorphisms of W that are homogeneous of degree 0. Let $\mathrm{GL}(W)_0$ denote the set of F-linear automorphisms of W that are homogeneous of degree 0, that is, the invertible elements of $\mathrm{End}_F(W)_0$.

We next describe a geometric structure on $\mathrm{Mod}^U(W)$, as in Example 5.17.

Discussion 8.2. We work in the setting of Notation 8.1.

A differential ∂ on W is an element of the graded vector space $\mathrm{Hom}_F(W, W)_{-1} = \bigoplus_{i=0}^{s} \mathrm{Hom}_F(W_i, W_{i-1})$ such that $\partial\partial = 0$. The vector space $\mathrm{Hom}_F(W_i, W_{i-1})$ has dimension $r_i r_{i-1}$, so the map ∂ corresponds to an element of the affine space $\mathbb{A}_F^d$ where $d := \sum_i r_i r_{i-1}$. The vanishing condition $\partial\partial = 0$ is equivalent to the entries of the matrices representing ∂ satisfying certain fixed homogeneous quadratic polynomial equations over F. Hence, the set of all differentials on W is a Zariski-closed subset of $\mathbb{A}_F^d$.

Once the differential ∂ has been chosen, a scalar multiplication μ is in particular a cycle in $\mathrm{Hom}_F(U \otimes_F W, W)_0 = \bigoplus_{i,j} \mathrm{Hom}_F(U_i \otimes_F W_j, W_{i+j})$. For all i, j, the vector space $\mathrm{Hom}_F(U_i \otimes_F W_j, W_{i+j})$ has dimension $n_i r_j r_{i+j}$, so the map μ corresponds to an element of the affine space $\mathbb{A}_F^{d'}$ where $d' := \sum_{i,j} n_i r_j r_{i+j}$. The condition that μ be an associative, unital cycle is equivalent to the entries of the matrices representing ∂ and μ satisfying certain fixed polynomials over F. Thus, the set $\mathrm{Mod}^U(W)$ is a Zariski-closed subset of $\mathbb{A}_F^d \times \mathbb{A}_F^{d'} \cong \mathbb{A}_F^{d+d'}$.

Example 8.3. We continue with the notation of Example 5.17. In this example, we have $\mathrm{Mod}^U(W) = \{1\}$ (representing the nontrivial scalar multiplication by 1) and $\mathrm{Mod}^U(W') = \{(x_1, x_0) \in F^2 \mid x_1 x_0 = 0\}$. Re-writing F^n as $\mathbb{A}_F^n$, we see that $\mathrm{Mod}^U(W)$ is a single point in $\mathbb{A}_F^1$ and $\mathrm{Mod}^U(W')$ is the union of the two coordinate axes $V(x_1 x_0) = V(x_0) \cup V(x_1)$.

Exercise 8.4. Continue with the notation of Example 5.17. Write out the coordinates and equations describing $\mathrm{Mod}^U(W'')$ and $\mathrm{Mod}^U(W''')$ where

$$\begin{aligned} W'' &= 0 \bigoplus F w_2 \bigoplus F w_1 \bigoplus F w_0 \bigoplus 0 \\ W''' &= 0 \bigoplus F z_2 \bigoplus (F z_{1,1} \bigoplus F z_{1,2}) \bigoplus F z_0 \bigoplus 0. \end{aligned}$$

For scalar multiplication, note that since multiplication by 1 is already determined by the F-vector space structure, we only need to worry about multiplication by e which maps $W''_i \to W''_{i+1}$ and $W'''_i \to W'''_{i+1}$ for $i = 0, 1, 2$.

We next describe a geometric structure on the set $\mathrm{GL}(W)_0$.

Discussion 8.5. We work in the setting of Notation 8.1.

A map $\alpha \in \mathrm{GL}(W)_0$ is an element of the graded vector space $\mathrm{Hom}_F(W, W)_0 = \bigoplus_{i=0}^s \mathrm{Hom}_F(W_i, W_i)$ with a multiplicative inverse. The vector space $\mathrm{Hom}_F(W_i, W_i)$ has dimension r_i^2, so the map α corresponds to an element of the affine space $\mathbb{A}_F^e$ where $e := \sum_i r_i^2$. The invertibility of α is equivalent to the invertibility of each "block" $\alpha_i \in \mathrm{Hom}_F(W_i, W_i)$, which is an open condition defined by the nonvanishing of the determinant polynomial. Thus, the set $\mathrm{GL}(W)_0$ is a Zariski-open subset of $\mathbb{A}_F^e$, so it is smooth over F.

Alternately, one can view $\mathrm{GL}(W)_0$ as the product $\mathrm{GL}(W_0) \times \cdots \times \mathrm{GL}(W_s)$. Since each $\mathrm{GL}(W_i)$ is an algebraic group smooth over F, it follows that $\mathrm{GL}(W)_0$ is also an algebraic group that is smooth over F.

Example 8.6. We continue with the notation of Example 5.17. It is straightforward to show that

$$\begin{aligned}
\mathrm{End}_F(W)_0 &= \mathrm{Hom}_F(F\nu_0, F\nu_0) \cong F = \mathbb{A}_F^1 \\
\mathrm{GL}_F(W)_0 &= \mathrm{Aut}_F(F\nu_0) \cong F^\times = U_x \subset \mathbb{A}_F^1 \\
\mathrm{End}_F(W')_0 &= \mathrm{Hom}_F(F\eta_1, F\eta_1) \bigoplus \mathrm{Hom}_F(F\eta_0, F\eta_0) \cong F \times F = \mathbb{A}_F^2 \\
\mathrm{GL}_F(W')_0 &= \mathrm{Aut}_F(F\eta_1) \bigoplus \mathrm{Aut}_F(F\eta_0) \cong F^\times \times F^\times = U_{x_1x_0} \subset \mathbb{A}_F^2.
\end{aligned}$$

Here U_x is the subset $\mathbb{A}_F^1 \smallsetminus V(x)$, and $U_{x_1x_0} = \mathbb{A}_F^2 \smallsetminus V(x_1x_0)$.

Exercise 8.7. With the notation of Example 5.17. Give coordinates and equations describing $\mathrm{GL}^U(W'')_0$ and $\mathrm{GL}^U(W''')_0$ where W'' and W''' are from Exercise 8.4.

Next, we describe an action of $\mathrm{GL}(W)_0$ on $\mathrm{Mod}^U(W)$.

Discussion 8.8. We work in the setting of Notation 8.1.

Let $\alpha \in \mathrm{GL}(W)_0$. For every $(\partial, \mu) \in \mathrm{Mod}^U(W)$, we define $\alpha \cdot (\partial, \mu) := (\tilde{\partial}, \tilde{\mu})$, where $\tilde{\partial}_i := \alpha_{i-1} \circ \partial_i \circ \alpha_i^{-1}$ and $\tilde{\mu}_{i+j} := \alpha_{i+j} \circ \mu_{i+j} \circ (U \otimes_F \alpha_j^{-1})$. For the multiplication, this defines a new multiplication

$$u_i \cdot_\alpha w_j := \alpha_{i+j}(u_i \cdot \alpha_j^{-1}(w_j))$$

where $\cdot$ is multiplication given by μ, as in Discussion 8.2: $u_i \cdot w_j := \mu_{i+j}(u_i \otimes w_j)$. Note that this leaves multiplication by 1_A unaffected:

$$1_A \cdot_\alpha w_j = \alpha_j(1_A \cdot \alpha_j^{-1}(w_j)) = \alpha_j(\alpha_j^{-1}(w_j)) = w_j.$$

It is routine to show that the ordered pair $(\tilde{\partial}, \tilde{\mu})$ describes a DG U-module structure for W, that is, we have $\alpha \cdot (\partial, \mu) := (\tilde{\partial}, \tilde{\mu}) \in \operatorname{Mod}^U(W)$. From the definition of $\alpha \cdot (\partial, \mu)$, it follows readily that this describes a $\mathrm{GL}(W)_0$-action on $\operatorname{Mod}^U(W)$.

Example 8.9. Continue with the notation of Example 5.17.

In this case, the only DG U-module structure on W is the trivial one $(\partial, \mu) = (0, 0)$, so we have $\alpha \cdot (\partial, \mu) = (\partial, \mu)$ for all $\alpha \in \mathrm{GL}(W)_0$.

The action of $\mathrm{GL}(W')_0$ on $\operatorname{Mod}^U(W')$ is a bit more interesting. Let $x_0, x_1 \in F$ such that $x_0 x_1 = 0$, as in Example 5.17. Identify $\mathrm{GL}_F(W')_0$ with $F^\times \times F^\times$, as in Example 8.3, and let $\alpha \in \mathrm{GL}_F(W')_0$ be given by the ordered pair $(y_1, y_0) \in F^\times \times F^\times$. The differential $\tilde{\partial}$ is defined so that the following diagram commutes.

$$\begin{array}{lccccccc}
\partial: & 0 & \longrightarrow & F\eta_1 & \xrightarrow{x_1} & F\eta_0 & \longrightarrow & 0 \\
 & & & \big\downarrow{\scriptstyle y_1} & & \big\downarrow{\scriptstyle y_0} & & \\
\tilde{\partial}: & 0 & \longrightarrow & F\tilde{\eta}_1 & \xrightarrow{\tilde{x}_1} & F\tilde{\eta}_0 & \longrightarrow & 0
\end{array}$$

so we have $\tilde{\partial}_1(\tilde{\eta}_1) = y_0 x_1 y_1^{-1} \tilde{\eta}_0$, i.e., $\tilde{x}_1 = y_0 x_1 y_1^{-1}$.

Since multiplication by 1 is already determined, and we have $e \cdot_\alpha \tilde{\eta}_1 = 0$ because of degree considerations, we only need to understand $e \cdot_\alpha \tilde{\eta}_0$. From Discussion 8.8, this is given by

$$\begin{aligned}
e \cdot_\alpha \tilde{\eta}_0 &= \alpha_1(e \cdot \alpha_0^{-1}(\tilde{\eta}_0)) = \alpha_1(e \cdot y_0^{-1}\eta_0) = y_0^{-1}\alpha_1(e \cdot \eta_0) \\
&= y_0^{-1}\alpha_1(x_0\eta_1) = y_0^{-1} y_1 x_0 \tilde{\eta}_1.
\end{aligned}$$

Exercise 8.10. Continue with the notation of Example 5.17. Using the solutions to Exercises 8.4 and 8.7 describe the actions of $\mathrm{GL}(W'')_0$ and $\mathrm{GL}(W''')_0$ on $\operatorname{Mod}^U(W'')$ and $\operatorname{Mod}^U(W''')$, respectively, as in the previous example.

Next, we describe some properties of the action from Discussion 8.8 that indicate a deeper connection between the algebra and geometry.

Discussion 8.11. We work in the setting of Notation 8.1.

Let $\alpha \in \mathrm{GL}(W)_0$. For every $(\partial, \mu) \in \operatorname{Mod}^U(W)$, let $\alpha \cdot (\partial, \mu) := (\tilde{\partial}, \tilde{\mu})$ be as in Discussion 8.8. It is straightforward to show that a map α gives a DG U-module isomorphism $(W, \partial, \mu) \xrightarrow{\cong} (W, \tilde{\partial}, \tilde{\mu})$. Conversely, given another element $(\partial', \mu') \in \operatorname{Mod}^U(W)$, if there is a DG U-module isomorphism $\beta\colon (W, \partial, \mu) \xrightarrow{\cong} (W, \partial', \mu')$, then $\beta \in \mathrm{GL}(W)_0$ and $(\partial', \mu') = \beta \cdot (\partial, \mu)$. In other words, the orbits in $\operatorname{Mod}^U(W)$ under the action of $\mathrm{GL}(W)_0$ are the isomorphism classes of DG U-module structures on W. Given an element $M = (\partial, \mu) \in \operatorname{Mod}^U(W)$, the orbit $\mathrm{GL}(W)_0 \cdot M$ is locally closed in $\operatorname{Mod}^U(W)$; see [18, Chap. II, Sect. 5.3].

Note that the maps defining the action of $\mathrm{GL}(W)_0$ on $\mathrm{Mod}^U(W)$ are regular, that is, determined by polynomial functions. This is because the inversion map $\alpha \mapsto \alpha^{-1}$ on $\mathrm{GL}(W)_0$ is regular, as is the multiplication of matrices corresponding to the compositions defining $\tilde{\partial}$ and $\tilde{\mu}$.

Next, we consider even more geometry by identifying tangent spaces to two of our objects of study.

Notation 8.12. We work in the setting of Notation 8.1. Let $F[\epsilon] := F\epsilon \bigoplus F$ be the algebra of dual numbers, where $\epsilon^2 = 0$ and $|\epsilon| = 0$. For our convenience, we write elements of $F[\epsilon]$ as column vectors: $a\epsilon + b = \left[\begin{smallmatrix} a \\ b \end{smallmatrix}\right]$. We identify $U[\epsilon] := F[\epsilon] \otimes_F U$ with $U\epsilon \bigoplus U \cong U \bigoplus U$, and $W[\epsilon] := F[\epsilon] \otimes_F W$ with $W\epsilon \bigoplus W \cong W \bigoplus W$. Using this protocol, we have $\partial_i^{U[\epsilon]} = \left[\begin{smallmatrix} \partial_i^U & 0 \\ 0 & \partial_i^U \end{smallmatrix}\right]$.

Let $\mathrm{Mod}^{U[\epsilon]}(W[\epsilon])$ denote the set of all ordered pairs (∂, μ) making $W[\epsilon]$ into a DG $U[\epsilon]$-module. Let $\mathrm{End}_{F[\epsilon]}(W[\epsilon])_0$ denote the set of $F[\epsilon]$-linear endomorphisms of $W[\epsilon]$ that are homogeneous of degree 0. Let $\mathrm{GL}(W[\epsilon])_0$ denote the set of $F[\epsilon]$-linear automorphisms of $W[\epsilon]$ that are homogeneous of degree 0, that is, the invertible elements of $\mathrm{End}_{F[\epsilon]}(W[\epsilon])_0$.

Given an element $M = (\partial, \mu) \in \mathrm{Mod}^U(W)$, the tangent space $\mathsf{T}_M^{\mathrm{Mod}^U(W)}$ is the set of all ordered pairs $(\overline{\partial}, \overline{\mu}) \in \mathrm{Mod}^{U[\epsilon]}(W[\epsilon])$ that give rise to M modulo ϵ. The tangent space $\mathsf{T}_{\mathrm{id}_W}^{\mathrm{GL}(W)_0}$ is the set of all elements of $\mathrm{GL}(W[\epsilon])_0$ that give rise to id_W modulo ϵ.

Remark 8.13. Alternate descriptions of the tangent spaces from Notation 8.12 are contained in [32, Lemmas 4.8 and 4.10]. Because of smoothness conditions, the map $\mathrm{GL}(W)_0 \xrightarrow{\cdot M} \mathrm{Mod}^U(W)$ induces a linear transformation $\mathsf{T}_{\mathrm{id}_W}^{\mathrm{GL}(W)_0} \to \mathsf{T}_M^{\mathrm{Mod}^U(W)}$ whose image is $\mathsf{T}_M^{\mathrm{GL}(W)_0 \cdot M}$; see [32, 4.11. Proof of Theorem B].

The next two results show some profound connections between the algebra and the geometry of the objects under consideration. The ideas behind these results are due to Voigt [42] and Gabriel [22, 1.2 Corollary].

Theorem 8.14 ([32, 4.11. Proof of Theorem B]). *We work in the setting of Notation 8.1. Given an element $M = (\partial, \mu) \in \mathrm{Mod}^U(W)$, there is an isomorphism of abelian groups*

$$T_M^{\mathrm{Mod}^U(W)} \,/\, T_M^{\mathrm{GL}(W)_0 \cdot M} \cong \mathrm{YExt}_U^1(M, M).$$

Proof (Sketch of Proof). Using Notation 8.12, let $N = (\overline{\partial}, \overline{\mu})$ be an element of $\mathsf{T}_M^{\mathrm{Mod}^U(W)}$. Since N is a DG $U[\epsilon]$-module, restriction of scalars along the natural inclusion $U \to U[\epsilon]$ makes N a DG U-module.

Define $\rho\colon M \to N$ and $\pi\colon N \to M$ by the formulas $\rho(w) := \left[\begin{smallmatrix} w \\ 0 \end{smallmatrix}\right]$ and $\pi\left(\left[\begin{smallmatrix} w' \\ w \end{smallmatrix}\right]\right) := w$. With [32, Lemmas 4.8 and 4.10], one shows that ρ and π are chain maps and that ρ and π are U-linear. In other words, we have an exact sequence

$$0 \to M \xrightarrow{\rho} N \xrightarrow{\pi} M \to 0$$

of DG U-module morphisms. So, we obtain a map $\tau\colon \mathsf{T}_M^{\mathrm{Mod}^U(W)} \to \mathrm{YExt}_U^1(M,M)$ where $\tau(N)$ is the equivalence class of the displayed sequence in $\mathrm{YExt}_U^1(M,M)$. One shows that τ is a surjective abelian group homomorphism with $\mathrm{Ker}(\tau) = \mathsf{T}_M^{\mathrm{GL}(W)_0\cdot M}$, and the result follows from the First Isomorphism Theorem.

To show that τ is surjective, fix an arbitrary element $\zeta \in \mathrm{YExt}_U^1(M,M)$, represented by the sequence $0 \to M \xrightarrow{f} Z \xrightarrow{g} M \to 0$. In particular, this is an exact sequence of F-complexes, so it is degree-wise split. This implies that we have a commutative diagram of graded vector spaces:

$$\begin{array}{ccccccccc} 0 & \longrightarrow & M & \xrightarrow{f} & Z & \xrightarrow{g} & M & \longrightarrow & 0 \\ & & {\scriptstyle =}\downarrow & & \downarrow{\scriptstyle \vartheta} & & \downarrow{\scriptstyle =} & & \\ 0 & \longrightarrow & M & \xrightarrow{\rho} & W[\epsilon] & \xrightarrow{\pi} & M & \longrightarrow & 0 \end{array}$$

where $\rho(w) = \left[\begin{smallmatrix} w \\ 0 \end{smallmatrix}\right]$, $\pi\left(\left[\begin{smallmatrix} w' \\ w \end{smallmatrix}\right]\right) = w$, and ϑ is an isomorphism of graded F-vector spaces. The map ϑ allows us to endow $W[\epsilon]$ with a DG $U[\epsilon]$-module structure $(\overline{\partial}, \overline{\mu})$ that gives rise to M modulo ϵ. So we have $N := (\overline{\partial}, \overline{\mu}) \in \mathsf{T}_M^{\mathrm{Mod}^U(W)}$. Furthermore, we have $\tau(N) = \zeta$, so τ is surjective.

See [32, 4.11. Proof of Theorem B] for more details. □

Corollary 8.15 ([32, Corollary 4.12]). *We work in the setting of Notation 8.1. Let C be a semidualizing DG U-module, and let $s \geqslant \sup(C)$. Set $M = \tau(C)_{(\leqslant s)}$ and $W = M^\natural$. Then the orbit $\mathrm{GL}(W)_0 \cdot M$ is open in $\mathrm{Mod}^U(W)$.*

Proof. Fact 7.37 implies that $\mathrm{YExt}_U^1(M,M) = 0$, so $\mathsf{T}_M^{\mathrm{Mod}^U(W)} = \mathsf{T}_M^{\mathrm{GL}(W)_0\cdot M}$ by Theorem 8.14. As the orbit $\underline{\mathrm{GL}}(W)_0 \cdot M$ is smooth and locally closed, this implies that $\underline{\mathrm{GL}}(W)_0 \cdot M$ is open in $\mathrm{Mod}^U(W)$. See [32, Corollary 4.12] for more details. □

We are now in a position to state and prove our version of Happel's result [25, proof of first proposition in Sect. 3] that was used in the proof of Theorem 2.13.

Lemma 8.16 ([32, Lemma 5.1]). *We work in the setting of Notation 8.1. The set $\mathfrak{S}_W(U)$ of quasiisomorphism classes of semi-free semidualizing DG U-modules C such that $s \geqslant \sup(C)$, $C_i = 0$ for all $i < 0$, and $(\tau(C)_{(\leqslant s)})^\natural \cong W$ is finite.*

Proof. Fix a representative C for each quasiisomorphism class in $\mathfrak{S}_W(U)$, and write $[C] \in \mathfrak{S}_W(U)$ and $M_C = \tau(C)_{(\leqslant s)}$.

Let $[C], [C'] \in \mathfrak{S}_W(U)$. If $\mathrm{GL}(W)_0 \cdot M_C = \mathrm{GL}(W)_0 \cdot M_{C'}$, then $[C] = [C']$: indeed, Discussion 8.8 explains the second step in the next display

$$C \simeq M_C \cong M_{C'} \simeq C'$$

and the remaining steps follow from the assumptions $s \geqslant \sup(C)$ and $s \geqslant \sup(C')$, by Exercise 5.27.

Now, each orbit $\mathrm{GL}(W)_0 \cdot M_C$ is open in $\mathrm{Mod}^U(W)$ by Corollary 8.15. Since $\mathrm{Mod}^U(W)$ is a subset of an affine space over F, it is quasi-compact, so it can only have finitely many open orbits. By the previous paragraph, this implies that there are only finitely many distinct elements $[C] \in \mathfrak{S}_W(U)$. □

We conclude this section with the third and final part of the proof of Theorem 1.4.

8.17 (Final Part of the Proof of Theorem 1.4). We need to prove that $\mathfrak{S}(U)$ is finite where $U = k \otimes_Q A$. Set $s = \dim(R) - \mathrm{depth}(R) + n$. One uses various accounting principles to prove that every semidualizing DG U-module is equivalent to a semidualizing DG U-module C' such that $\mathrm{H}_i(C') = 0$ for all $i < 0$ and for all $i > s$. Let $L \xrightarrow{\simeq} C'$ be a minimal semi-free resolution of C' over U. The conditions $\sup(L) = \sup(C') \leqslant s$ imply that L (and hence C') is quasiisomorphic to the truncation $\tilde{L} := \tau(L)_{\leqslant s}$. We set $W := \tilde{L}^{\natural}$ and work in the setting of Notation 8.1.

One then uses further accounting principles to prove that there is an integer $\lambda \geqslant 0$, depending only on R and U, such that $\sum_{i=0}^{s} r_i \leqslant \lambda$. Compare this with Lemma 2.9. (Recall that r_i and other quantities are fixed in Notation 8.1.) Then, because there are only finitely many $(r_0, \ldots, r_s) \in \mathbb{N}^{s+1}$ with $\sum_{i=0}^{s} r_i \leqslant \lambda$, there are only finitely many W that occur from this construction, say $W^{(1)}, \ldots, W^{(b)}$. Lemma 8.16 implies that $\mathfrak{S}(U) = \mathfrak{S}_{W^{(1)}}(U) \cup \cdots \cup \mathfrak{S}_{W^{(b)}}(U) \cup \{[U]\}$ is finite. □

9 Applications of Semidualizing Modules

This section contains three applications of semidualizing modules, to indicate why Theorem 1.4 might be interesting.

Assumption 9.1. Throughout this section, $(R, \mathfrak{m}, k)$ is local.

9.1 Application I: Asymptotic Behavior of Bass Numbers

Our first application shows that the existence of nontrivial semidualizing modules forces the sequence of Bass numbers of a local ring to be unbounded. This partially answers a question of Huneke.

Definition 9.2. The *ith Bass number* of R is $\mu_R^i := \mathrm{rank}_k(\mathrm{Ext}_R^i(k, R))$. The *Bass series* of R is the formal power series $I_R(t) = \sum_{i=1}^{\infty} \mu_R^i t^i$.

Remark 9.3. The Bass numbers of R contain important structural information about the minimal injective resolution J of R. They also keep track of the depth and injective dimension of R:

$$\mathrm{depth}(R) = \min\{i \geqslant 0 \mid \mu_R^i \neq 0\}$$
$$\mathrm{id}_R(R) = \sup\{i \geqslant 0 \mid \mu_R^i \neq 0\}.$$

In particular, R is Gorenstein if and only if the sequence $\{\mu_R^i\}$ is eventually 0. If R has a dualizing module D, then the Bass numbers of R are related to the Betti numbers of D by the formula

$$\mu_R^{i+\mathrm{depth}(R)} = \beta_i^R(D) := \mathrm{rank}_k(\mathrm{Ext}_R^i(D,k)).$$

Viewed in the context of the characterization of Gorenstein rings in Remark 9.3, the next question is natural, even if it is a bit bold.[11]

Question 9.4 (Huneke). If the sequence $\{\mu_R^i\}$ is bounded, must R be Gorenstein? Equivalently, if R is not Gorenstein, must the sequence $\{\mu_R^i\}$ be unbounded?

The connection between semidualizing modules and Huneke's question is found in the following result. It shows that Huneke's question reduces to the case where R has only trivial semidualizing modules. It is worth noting that the same conclusion holds when R is not assumed to be Cohen–Macaulay and C is a semidualizing *DG* module that is neither free nor dualizing. However, since we have not talked about dualizing DG modules, we only state the module case.

Theorem 9.5 ([37, Theorem B]). *If R is Cohen–Macaulay and has a semidualizing module C that is neither free nor dualizing, then the sequence $\{\mu_R^i\}$ is unbounded.*

Proof (Sketch of Proof). Pass to the completion of R to assume that R is complete. This does not change the sequence $\{\mu_R^i\}$ nor the assumptions on C. As R is complete, it has a dualizing module D, and one has $\mu_R^i = \mathrm{rank}_k(\mathrm{Ext}_R^{i-d}(D,k))$ for all i, where $d = \mathrm{depth}(R)$. Thus, it suffices to show that the sequence $\{\mathrm{rank}_k(\mathrm{Ext}_R^i(D,k))\}$ is unbounded. With $C' = \mathrm{Hom}_R(C,D)$, one has $\mathrm{rank}_k(\mathrm{Ext}_R^i(C,k)), \mathrm{rank}_k(\mathrm{Ext}_R^i(C',k)) \geqslant 1$ for all $i \geqslant 0$. Hence, the computation

$$\mathrm{rank}_k(\mathrm{Ext}_R^i(D,k)) = \sum_{p=0}^{i} \mathrm{rank}_k(\mathrm{Ext}_R^p(C,k))\,\mathrm{rank}_k(\mathrm{Ext}_R^{p-i}(C',k)) \geqslant \sum_{p=0}^{i} 1 = p+1$$

gives the desired unboundedness. □

9.2 Application II: Structure of Quasi-Deformations

Our second application shows how semidualizing modules can be used to improve given structures. Specifically, one can use a particular semidualizing module to improve the closed fibre of a given quasi-deformation.

[11] As best we know, Huneke has not posed this question in print.

Definition 9.6 ([9, (1.1) and (1.2)]). A *quasi-deformation* of R is a diagram of local ring homomorphisms $R \xrightarrow{\varphi} R' \xleftarrow{\tau} Q$ such that φ is flat and τ is surjective with kernel generated by a Q-regular sequence. A finitely generated R-module M has *finite CI-dimension* if there is a quasideformation $R \to R' \leftarrow Q$ such that $\mathrm{pd}_Q(R' \otimes_R M) < \infty$.

Remark 9.7. A straightforward localization and completion argument shows that if M is an R-module of finite CI-dimension, then there is a quasideformation $R \to R' \leftarrow Q$ such that $\mathrm{pd}_Q(R' \otimes_R M)$ is finite, Q is complete, and $R'/\mathfrak{m}R'$ is artinian, hence Cohen–Macaulay.

The next result is a souped-up version of the previous remark. In contrast to the application of semidualizing modules in Theorem 9.5, this one does not refer to any semidualizing modules in the statement. Instead, in its proof, one uses a semidualizing module to improve the quasideformation given by definition to one satisfying the desired conclusions.

Theorem 9.8 ([36, Theorem F]). *If M is an R-module of finite CI-dimension, then there is a quasideformation $R \to R' \leftarrow Q$ such that $\mathrm{pd}_Q(R' \otimes_R M) < \infty$ and such that $R'/\mathfrak{m}R'$ is artinian and Gorenstein.*

Proof (Sketch of Proof). By Remark 9.7, there is a quasideformation $R \xrightarrow{\varphi} R' \leftarrow Q$ such that $\mathrm{pd}_Q(R' \otimes_R M)$ is finite, Q is complete, and $R'/\mathfrak{m}R'$ is artinian. We work to improve this quasi-deformation.

A relative version of Cohen's Structure Theorem due to Avramov et al. [8, (1.1) Theorem] provides "Cohen factorization" of φ, that is a commutative diagram of local ring homomorphisms

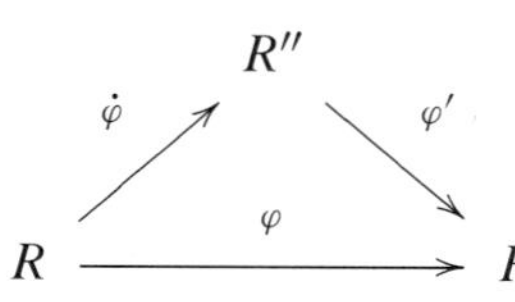

such that $\dot{\varphi}$ is flat, $R''/\mathfrak{m}R''$ is regular, and φ' is surjective. Since φ is flat and $R'/\mathfrak{m}R'$ is Cohen–Macaulay, it follows that R' is perfect over R''. From this, we conclude that $\mathrm{Ext}^i_{R''}(R', R'') = 0$ for all $i \neq c$ where $c = \mathrm{depth}(R'') - \mathrm{depth}(R')$, and that $D^\varphi := \mathrm{Ext}^c_{R''}(R', R'')$ is a semidualizing R'-module. (This is the "relative dualizing module" of Avramov and Foxby [5].) This implies that $\mathrm{Ext}^2_{R'}(D^\varphi, D^\varphi) = 0$, so there is a semidualizing Q-module B such that $D^\varphi \cong R' \otimes_Q B$. (This essentially follows from a lifting theorem of Auslander et al. [3, Proposition 1.7], since Q is complete.)

The desired quasi-deformation is in the bottom row of the following diagram

$$\begin{array}{ccccc} & & R' & \longleftarrow & Q \\ & \overset{\varphi}{\nearrow} & \downarrow & & \downarrow \\ R & \longrightarrow & R' \ltimes D^{\varphi} & \longleftarrow & Q \ltimes B \end{array}$$

where $Q \ltimes B$ and $R' \ltimes D^{\varphi}$ are "trivial extensions," i.e., Nagata's "idealizations." □

9.3 Application III: Bass Series of Local Ring Homomorphisms

Our third application of semidualizing modules is a version of Theorem 1.3 from [5] where φ is only assumed to have finite G-dimension, defined in the next few items.

Definition 9.9. A finitely generated R-module G is *totally reflexive* if one has $G \cong \mathrm{Hom}_R(\mathrm{Hom}_R(G, R), R)$ and $\mathrm{Ext}^i_R(G, R) = 0 = \mathrm{Ext}^i_R(\mathrm{Hom}_R(G, R), R)$ for all $i \geqslant 1$. A finitely generated R-module M has *finite G-dimension* if there is an exact sequence $0 \to G_n \to \cdots \to G_0 \to M \to 0$ such that each G_i is totally reflexive.

Remark 9.10. If M is an R-module of finite G-dimension, then $\mathrm{Ext}^i_R(M, R) = 0$ for $i \gg 0$ and $\mathrm{Ext}_R^{\mathrm{depth}(R)-\mathrm{depth}_R(M)}(M, R) \neq 0$.

Definition 9.11. Let $R \xrightarrow{\varphi} S$ be a local ring homomorphism, and let $S \xrightarrow{\psi} \hat{S}$ be the natural map where $\hat{S}$ is the completion of S. Fix a Cohen factorization $R \xrightarrow{\dot{\varphi}} R' \xrightarrow{\varphi'} \hat{S}$ of the "semi-completion of φ," i.e., the composition $R \xrightarrow{\psi\varphi} \hat{S}$ (see the proof of Theorem 9.8). We say that φ has *finite G-dimension* if $\hat{S}$ has finite G-dimension over R'. Moreover, the map φ is *quasi-Gorenstein* if it has finite G-dimension and $\mathrm{Ext}^i_R(R', \hat{S}) = 0$ for all $i \neq \mathrm{depth}(R') - \mathrm{depth}(\hat{S})$.

The next result is the aforementioned improvement of Theorem 1.3. As with Theorem 9.8, note that the statement does not involve semidualizing modules.

Theorem 9.12 ([5, (7.1) Theorem]). *Let $(R, \mathfrak{m}) \to (S, \mathfrak{n})$ be a local ring homomorphism of finite G-dimension. Then there is a formal Laurent series $I_\varphi(t)$ with nonnegative integer coefficients such that $I_S(t) = I_R(t) I_\varphi(t)$. In particular, if S is Gorenstein, then so is R.*

Proof (Sketch of Proof When φ is Quasi-Gorenstein). Fix a Cohen factorization $R \xrightarrow{\dot{\varphi}} R' \xrightarrow{\varphi'} \hat{S}$ of the semi-completion of φ, and set $d = \mathrm{depth}(R') - \mathrm{depth}(\hat{S})$. Since φ is quasi-Gorenstein, the $\hat{S}$-module $D' = \mathrm{Ext}^d_R(R', \hat{S})$ is semidualizing. (Again, this is Avramov and Foxby's relative dualizing module [5].) If l denotes the residue field of S, then the series $I_\varphi := \sum_i \mathrm{rank}_l(\mathrm{Ext}^{i+d'}_{\hat{S}}(D', l))t^i$ satisfies the desired conditions where $d' = \mathrm{depth}(R) - \mathrm{depth}(S)$. □

10 Sketches of Solutions to Exercises

10.1 (Sketch of Solution to Exercise 2.8). As S is R-flat, there is an isomorphism

$$\operatorname{Ext}^i_S(S \otimes_R C, S \otimes_R C) \cong S \otimes_R \operatorname{Ext}^i_R(C, C)$$

for each i. It follows that (1) if $\operatorname{Ext}^i_R(C, C) = 0$, then $\operatorname{Ext}^i_S(S \otimes_R C, S \otimes_R C) = 0$, and (2) if φ is faithfully flat and $\operatorname{Ext}^i_S(S \otimes_R C, S \otimes_R C) = 0$, then $\operatorname{Ext}^i_R(C, C) = 0$. Similarly, there is a commutative diagram

$$\begin{array}{ccc} S & \xrightarrow{S \otimes_R \chi^R_C} & S \otimes_R \operatorname{Hom}_R(C, C) \\ {\scriptstyle \chi^S_{S \otimes_R C}} \downarrow & \swarrow {\scriptstyle \cong} & \\ \operatorname{Hom}_S(S \otimes_R C, S \otimes_R C) & & \end{array}$$

so (1) if χ^R_C is an isomorphism, then so is $\chi^S_{S \otimes_R C}$, and (2) if φ is faithfully flat and $\chi^S_{S \otimes_R C}$ is an isomorphism, then so is χ^R_C. □

10.2 (Sketch of Solution to Exercise 3.4). (a) It is routine to show that $\partial_n^{\operatorname{Hom}_R(X,Y)}$ is R-linear and maps $\operatorname{Hom}_R(X, Y)_n$ to $\operatorname{Hom}_R(X, Y)_{n-1}$. To show that $\operatorname{Hom}_R(X, Y)$ is an R-complex, we compute:

$$\begin{aligned}
&\partial_{n-1}^{\operatorname{Hom}_R(X,Y)}(\partial_n^{\operatorname{Hom}_R(X,Y)}(\{f_p\})) \\
&= \partial_{n-1}^{\operatorname{Hom}_R(X,Y)}(\{\partial^Y_{p+n} f_p - (-1)^n f_{p-1} \partial^X_p\}) \\
&= \{\partial^Y_{p+n-1}[\partial^Y_{p+n} f_p - (-1)^n f_{p-1}\partial^X_p] - (-1)^{n-1}[\partial^Y_{p+n-1} f_{p-1} - (-1)^n f_{p-2}\partial^X_{p-1}]\partial^X_p\} \\
&= \{\underbrace{\partial^Y_{p+n-1}\partial^Y_{p+n}}_{=0} f_p \underbrace{-(-1)^n \partial^Y_{p+n} f_{p-1}\partial^X_p - (-1)^{n-1}\partial^Y_{p+n-1} f_{p-1}\partial^X_p}_{=0} + f_{p-2} \underbrace{\partial^X_{p-1}\partial^X_p}_{=0}\} \\
&= 0.
\end{aligned}$$

(b) For $f = \{f_p\} \in \operatorname{Hom}_R(X, Y)_0$, we have

$$\partial_0^{\operatorname{Hom}_R(X,Y)}(\{f_p\}) = \{\partial^Y_p f_p - f_{p-1}\partial^X_p\}.$$

Hence, f is a chain map if and only if $\partial^Y_p f_p - f_{p-1}\partial^X_p = 0$ for all p, which is equivalent to the commutativity of the given diagram.
(c) This follows by the fact $\partial_0^{\operatorname{Hom}_R(X,Y)} \circ \partial_1^{\operatorname{Hom}_R(X,Y)} = 0$, from part (a).
(d) Since $\partial_1^{\operatorname{Hom}_R(X,Y)}(\{s_p\}) = \{\partial^Y_{p+1} s_p + s_{p-1}\partial^X_p\}$, this follows by definition. □

10.3 (Sketch of Solution to Exercise 3.5). Let $\tau\colon \mathrm{Hom}_R(R,X)\to X$ be given by $\tau_n(\{f_p\}) = f_n(1)$. We consider R as a complex concentrated in degree 0, so for all $p\neq 0$ we have $R_p = 0$, and for all n we have $\partial_n^R = 0$. It follows that for all n and all $f = \{f_p\}\in \mathrm{Hom}_R(R,X)_n$ and all $p\neq 0$, we have $f_p = 0$. Thus, for each n the natural maps $\mathrm{Hom}_R(R,X)_n \xrightarrow{\cong} \mathrm{Hom}_R(R,X_n) \xrightarrow{\cong} X_n$ are R-module isomorphisms, the composition of which is τ_n. To show that τ is a chain map, we compute:

$$\begin{aligned}
\tau_{n-1}(\partial_n^{\mathrm{Hom}_R(R,X)}(\{f_p\})) &= \tau_{n-1}(\partial_n^{\mathrm{Hom}_R(R,X)}(\ldots,0,f_0,0,\ldots))\\
&= \tau_{n-1}(\ldots,0,\partial_n^X f_0, -(-1)^n f_0\partial_1^R,0,\ldots)\\
&= \tau_{n-1}(\ldots,0,\partial_n^X f_0,0,0,\ldots)\\
&= \partial_n^X(f_0(1))\\
&= \partial_n^X(\tau_n(\{f_p\})).
\end{aligned}$$

(Note that these steps are optimal for presentation, in some sense, but they do not exactly represent the thought process we used to find the solution. Instead, we computed and simplified $\tau_{n-1}(\partial_n^{\mathrm{Hom}_R(R,X)}(\{f_p\}))$ and $\partial_n^X(\tau_n(\{f_p\}))$ separately and checked that the resulting expression was the same for both. Similar comments apply to many solutions.) □

10.4 (Sketch of Solution to Exercise 3.6). Let X be an R-complex, and let M be an R-module.

(a) Write X_* for the complex

$$X_* = \cdots \xrightarrow{(\partial_{n+1}^X)_*} (X_n)_* \xrightarrow{(\partial_n^X)_*} (X_{n-1})_* \xrightarrow{(\partial_{n-1}^X)_*} \cdots.$$

We consider M as a complex concentrated in degree 0, so for all $p\neq 0$ we have $M_p = 0$, and for all n we have $\partial_n^M = 0$. It follows that for all n and all $f = \{f_p\}\in \mathrm{Hom}_R(M,X)_n$ and all $p\neq 0$, we have $f_p = 0$. Thus, for each n the natural map $\tau_n\colon \mathrm{Hom}_R(M,X)_n \xrightarrow{\cong} \mathrm{Hom}_R(M,X_n) = (X_*)_n$ is an R-module isomorphism. Thus, it remains to show that the map $\tau\colon \mathrm{Hom}_R(M,X)\to X_*$ given by $f = \{f_p\}\mapsto f_0$ is a chain map. We compute:

$$\begin{aligned}
\tau_{n-1}(\partial_n^{\mathrm{Hom}_R(M,X)}(\{f_p\})) &= \tau_{n-1}(\partial_n^{\mathrm{Hom}_R(M,X)}(\ldots,0,f_0,0,\ldots))\\
&= \tau_{n-1}(\ldots,0,\partial_n^X f_0, -(-1)^n f_0\partial_1^M,0,\ldots)\\
&= \tau_{n-1}(\ldots,0,\partial_n^X f_0,0,0,\ldots)\\
&= \partial_n^X f_0.
\end{aligned}$$

This yields

$$\tau_{n-1}(\partial_n^{\mathrm{Hom}_R(M,X)}(\{f_p\})) = \partial_n^X f_0 = (\partial_n^X)_*(f_0) = \partial_n^{X_*}(f_0) = \partial_n^{X_*}(\tau_n(\{f_p\}))$$

as desired

(b) Write X^* and $X^\dagger$ for the complexes

$$X^* = \cdots \xrightarrow{(-1)^n(\partial_n^X)^*} X_n^* \xrightarrow{(-1)^{n+1}(\partial_{n+1}^X)^*} X_{n+1}^* \xrightarrow{(-1)^{n+2}(\partial_{n+2}^X)^*} \cdots$$

$$X^\dagger = \cdots \xrightarrow{(\partial_n^X)^*} X_n^* \xrightarrow{(\partial_{n+1}^X)^*} X_{n+1}^* \xrightarrow{(\partial_{n+2}^X)^*} \cdots .$$

Note that the displayed pieces for X^* are in degree $-n$, $1-n$, and similarly for $X^\dagger$. We prove that $\mathrm{Hom}_R(X,M) \cong X^* \cong X^\dagger$.

As in part (a), for all $p \neq 0$ we have $M_p = 0$, and for all n we have $\partial_n^M = 0$. It follows that, for all n and all $f = \{f_p\} \in \mathrm{Hom}_R(X,M)_n$ and all $p \neq -n$, we have $f_p = 0$. Thus, for each n the map $\tau_n : \mathrm{Hom}_R(X,M)_n \xrightarrow{\cong} \mathrm{Hom}_R(X_{-n},M) = (X^*)_n$ given by $\{f_p\} \mapsto f_{-n}$ is an R-module isomorphism. Thus, for the isomorphism $\mathrm{Hom}_R(M,X) \cong X^*$, it remains to show that $\tau : \mathrm{Hom}_R(X,M) \to X^*$ is a chain map. We compute:

$$\begin{aligned}
\tau_{n-1}(\partial_n^{\mathrm{Hom}_R(X,M)}(\{f_p\})) &= \tau_{n-1}(\partial_n^{\mathrm{Hom}_R(X,M)}(\ldots,0,f_{-n},0,\ldots)) \\
&= \tau_{n-1}(\ldots,0,\partial_0^M f_{-n}, -(-1)^n f_{-n}\partial_{1-n}^X, 0,\ldots) \\
&= \tau_{n-1}(\ldots,0,(-1)^{n-1} f_{-n}\partial_{1-n}^X, 0,0,\ldots) \\
&= (-1)^{n-1} f_{-n}\partial_{1-n}^X \\
&= (-1)^{1-n}(\partial_{1-n}^X)^*(f_{-n}) \\
&= \partial_n^{X^*}(f_{-n}) \\
&= \partial_n^{X^*}(\tau_n(\{f_p\})).
\end{aligned}$$

For the isomorphism $X^* \cong X^\dagger$, we first observe that $X^\dagger$ is an R-complex. Next, note the following: given an R-complex Y, the following diagram describes an isomorphism of R-complexes.

$$\begin{array}{ccccccccc}
Y_{4n+2} & \xrightarrow{(-1)^{-4n-1}\partial_{4n+2}^Y} & Y_{4n+1} & \xrightarrow{(-1)^{-4n}\partial_{4n+1}^Y} & Y_{4n} & \xrightarrow{(-1)^{-4n+1}\partial_{4n}^Y} & Y_{4n-1} & \xrightarrow{(-1)^{-4n+2}\partial_{4n-1}^Y} & Y_{4n-2} \\
\downarrow{\scriptstyle (-1)^{4n+2}} & & {\scriptstyle (-1)^{4n+1}}\downarrow & & {\scriptstyle (-1)^{4n+1}}\downarrow & & {\scriptstyle (-1)^{4n}}\downarrow & & {\scriptstyle (-1)^{4n-2}}\downarrow \\
Y_{4n+2} & \xrightarrow[\partial_{4n+2}^Y]{} & Y_{4n+1} & \xrightarrow[\partial_{4n+1}^Y]{} & Y_{4n} & \xrightarrow[\partial_{4n}^Y]{} & Y_{4n-1} & \xrightarrow[\partial_{4n-1}^Y]{} & Y_{4n-2}
\end{array}$$

Now, apply this observation to $X^\dagger$. □

10.5 (Sketch of Solution to Exercise 3.7). (a) Let $f\colon X \to Y$ be a chain map. By definition, we have $f_i\partial^X_{i+1} = \partial^Y_{i+1}f_{i+1}$. It follows readily that $f_i(\operatorname{Im}(\partial^X_{i+1})) \subseteq \operatorname{Im}(\partial^Y_{i+1})$ and $f_i(\operatorname{Ker}(\partial^X_i)) \subseteq \operatorname{Ker}(\partial^Y_i)$. From this, it is straightforward to show that the map $\operatorname{Ker}(\partial^X_i)/\operatorname{Im}(\partial^X_{i+1}) \to \operatorname{Ker}(\partial^Y_i)/\operatorname{Im}(\partial^Y_{i+1})$ given by $\overline{x} \mapsto \overline{f_i(x)}$ is a well-defined R-module homomorphism, as desired.

(b) Assume now that f is null-homotopic. By definition, there is an element $s = \{s_p\} \in \operatorname{Hom}_R(X,Y)_1$ such that $\{f_p\} = \partial_1^{\operatorname{Hom}_R(X,Y)}(\{s_p\}) = \{\partial^Y_{p+1}s_p + s_{p-1}\partial^X_p\}$. It follows that for each i and each $\overline{x} \in \mathrm{H}_i(X)$, one has

$$\mathrm{H}_i(f)(\overline{x}) = \overline{f_i(x)} = \overline{\underbrace{\partial^Y_{i+1}(s_i(x))}_{\in\operatorname{Im}(\partial^Y_{i+1})} + s_{i-1}\underbrace{(\partial^X_i(x))}_{=0}} = 0$$

in $\mathrm{H}_i(Y)$. □

10.6 (Sketch of Solution to Exercise 3.9). Let $f\colon X \xrightarrow{\cong} Y$ be an isomorphism between the R-complexes X and Y. Then for each i the map f_i induces isomorphisms $\operatorname{Ker}(\partial^X_i) \xrightarrow{\cong} \operatorname{Ker}(\partial^Y_i)$ and $\operatorname{Im}(\partial^X_{i+1}) \xrightarrow{\cong} \operatorname{Im}(\partial^Y_{i+1})$. It follows that f_i induces an isomorphism $\operatorname{Ker}(\partial^X_i)/\operatorname{Im}(\partial^X_{i+1}) \xrightarrow{\cong} \operatorname{Ker}(\partial^Y_i)/\operatorname{Im}(\partial^Y_{i+1})$, as desired. □

10.7 (Sketch of Solution to Exercise 3.10). Let M be an R-module with augmented projective resolution P^+.

$$P^+ = \cdots \xrightarrow{\partial^P_2} P_1 \xrightarrow{\partial^P_1} P_0 \xrightarrow{\tau} M \to 0.$$

It is straightforward to check that the following diagram commutes

$$\begin{array}{ccccccc}
P & & \cdots \xrightarrow{\partial^P_2} & P_1 & \xrightarrow{\partial^P_1} & P_0 & \longrightarrow & 0 \\
{\scriptstyle t}\downarrow & & & \downarrow & & {\scriptstyle \tau}\downarrow & & \downarrow \\
M & & & 0 & \longrightarrow & M & \longrightarrow & 0.
\end{array}$$

The exactness of P^+ and the definition of P implies that $\mathrm{H}_i(P) = 0 = \mathrm{H}_i(M)$ for $i \neq 0$. Thus, to show that t is a quasiisomorphism, it suffices to show that $\mathrm{H}_0(t)\colon \mathrm{H}_0(P) \to \mathrm{H}_0(M)$ is an isomorphism. Notice that this can be identified with the map $\tau'\colon \operatorname{Coker}(\partial^P_1) \to M$ induced by τ. Since τ is surjective, it is straightforward to show that τ' is surjective. Using the fact that $\operatorname{Ker}(\tau) = \operatorname{Im}(\partial^P_1)$, one shows readily that τ' is injective, as desired.

The case of an injective resolution result is handled similarly. □

10.8 (Sketch of Solution to Exercise 3.14). Fix a chain map $f\colon X \to Y$ and an R-complex Z. Consider a sequence $\{g_p\} \in \operatorname{Hom}_R(Y,Z)_n$. Note that the sequence

$\mathrm{Hom}_R(f,Z)_n(\{g_p\}) = \{g_p f_p\}$ is in $\mathrm{Hom}_R(X,Z)_n$, so the map $\mathrm{Hom}_R(f,Z)$ is well-defined and of degree 0. Also, it is straightforward to show that $\mathrm{Hom}_R(f,Z)$ is R-linear. To verify that $\mathrm{Hom}_R(f,Z)$ is a chain map, we compute:

$$\begin{aligned}
\partial_n^{\mathrm{Hom}_R(X,Z)}(\mathrm{Hom}_R(f,Z)_n(\{g_p\})) &= \partial_n^{\mathrm{Hom}_R(X,Z)}(\{g_p f_p\}) \\
&= \{\partial_{p+n}^Z g_p f_p - (-1)^n g_{p-1} f_{p-1} \partial_p^X\} \\
&= \{\partial_{p+n}^Z g_p f_p - (-1)^n g_{p-1} \partial_p^Y f_p\} \\
&= \mathrm{Hom}_R(f,Z)_n(\{\partial_{p+n}^Z g_p - (-1)^n g_{p-1} \partial_p^Y\}) \\
&= \mathrm{Hom}_R(f,Z)_n(\partial_n^{\mathrm{Hom}_R(Y,Z)}(\{g_p\})).
\end{aligned}$$

Note that the third equality follows from the fact that f is a chain map.

The computation for $\mathrm{Hom}_R(Z,f)\colon \mathrm{Hom}_R(Z,X) \to \mathrm{Hom}_R(Z,Y)$ is similar.

□

10.9 (Sketch of Solution to Exercise 3.16). It is straightforward to show that $\mu^{X,r}$ is R-linear. (Note that this uses the fact that R is commutative.) To show that $\mu^{X,r}$ is a chain map, we compute:

$$\partial_i^X(\mu_i^{X,r}(x)) = \partial_i^X(rx) = r\partial_i^X(x) = \mu_{i-1}^{X,r}(\partial_i^X(x)).$$

For the induced map $\mathrm{H}_i(\mu^{X,r})$, we have $\mathrm{H}_i(\mu^{X,r})(\overline{x}) = \overline{rx} = r\overline{x}$, as desired. □

10.10 (Sketch of Solution to Exercise 3.17). Exercise 3.16 shows that for all $r \in R$, the sequence $\chi_0^X(r) = \mu^{X,r}$ is a chain map. That is, $\chi_0^X(r)$ is a cycle in $\mathrm{Hom}_R(X,X)_0$. It follows that the next diagram commutes

$$\begin{array}{ccccc}
0 & \longrightarrow & R & \longrightarrow & 0 \\
\downarrow & & \downarrow{\scriptstyle \chi_0^X} & & \downarrow \\
\mathrm{Hom}_R(X,X)_1 & \xrightarrow{\partial_1^{\mathrm{Hom}_R(X,X)}} & \mathrm{Hom}_R(X,X)_0 & \xrightarrow{\partial_0^{\mathrm{Hom}_R(X,X)}} & \mathrm{Hom}_R(X,X)_{-1}
\end{array}$$

so χ^X is a chain map. □

10.11 (Sketch of Solution to Exercise 4.2). Let X, Y, and Z be R-complexes.

(a) It is routine to show that $\partial_n^{X\otimes_R Y}$ is R-linear and maps from $(X \otimes_R Y)_n$ to $(X \otimes_R Y)_{n-1}$. To show that $X \otimes_R Y$ is an R-complex, we compute:

$$\begin{aligned}
&\partial_{n-1}^{X\otimes_R Y}(\partial_n^{X\otimes_R Y}(\ldots,0,x_p \otimes y_{n-p},0,\ldots)) \\
&= \partial_{n-1}^{X\otimes_R Y}(\ldots,0,\partial_p^X(x_p)\otimes y_{n-p},(-1)^p x_p \otimes \partial_{n-p}^Y(y_{n-p}),0,\ldots) \\
&= \partial_{n-1}^{X\otimes_R Y}(\ldots,0,\partial_p^X(x_p)\otimes y_{n-p},0,0,\ldots) \\
&\quad + \partial_{n-1}^{X\otimes_R Y}(\ldots,0,0,(-1)^p x_p \otimes \partial_{n-p}^Y(y_{n-p}),0,\ldots)
\end{aligned}$$

$$= (\ldots, 0, \underbrace{\partial_{p-1}^X(\partial_p^X(x_p))}_{=0} \otimes y_{n-p}, (-1)^{p-1}\partial_p^X(x_p) \otimes \partial_{n-p}^Y(y_{n-p}), 0, \ldots)$$

$$+ (\ldots, 0, (-1)^p \partial_p^X(x_p) \otimes \partial_{n-p}^Y(y_{n-p}), (-1)^p x_p \otimes \underbrace{\partial_{n-p-1}^Y(\partial_{n-p}^Y(y_{n-p}))}_{=0}, 0, \ldots).$$

The only possibly nontrivial entry in this sum is

$$\begin{aligned}(-1)^{p-1}\partial_p^X(x_p) \otimes \partial_{n-p}^Y(y_{n-p}) &+ (-1)^p \partial_p^X(x_p) \otimes \partial_{n-p}^Y(y_{n-p}) \\ &= (-1)^{p-1}[\partial_p^X(x_p) \otimes \partial_{n-p}^Y(y_{n-p}) - \partial_p^X(x_p) \otimes \partial_{n-p}^Y(y_{n-p})] = 0\end{aligned}$$

so we have $\partial_{n-1}^{X \otimes_R Y} \partial_n^{X \otimes_R Y} = 0$.

(b) The isomorphism $X \xrightarrow{\cong} R \otimes_R X$ is given by $x \mapsto (\ldots, 0, 1 \otimes x, 0, \ldots)$. Check that this is an isomorphism as in 10.3, using that R is concentrated in degree 0.

(c) As the hint suggests, the isomorphism $g\colon X \otimes_R Y \xrightarrow{\cong} Y \otimes_R X$ requires a sign-change: we define

$$g_n(\ldots, 0, x_p \otimes y_{n-p}, 0, \ldots) := (\ldots, 0, (-1)^{p(n-p)} y_{n-p} \otimes x_p, 0, \ldots).$$

It is routine to show that $g = \{g_n\}$ is R-linear and maps $(X \otimes_R Y)_n$ to $(Y \otimes_R X)_n$. The following computation shows that g is a chain map:

$$\begin{aligned}
&g_{n-1}(\partial_n^{X \otimes_R Y}(\ldots, 0, x_p \otimes y_{n-p}, 0, \ldots)) \\
&= g_{n-1}(\ldots, 0, \partial_p^X(x_p) \otimes y_{n-p}, (-1)^p x_p \otimes \partial_{n-p}^Y(y_{n-p}), 0, \ldots) \\
&= g_{n-1}(\ldots, 0, \partial_p^X(x_p) \otimes y_{n-p}, 0, \ldots) \\
&\quad + g_{n-1}(\ldots, 0, (-1)^p x_p \otimes \partial_{n-p}^Y(y_{n-p}), 0, \ldots) \\
&= (\ldots, 0, (-1)^{(p-1)(n-p)} y_{n-p} \otimes \partial_p^X(x_p), 0, \ldots) \\
&\quad + (\ldots, 0, (-1)^{p+p(n-p-1)} \partial_{n-p}^Y(y_{n-p}) \otimes x_p, 0, \ldots) \\
&= (\ldots, 0, (-1)^{(p+1)(n-p)} y_{n-p} \otimes \partial_p^X(x_p), 0, \ldots) \\
&\quad + (\ldots, 0, (-1)^{p(n-p)} \partial_{n-p}^Y(y_{n-p}) \otimes x_p, 0, \ldots) \\
&= (\ldots, 0, (-1)^{p(n-p)} \partial_{n-p}^Y(y_{n-p}) \otimes x_p, (-1)^{(p+1)(n-p)} y_{n-p} \otimes \partial_p^X(x_p), 0, \ldots) \\
&= (\ldots, 0, (-1)^{p(n-p)} \partial_{n-p}^Y(y_{n-p}) \otimes x_p, (-1)^{p(n-p)+(n-p)} y_{n-p} \otimes \partial_p^X(x_p), 0, \ldots) \\
&= \partial_n^{Y \otimes_R X}(\ldots, 0, (-1)^{p(n-p)} y_{n-p} \otimes x_p, 0, \ldots) \\
&= \partial_n^{Y \otimes_R X}(g_n(\ldots, 0, x_p \otimes y_{n-p}, 0, \ldots))
\end{aligned}$$

Similarly, one shows that the map $h: Y \otimes_R X \xrightarrow{\cong} X \otimes_R Y$ defined as

$$h_n(\dots, 0, y_p \otimes x_{n-p}, 0, \dots) := (\dots, 0, (-1)^{p(n-p)} x_{n-p} \otimes y_p, 0, \dots)$$

is a chain map. Moreover, one has

$$\begin{aligned} g_n(h_n(\dots, 0, y_p \otimes x_{n-p}, 0, \dots)) &= g_n(\dots, 0, (-1)^{p(n-p)} x_{n-p} \otimes y_p, 0, \dots) \\ &= (\dots, 0, (-1)^{p(n-p)}(-1)^{(n-p)p} y_p \otimes x_{n-p}, 0, \dots) \\ &= (\dots, 0, ((-1)^{p(n-p)})^2 y_p \otimes x_{n-p}, 0, \dots) \\ &= (\dots, 0, y_p \otimes x_{n-p}, 0, \dots) \end{aligned}$$

so hg is the identity on $Y \otimes_R X$. Similarly, gh is the identity on $X \otimes_R Y$. It follows that h is a two-sided inverse for g, so g is an isomorphism.

(d) For the isomorphism $X \otimes_R (Y \otimes_R Z) \to (X \otimes_R Y) \otimes_R Z$, we change notation. The point is that elements of $X \otimes_R (Y \otimes_R Z)$ are sequences where each entry is itself a sequence. For instance, a generator in $X \otimes_R (Y \otimes_R Z)_n$ is of the form

$$\big(\dots, (\dots, 0, 0, \dots), (\dots, 0, x_p \otimes (y_q \otimes z_{n-p-q}), 0, \dots), (\dots, 0, 0, \dots), \dots\big)$$

which we simply write as $x_p \otimes (y_q \otimes z_{n-p-q})$. One has to be careful here not to combine elements illegally: the elements $x_a \otimes (y_b \otimes z_{n-a-b})$ and $x_p \otimes (y_q \otimes z_{n-p-q})$ are only in the same summand of $X \otimes_R (Y \otimes_R Z)_n$ if $a = p$ and $b = q$.

Using this protocol, define $f: X \otimes_R (Y \otimes_R Z) \to (X \otimes_R Y) \otimes_R Z$ as

$$f_{p+q+r}(x_p \otimes (y_q \otimes z_r)) = (x_p \otimes y_q) \otimes z_r.$$

(This has no sign-change since no factors are commuted.) As in the previous case, showing that f is an isomorphism reduces to showing that it is a chain map:

$$\begin{aligned} &\partial^{(X \otimes_R Y) \otimes_R Z}_{p+q+r}(f_{p+q+r}(x_p \otimes (y_q \otimes z_r))) \\ &\quad = \partial^{(X \otimes_R Y) \otimes_R Z}_{p+q+r}((x_p \otimes y_q) \otimes z_r) \\ &\quad = \partial^{X \otimes_R Y}_{p+q}(x_p \otimes y_q) \otimes z_r + (-1)^{p+q}(x_p \otimes y_q) \otimes \partial^Z_r(z_r) \\ &\quad = (\partial^X_p(x_p) \otimes y_q) \otimes z_r + (-1)^p (x_p \otimes \partial^Y_q(y_q)) \otimes z_r + (-1)^{p+q}(x_p \otimes y_q) \otimes \partial^Z_r(z_r) \\ &\quad = f_{p+q+r-1}(\partial^X_p(x_p) \otimes (y_q \otimes z_r)) + (-1)^p x_p \otimes (\partial^Y_q(y_q) \otimes z_r) \\ &\qquad\quad + (-1)^{p+q} x_p \otimes (y_q \otimes \partial^Z_r(z_r)) \\ &\quad = f_{p+q+r-1}(\partial^{X \otimes_R (Y \otimes_R Z)}_{p+q+r}(x_p \otimes (y_q \otimes z_r))). \end{aligned}$$

The last equality in the sequence follows like the first two. □

10.12 (Sketch of Solution to Exercise 4.8). Fix a chain map $f: X \to Y$ and an R-complex Z. For each element $z_p \otimes x_q \in (Z \otimes_R X)_n$, the output $(Z \otimes_R f)_n(z_p \otimes x_q) = z_p \otimes f_q(x_q)$ is in $(X \otimes_R Z)_n$, so the map $Z \otimes_R f: Z \otimes_R X \to Z \otimes_R Y$ is well-defined and of degree 0. Also, it is straightforward to show that $Z \otimes_R f$ is R-linear. To show that $Z \otimes_R f$ is a chain map, we compute:

$$\begin{aligned}
\partial_{p+q}^{Z\otimes_R Y}((Z \otimes_R f)_{p+q}(z_p \otimes x_q)) &= \partial_{p+q}^{Z\otimes_R Y}(z_p \otimes f_q(x_q)) \\
&= \partial_p^Z(z_p) \otimes f_q(x_q) + (-1)^p z_p \otimes \partial_q^Y(f_q(x_q)) \\
&= \partial_p^Z(z_p) \otimes f_q(x_q) + (-1)^p z_p \otimes f_{q-1}(\partial_q^X(x_q)) \\
&= (Z \otimes_R f)_{p+q-1}(\partial_p^Z(z_p) \otimes x_q + (-1)^p z_p \otimes \partial_q^X(x_q)) \\
&= (Z \otimes_R f)_{p+q-1}(\partial_{p+q}^{Z\otimes_R Y}(z_p \otimes x_q)).
\end{aligned}$$

The proof for $f \otimes_R Z$ is similar. □

10.13 (Sketch of Solution to Exercise 4.11). We briefly describe our linear algebra protocols before sketching this solution. For each $n \in \mathbb{N}$, the module R^n consists of the column vectors of size n with entries from R. Under this convention, the R-module homomorphisms $R^m \xrightarrow{f} R^n$ are in bijection with the $n \times m$ matrices with entries from R. Moreover, the matrix representing f has columns $f(e_1), \ldots, f(e_m)$ where $e_1, \ldots, e_m$ is the standard basis for R^m.

More generally, given free R-modules V and W with ordered bases $v_1, \ldots, v_m$ and $w_1, \ldots, w_n$ respectively, the R-module homomorphisms $V \xrightarrow{f} W$ are in bijection with the $n \times m$ matrices with entries from R. Moreover, the jth column of the matrix representing f with these bases is $\begin{pmatrix} a_{1,j} \\ a_{2,j} \\ \vdots \\ a_{n,j} \end{pmatrix}$ where $f(v_j) = \sum_{i=1}^n a_{i,j} w_i$.

Note that these protocols allow for function composition to be represented by matrix multiplication in the same order: if f and g are represented by matrices A and B, respectively, then gf is represented by BA, using the same ordered bases.

Now for the sketch of the solution. By definition, we have

$$\begin{aligned}
K^R(x) = &\quad 0 \to Re \xrightarrow{x} R1 \to 0 \\
K^R(y) = &\quad 0 \to Rf \xrightarrow{y} R1 \to 0 \\
K^R(z) = &\quad 0 \to Rg \xrightarrow{z} R1 \to 0.
\end{aligned}$$

The notation indicates that, in each case, the basis vector in degree 0 is 1 and the basis vectors in degree 1 are e, f, and g, respectively.

We have $K^R(x, y) = K^R(x) \otimes_R K^R(y)$, by definition. It follows that $K^R(x, y)_n = \bigoplus_{i+j=n} K^R(x)_i \otimes K^R(y)_j$. Since $K^R(x)_i = 0 = K^R(y)_i$ for all $i \neq 0, 1$, it follows readily that $K^R(x, y)_n = 0$ for all $n \neq 0, 1, 2$. In degree 0, we have

$$K^R(x, y)_0 = K^R(x)_0 \otimes_R K^R(y)_0 = R1 \otimes_R R1 \cong R$$

with basis vector $1 \otimes 1$. Similarly, we have

$$K^R(x, y)_1 = (Re \otimes_R R1) \bigoplus (R1 \otimes_R Rf) \cong R^2$$

with ordered basis $e \otimes 1, 1 \otimes f$. In degree 2, we have

$$K^R(x, y)_2 = Re \otimes_R Rf \cong R$$

with basis vector $e \otimes f$. In particular, $K^R(x, y)$ has the following shape:

$$K^R(x, y) = \quad 0 \to R \xrightarrow{d_2} R^2 \xrightarrow{d_1} R \to 0.$$

Using the linear algebra protocols described above, we identify an element $re \otimes 1 + s1 \otimes f \in K^R(x, y)_1 \cong R^2$ with the column vector $\binom{r}{s}$. It follows that d_2 is a 2×1 matrix, and d_1 is a 1×2 matrix, which we identify from the images of the corresponding basis vectors. First, for d_2:

$$d_2(e \otimes f) = \partial_1^{K^R(x)}(e) \otimes f + (-1)^{|e|} e \otimes \partial_1^{K^R(y)}(f) = x1 \otimes f - ye \otimes 1.$$

This corresponds to the column vector $\binom{-y}{x}$, and it follows that d_2 is represented by the matrix $\binom{-y}{x}$. For d_1, we have two basis vectors to consider, in order:

$$d_1(e \otimes 1) = \partial_1^{K^R(x)}(e) \otimes 1 + (-1)^{|e|} e \otimes \partial_1^{K^R(y)}(1) = x1 \otimes 1$$
$$d_1(1 \otimes f) = \partial_1^{K^R(x)}(1) \otimes f + (-1)^{|1|} 1 \otimes \partial_1^{K^R(y)}(f) = y1 \otimes 1.$$

It follows that the map d_1 is represented by the row matrix $(x \;\; y)$, so in summary:

$$K^R(x, y) = \quad 0 \to R \xrightarrow{\binom{-y}{x}} R^2 \xrightarrow{(x \;\; y)} R \to 0.$$

The condition $d_1 d_2 = 0$ follows from the fact that $xy = yx$, that is, from the commutativity of R. Sometimes, relations of this form are called "Koszul relations."

We repeat the process for $K^R(x, y, z) = K^R(x, y) \otimes_R K^R(z)$, showing

$$K^R(x, y, z) = \quad 0 \to R \xrightarrow{\begin{pmatrix} z \\ -y \\ x \end{pmatrix}} R^3 \xrightarrow{\begin{pmatrix} -y & -z & 0 \\ x & 0 & -z \\ 0 & x & y \end{pmatrix}} R^3 \xrightarrow{(x \;\; y \;\; z)} R \to 0.$$

For all n, we have $K^R(x, y, z)_n = \bigoplus_{i+j=n} K^R(x, y)_i \otimes K^R(z)_j$. As $K^R(x, y)_i = 0 = K^R(z)_j$ for all $i \neq 0, 1, 2$ and all $j \neq 0, 1$, we have $K^R(x, y, z)_n = 0$ for $n \neq 0, 1, 2, 3$. In degree 0,

$$K^R(x, y, z)_0 = K^R(x, y)_0 \otimes_R K^R(z)_0 = R(1 \otimes 1) \otimes_R R1 \cong R$$

with basis vector $1 \otimes 1 \otimes 1$. The other modules and bases are computed similarly. We summarize in the following table:

i	$K^R(x,y,z)_i$	Ordered basis
3	R^1	$e \otimes f \otimes g$
2	R^3	$e \otimes f \otimes 1, e \otimes 1 \otimes g, 1 \otimes f \otimes g$
1	R^3	$e \otimes 1 \otimes 1, 1 \otimes f \otimes 1, 1 \otimes 1 \otimes g$
0	R^1	$1 \otimes 1 \otimes 1$

In particular, $K^R(x,y,z)$ has the following shape:

$$K^R(x,y,z) = \quad 0 \to R \xrightarrow{d_3} R^3 \xrightarrow{d_2} R^3 \xrightarrow{d_1} R \to 0.$$

It follows that d_3 is a 3×1 matrix, d_2 is a 3×3 matrix, and d_1 is a 1×3 matrix, which we identify from the images of the corresponding basis vectors. First, for d_3:

$$\begin{aligned} d_3(e \otimes f \otimes g) &= \partial_1^{K^R(x,y)}(e \otimes f) \otimes g + (-1)^{|e \otimes f|} e \otimes f \otimes \partial_1^{K^R(z)}(g) \\ &= x1 \otimes f \otimes g - ye \otimes 1 \otimes g + ze \otimes f \otimes 1. \end{aligned}$$

This corresponds to the column vector $\begin{pmatrix} z \\ -y \\ x \end{pmatrix}$, and it follows that d_2 is represented by the matrix $\begin{pmatrix} z \\ -y \\ x \end{pmatrix}$. Next, for d_2:

$$\begin{aligned} d_2(e \otimes f \otimes 1) &= \partial_1^{K^R(x,y)}(e \otimes f) \otimes 1 + (-1)^{|e \otimes f|} e \otimes f \otimes \partial_1^{K^R(z)}(1) \\ &= x1 \otimes f \otimes 1 - ye \otimes 1 \otimes 1 \\ d_2(e \otimes 1 \otimes g) &= \partial_1^{K^R(x,y)}(e \otimes 1) \otimes g + (-1)^{|e \otimes 1|} e \otimes 1 \otimes \partial_1^{K^R(z)}(g) \\ &= x1 \otimes 1 \otimes g - ze \otimes 1 \otimes 1 \\ d_2(1 \otimes f \otimes g) &= \partial_1^{K^R(x,y)}(1 \otimes f) \otimes g + (-1)^{|1 \otimes f|} 1 \otimes f \otimes \partial_1^{K^R(z)}(g) \\ &= y1 \otimes 1 \otimes g - z1 \otimes f \otimes 1. \end{aligned}$$

It follows that d_2 is represented by the following matrix $d_2 = \begin{pmatrix} -y & -z & 0 \\ x & 0 & -z \\ 0 & x & y \end{pmatrix}$. For d_1, we have three basis vectors to consider, in order:

$$\begin{aligned} d_1(e \otimes 1 \otimes 1) &= \partial_1^{K^R(x,y)}(e \otimes 1) \otimes 1 + (-1)^{|e \otimes 1|} e \otimes 1 \otimes \partial_1^{K^R(z)}(1) \\ &= x1 \otimes 1 \otimes 1 \\ d_1(1 \otimes f \otimes 1) &= \partial_1^{K^R(x,y)}(1 \otimes f) \otimes 1 + (-1)^{|1 \otimes f|} 1 \otimes f \otimes \partial_1^{K^R(z)}(1) \\ &= y1 \otimes 1 \otimes 1 \end{aligned}$$

$$d_1(1 \otimes 1 \otimes g) = \partial_1^{K^R(x,y)}(1 \otimes 1) \otimes g + (-1)^{|1 \otimes 1|} 1 \otimes 1 \otimes \partial_1^{K^R(z)}(g)$$
$$= z1 \otimes 1 \otimes 1$$

It follows that the map d_1 is represented by the row matrix $(x \ \ y \ \ z)$, as desired. □

10.14 (Sketch of Solution to Exercise 4.12). This is essentially a manifestation of Pascal's Triangle. We proceed by induction on n, where $\mathbf{x} = x_1, \ldots, x_n$. The base case $n = 1$ follows from directly from Definition 4.10. Assume now that the result holds for sequences of length $t \geqslant 1$. We prove the it holds for sequences $x_1, \ldots, x_t, x_{t+1}$. By definition, we have

$$K^R(x_1, \ldots, x_{t+1}) = K^R(x_1, \ldots, x_t) \otimes_R K^R(x_{t+1})$$

Since $K^R(x_{t+1})_i = 0$ for all $i \neq 0, 1$, it follows that

$$\begin{aligned} K^R(x_1, \ldots, x_{t+1})_m &= K^R(x_1, \ldots, x_t) \otimes_R K^R(x_{t+1}) \\ &= (K^R(x_1, \ldots, x_t)_m \otimes_R K^R(x_{t+1})_0) \\ &\bigoplus (K^R(x_1, \ldots, x_t)_{m-1} \otimes_R K^R(x_{t+1})_1) \\ &\cong (R^{\binom{t}{m}} \otimes_R R) \bigoplus (R^{\binom{t}{m-1}} \otimes_R R) \\ &\cong R^{\binom{t}{m} + \binom{t}{m-1}} \\ &= R^{\binom{t+1}{m}} \end{aligned}$$

which is the desired conclusion. □

10.15 (Sketch of Solution to Exercise 4.13). This follows from the commutativity of tensor product

$$K^R(\mathbf{x}) = K^R(x_1) \otimes_R \cdots \otimes_R K^R(x_n) \cong K^R(x_{\sigma(1)}) \otimes_R \cdots \otimes_R K^R(x_{\sigma(n)}) = K^R(\mathbf{x}')$$

which is the desired conclusion. □

10.16 (Sketch of Solution to Exercise 4.17). Set $(-)^* = \operatorname{Hom}_R(-, R)$. We use the following linear algebra facts freely. First, there is an isomorphism $(R^n)^* \cong R^n$. Second, given an R-module homomorphism $R^m \to R^n$ represented by a matrix A, the dual map $(R^n)^* \to (R^m)^*$ is represented by the transpose A^{T}.

Let $x, y, z \in R$. First, we verify that $\operatorname{Hom}_R(K^R(x), R) \cong \Sigma K^R(x)$, by the above linear algebra facts with Exercise 3.6. The complex $K^R(x)$ has the form

$$K^R(x) = \quad 0 \to R \xrightarrow{x} R \to 0$$

concentrated in degrees 0 and 1. Thus, the shifted complex $\Sigma K^R(x)$ is of the form

$$\Sigma K^R(x) = \quad 0 \to R \xrightarrow{-x} R \to 0$$

concentrated in degrees 0 and -1. By Exercise 3.6, the dual $\operatorname{Hom}_R(K^R(x), R)$ is concentrated in degrees 0 and -1, and has the form

$$\operatorname{Hom}_R(K^R(x), R) = \quad 0 \to R \xrightarrow{x} R \to 0.$$

The following diagram shows that these complexes are isomorphic.

$$\begin{array}{ccccccccc}
\Sigma K^R(x) & & 0 & \longrightarrow & R & \xrightarrow{-x} & R & \longrightarrow & 0 \\
\cong \downarrow & & & & \downarrow {\scriptstyle -1} & & \downarrow {\scriptstyle 1} & & \\
\operatorname{Hom}_R(K^R(x), R) & & 0 & \longrightarrow & R & \xrightarrow{x} & R & \longrightarrow & 0
\end{array}$$

Next, we verify the isomorphism $\operatorname{Hom}_R(K^R(x, y), R) \cong \Sigma^2 K^R(x, y)$. By Exercise 4.11, the complex $K^R(x, y)$ is of the form

$$K^R(x, y) = \quad 0 \to R \xrightarrow{\left(\begin{smallmatrix} -y \\ x \end{smallmatrix}\right)} R^2 \xrightarrow{(x\ y)} R \to 0$$

concentrated in degrees 0, 1, 2. Thus, the shifted complex $\Sigma K^R(x)$ is of the form

$$\Sigma K^R(x, y) = \quad 0 \to R \xrightarrow{\left(\begin{smallmatrix} -y \\ x \end{smallmatrix}\right)} R^2 \xrightarrow{(x\ y)} R \to 0$$

concentrated in degrees $0, -1, -2$. By Exercise 3.6, the dual $\operatorname{Hom}_R(K^R(x, y), R)$ has the form and is concentrated in degrees 0 and -1

$$\operatorname{Hom}_R(K^R(x, y), R) = \quad 0 \to R \xrightarrow{\left(\begin{smallmatrix} x \\ y \end{smallmatrix}\right)} R^2 \xrightarrow{(-y\ x)} R \to 0.$$

The following diagram shows that these complexes are isomorphic.

$$\begin{array}{ccccccccccc}
\Sigma K^R(x, y) & & 0 & \longrightarrow & R & \xrightarrow{\left(\begin{smallmatrix} -y \\ x \end{smallmatrix}\right)} & R^2 & \xrightarrow{(x\ y)} & R & \longrightarrow & 0 \\
\cong \downarrow & & & & \downarrow {\scriptstyle 1} & & \downarrow {\scriptstyle \left(\begin{smallmatrix} 0 & 1 \\ -1 & 0 \end{smallmatrix}\right)} & & \downarrow {\scriptstyle 1} & & \\
\operatorname{Hom}_R(K^R(x, y), R) & & 0 & \longrightarrow & R & \xrightarrow[\left(\begin{smallmatrix} x \\ y \end{smallmatrix}\right)]{} & R^2 & \xrightarrow{(-y\ x)} & R & \longrightarrow & 0
\end{array}$$

Note that we found this isomorphism, as follows. Use the identity in degree 1. Consider the matrix $\left(\begin{smallmatrix} a & b \\ c & d \end{smallmatrix}\right)$ in degree 1, and solve the linear equations needed to make the right-most square commute. Then check that the identity in degree 2 makes the left-most square commute.

Lastly, we verify the isomorphism $\mathrm{Hom}_R(K^R(x,y,z),R) \cong \Sigma^3 K^R(x,y,z)$. By Exercise 4.11, the complex $K^R(x,y,x)$ is of the form

$$K^R(x,y,z) = \quad 0 \to R \xrightarrow{\begin{pmatrix} z \\ -y \\ x \end{pmatrix}} R^3 \xrightarrow{\begin{pmatrix} -y & -z & 0 \\ x & 0 & -z \\ 0 & x & y \end{pmatrix}} R^3 \xrightarrow{(x\ y\ z)} R \to 0$$

concentrated in degrees $0, 1, 2, 3$. The shifted complex $\Sigma^3 K^R(x,y,z)$ is of the form

$$\Sigma^3 K^R(x,y,z) = \quad 0 \to R \xrightarrow{\begin{pmatrix} -z \\ y \\ -x \end{pmatrix}} R^3 \xrightarrow{\begin{pmatrix} y & z & 0 \\ -x & 0 & z \\ 0 & -x & -y \end{pmatrix}} R^3 \xrightarrow{(-x\ -y\ -z)} R \to 0$$

concentrated in degrees $0, -1, -2, -3$. By Exercise 3.6, the dual $\mathrm{Hom}_R(K^R(x),R)$ has the form and is concentrated in degrees $0, -1, -2, -3$

$$\mathrm{Hom}_R(K^R(x,y,z),R) = \quad 0 \to R \xrightarrow{\begin{pmatrix} x \\ y \\ z \end{pmatrix}} R^3 \xrightarrow{\begin{pmatrix} -y & x & 0 \\ -z & 0 & x \\ 0 & -z & y \end{pmatrix}} R^3 \xrightarrow{(z\ -y\ x)} R \to 0.$$

The following diagram

$$\begin{array}{ccccccccccc}
\Sigma^3 K^R(x,y,z) & & 0 \to & R & \xrightarrow{\begin{pmatrix} -z \\ y \\ -x \end{pmatrix}} & R^3 & \xrightarrow{\begin{pmatrix} y & z & 0 \\ -x & 0 & z \\ 0 & -x & -y \end{pmatrix}} & R^3 & \xrightarrow{(-x\ -y\ -z)} & R & \to 0 \\
\cong \downarrow & & & 1 \downarrow & & \downarrow \begin{pmatrix} 0 & 0 & -1 \\ 0 & 1 & 0 \\ -1 & 0 & 0 \end{pmatrix} & & \downarrow \begin{pmatrix} 0 & 0 & -1 \\ 0 & 1 & 0 \\ -1 & 0 & 0 \end{pmatrix} & & \downarrow 1 & \\
\mathrm{Hom}_R(K^R(x,y,z),R) & & 0 \to & R & \xrightarrow[\begin{pmatrix} x \\ y \\ z \end{pmatrix}]{} & R^3 & \xrightarrow[\begin{pmatrix} -y & x & 0 \\ -z & 0 & x \\ 0 & -z & y \end{pmatrix}]{} & R^3 & \xrightarrow[(z\ -y\ x)]{} & R & \to 0
\end{array}$$

shows that these complexes are isomorphic. □

10.17 (Sketch of Solution to Exercise 4.21). Let $\mathbf{x} = x_1, \ldots, x_n \in R$.

By definition of $\tilde{K}^R(\mathbf{x})$, the first differential is given by $d_1(e_i) = x_i$ for $i = 1, \ldots, n$, represented in matrix form by $R^n \xrightarrow{(x_1\ \cdots\ x_n)} R$. In particular, we have

$$\tilde{K}^R(x_1) = \quad 0 \to R \xrightarrow{x_1} R \to 0.$$

For $\tilde{K}^R(x_1, x_2)$ this says that we only have to compute d_2:

$$d_2(e_1 \wedge e_2) = x_1 e_2 - x_2 e_1$$

so we have

$$\tilde{K}^R(x_1,x_2) = \quad 0 \to R \xrightarrow{\left(\begin{smallmatrix}-x_2\\ x_1\end{smallmatrix}\right)} R^2 \xrightarrow{(x_1\ x_2)} R \to 0.$$

In comparison with Exercise 4.11, this says that

$$\tilde{K}^R(x,y) = \quad 0 \to R \xrightarrow{\left(\begin{smallmatrix}-y\\ x\end{smallmatrix}\right)} R^2 \xrightarrow{(x\ y)} R \to 0.$$

For $\tilde{K}^R(x,y,z)$, we specify an ordering on the basis for $\tilde{K}^R(x,y,z)_2$

$$e_1 \wedge e_2, e_1 \wedge e_3, e_2 \wedge e_3$$

and we compute:

$$\begin{aligned}
d_2(e_1 \wedge e_2) &= x_1e_2 - x_2e_1\\
d_2(e_1 \wedge e_3) &= x_1e_3 - x_3e_1\\
d_2(e_2 \wedge e_3) &= x_2e_3 - x_3e_2\\
d_3(e_1 \wedge e_2 \wedge e_3) &= x_1e_2 \wedge e_3 - x_2e_1 \wedge e_3 + x_3e_1 \wedge e_2.
\end{aligned}$$

In summary, we have the following:

$$K^R(x_1,x_2,x_3) = \quad 0 \to R \xrightarrow{\left(\begin{smallmatrix}x_3\\ -x_2\\ x_1\end{smallmatrix}\right)} R^3 \xrightarrow{\left(\begin{smallmatrix}-x_2 & -x_3 & 0\\ x_1 & 0 & -x_3\\ 0 & x_1 & x_2\end{smallmatrix}\right)} R^3 \xrightarrow{(x_1\ x_2\ x_3)} R \to 0.$$

Again, this compares directly with Exercise 4.11. □

10.18 (Sketch of Solution to Exercise 4.27). In the following multiplication tables, given an element x from the left column and an element y from the top row, the corresponding element in the table is the product xy. (See also the tables on the following page.)

$\bigwedge R^1$	1	e
1	1	e
e	e	0

$\bigwedge R^2$	1	e_1	e_2	$e_1 \wedge e_2$
1	1	e_1	e_2	$e_1 \wedge e_2$
e_1	e_1	0	$e_1 \wedge e_2$	0
e_2	e_2	$-e_1 \wedge e_2$	0	0
$e_1 \wedge e_2$	$e_1 \wedge e_2$	0	0	0

10.19 (Sketch of Solution to Exercise 4.29). We proceed by induction on t.

Base case: $t = 2$. The result is trivial if ι is the identity, so assume for the rest of this paragraph that ι is not the identity permutation. Because of our assumptions on ι, in this case it has the cycle-notation form $\iota = (j_1\ \ j_2)$. If $j_1 < j_2$, then by definition we have

$\bigwedge R^3$	1	e_1	e_2	e_3
1	1	e_1	e_2	e_3
e_1	e_1	0	$e_1 \wedge e_2$	$e_1 \wedge e_3$
e_2	e_2	$-e_1 \wedge e_2$	0	$e_2 \wedge e_3$
e_3	e_3	$-e_1 \wedge e_3$	$-e_2 \wedge e_3$	0
$e_1 \wedge e_2$	$e_1 \wedge e_2$	0	0	$e_1 \wedge e_2 \wedge e_3$
$e_1 \wedge e_3$	$e_1 \wedge e_3$	0	$-e_1 \wedge e_2 \wedge e_3$	0
$e_2 \wedge e_3$	$e_2 \wedge e_3$	$e_1 \wedge e_2 \wedge e_3$	0	0
$e_1 \wedge e_2 \wedge e_3$	$e_1 \wedge e_2 \wedge e_3$	0	0	0

$\bigwedge R^3$	$e_1 \wedge e_2$	$e_1 \wedge e_3$	$e_2 \wedge e_3$	$e_1 \wedge e_2 \wedge e_3$
1	$e_1 \wedge e_2$	$e_1 \wedge e_3$	$e_2 \wedge e_3$	$e_1 \wedge e_2 \wedge e_3$
e_1	0	0	$e_1 \wedge e_2 \wedge e_3$	0
e_2	0	$-e_1 \wedge e_2 \wedge e_3$	0	0
e_3	$e_1 \wedge e_2 \wedge e_3$	0	0	0
$e_1 \wedge e_2$	0	0	0	0
$e_1 \wedge e_3$	0	0	0	0
$e_2 \wedge e_3$	0	0	0	0
$e_1 \wedge e_2 \wedge e_3$	0	0	0	0

$$e_{\iota(j_1)} \wedge e_{\iota(j_2)} = e_{j_2} \wedge e_{j_1} = -e_{j_1} \wedge e_{j_2} = \operatorname{sgn}(\iota) e_{j_1} \wedge e_{j_2}.$$

The same logic applies when $j_i > j_2$.

Induction step: Assume that $t \geqslant 3$ and that the result holds for elements of the form $e_{i_1} \wedge \cdots \wedge e_{i_s}$ such that $2 \leqslant s < t$. We proceed by cases.

Case 1: $\iota(j_1) = j_1$. In this case, ι describes a permutation of $\{j_2, \ldots, j_t\}$ with the same signum as ι; thus, our induction hypothesis explains the third equality in the following display

$$\begin{aligned} e_{\iota(j_1)} \wedge e_{\iota(j_2)} \wedge \cdots \wedge e_{\iota(j_t)} &= e_{\iota(j_1)} \wedge (e_{\iota(j_2)} \wedge \cdots \wedge e_{\iota(j_t)}) \\ &= e_{j_1} \wedge (e_{\iota(j_2)} \wedge \cdots \wedge e_{\iota(j_t)}) \\ &= e_{j_1} \wedge (\operatorname{sgn}(\iota) e_{j_2} \wedge \cdots \wedge e_{j_t}) \\ &= \operatorname{sgn}(\iota) e_{j_1} \wedge (e_{j_2} \wedge \cdots \wedge e_{j_t}) \\ &= \operatorname{sgn}(\iota) e_{j_1} \wedge e_{j_2} \wedge \cdots \wedge e_{j_t} \end{aligned}$$

and the other equalities are by definition and the condition $\iota(j_1) = j_1$.

Case 2: ι has the cycle-notation form $\iota = (j_1 \;\; j_2)$ and $j_2 = \min\{j_2, j_3, \ldots, j_t\}$. In this case, we are trying to show that

$$e_{j_2} \wedge e_{j_1} \wedge e_{j_3} \wedge \cdots \wedge e_{j_t} = -e_{j_1} \wedge e_{j_2} \wedge e_{j_3} \wedge \cdots \wedge e_{j_t}. \tag{3}$$

Let α be the permutation of $\{j_3, \ldots, j_t\}$ such that $\alpha(j_3) < \cdots < \alpha(j_t)$. Our assumption on j_2 implies that $j_2 < \alpha(j_3) < \cdots < \alpha(j_t)$.

Case 2a: $j_1 < j_2 < \alpha(j_3) < \cdots < \alpha(j_t)$. Our induction hypothesis explains the second equality in the following sequence

$$\begin{aligned}
e_{j_2} \wedge e_{j_1} \wedge e_{j_3} \wedge \cdots \wedge e_{j_t} &= e_{j_2} \wedge (e_{j_1} \wedge (e_{j_3} \wedge \cdots \wedge e_{j_t}))\\
&= e_{j_2} \wedge (e_{j_1} \wedge (\operatorname{sgn}(\alpha) e_{\alpha(j_3)} \wedge \cdots \wedge e_{\alpha(j_t)}))\\
&= \operatorname{sgn}(\alpha) e_{j_2} \wedge (e_{j_1} \wedge (e_{\alpha(j_3)} \wedge \cdots \wedge e_{\alpha(j_t)}))\\
&= \operatorname{sgn}(\alpha) e_{j_2} \wedge (e_{j_1} \wedge e_{\alpha(j_3)} \wedge \cdots \wedge e_{\alpha(j_t)})\\
&= -\operatorname{sgn}(\alpha) e_{j_1} \wedge e_{j_2} \wedge e_{\alpha(j_3)} \wedge \cdots \wedge e_{\alpha(j_t)}
\end{aligned}$$

and the remaining equalities are by definition, where the fourth and fifth ones use our Case 2a assumption. Similar logic explains the next sequence:

$$\begin{aligned}
-e_{j_1} \wedge e_{j_2} \wedge e_{j_3} \wedge \cdots \wedge e_{j_t} &= -e_{j_1} \wedge (e_{j_2} \wedge (e_{j_3} \wedge \cdots \wedge e_{j_t}))\\
&= -e_{j_1} \wedge (e_{j_2} \wedge (\operatorname{sgn}(\alpha) e_{\alpha(j_3)} \wedge \cdots \wedge e_{\alpha(j_t)}))\\
&= -\operatorname{sgn}(\alpha) e_{j_1} \wedge (e_{j_2} \wedge (e_{\alpha(j_3)} \wedge \cdots \wedge e_{\alpha(j_t)}))\\
&= -\operatorname{sgn}(\alpha) e_{j_1} \wedge e_{j_2} \wedge e_{\alpha(j_3)} \wedge \cdots \wedge e_{\alpha(j_t)}.
\end{aligned}$$

This explains Eq. (3) in Case 2a.

Case 2a: $j_2 < j_1 < \alpha(j_3) < \cdots < \alpha(j_t)$. Our induction hypothesis explains the second equality in the following sequence

$$\begin{aligned}
e_{j_2} \wedge e_{j_1} \wedge e_{j_3} \wedge \cdots \wedge e_{j_t} &= e_{j_2} \wedge (e_{j_1} \wedge (e_{j_3} \wedge \cdots \wedge e_{j_t}))\\
&= e_{j_2} \wedge (e_{j_1} \wedge (\operatorname{sgn}(\alpha) e_{\alpha(j_3)} \wedge \cdots \wedge e_{\alpha(j_t)}))\\
&= \operatorname{sgn}(\alpha) e_{j_2} \wedge (e_{j_1} \wedge (e_{\alpha(j_3)} \wedge \cdots \wedge e_{\alpha(j_t)}))\\
&= \operatorname{sgn}(\alpha) e_{j_2} \wedge e_{j_1} \wedge e_{\alpha(j_3)} \wedge \cdots \wedge e_{\alpha(j_t)}
\end{aligned}$$

and the remaining equalities are by definition, where the fourth one uses our Case 2b assumption. Similar logic explains the next sequence:

$$\begin{aligned}
-e_{j_1} \wedge e_{j_2} \wedge e_{j_3} \wedge \cdots \wedge e_{j_t} &= -e_{j_1} \wedge (e_{j_2} \wedge (e_{j_3} \wedge \cdots \wedge e_{j_t}))\\
&= -e_{j_1} \wedge (e_{j_2} \wedge (\operatorname{sgn}(\alpha) e_{\alpha(j_3)} \wedge \cdots \wedge e_{\alpha(j_t)}))\\
&= -\operatorname{sgn}(\alpha) e_{j_1} \wedge (e_{j_2} \wedge (e_{\alpha(j_3)} \wedge \cdots \wedge e_{\alpha(j_t)}))\\
&= -\operatorname{sgn}(\alpha) e_{j_1} \wedge (e_{j_2} \wedge e_{\alpha(j_3)} \wedge \cdots \wedge e_{\alpha(j_t)})\\
&= \operatorname{sgn}(\alpha) e_{j_2} \wedge e_{j_1} \wedge e_{\alpha(j_3)} \wedge \cdots \wedge e_{\alpha(j_t)}.
\end{aligned}$$

This explains Eq. (3) in Case 2b.

Case 2c: $j_2 < \alpha(j_3) < \cdots < \alpha(j_p) < j_1 < \alpha(j_{p+1}) < \cdots < \alpha(j_t)$. Similar logic as in the previous cases explains the following sequences

$$\begin{aligned}
&e_{j_2} \wedge e_{j_1} \wedge e_{j_3} \wedge \cdots \wedge e_{j_t} \\
&\quad = e_{j_2} \wedge (e_{j_1} \wedge (e_{j_3} \wedge \cdots \wedge e_{j_t})) \\
&\quad = e_{j_2} \wedge (e_{j_1} \wedge (\operatorname{sgn}(\alpha) e_{\alpha(j_3)} \wedge \cdots \wedge e_{\alpha(j_t)})) \\
&\quad = \operatorname{sgn}(\alpha) e_{j_2} \wedge (e_{j_1} \wedge (e_{\alpha(j_3)} \wedge \cdots \wedge e_{\alpha(j_t)})) \\
&\quad = \operatorname{sgn}(\alpha) e_{j_2} \wedge ((-1)^{p-2} e_{\alpha(j_3)} \wedge \cdots \wedge e_{\alpha(j_p)} \wedge e_{j_1} \wedge e_{\alpha(j_{p+1})} \wedge \cdots e_{\alpha(j_t)}) \\
&\quad = (-1)^{p-2} \operatorname{sgn}(\alpha) e_{j_2} \wedge e_{\alpha(j_3)} \wedge \cdots \wedge e_{\alpha(j_p)} \wedge e_{j_1} \wedge e_{\alpha(j_{p+1})} \wedge \cdots e_{\alpha(j_t)}
\end{aligned}$$

$$\begin{aligned}
&-e_{j_1} \wedge e_{j_2} \wedge e_{j_3} \wedge \cdots \wedge e_{j_t} \\
&\quad = -e_{j_1} \wedge (e_{j_2} \wedge (e_{j_3} \wedge \cdots \wedge e_{j_t})) \\
&\quad = -e_{j_1} \wedge (e_{j_2} \wedge (\operatorname{sgn}(\alpha) e_{\alpha(j_3)} \wedge \cdots \wedge e_{\alpha(j_t)})) \\
&\quad = -\operatorname{sgn}(\alpha) e_{j_1} \wedge (e_{j_2} \wedge (e_{\alpha(j_3)} \wedge \cdots \wedge e_{\alpha(j_t)})) \\
&\quad = -\operatorname{sgn}(\alpha) e_{j_1} \wedge (e_{j_2} \wedge e_{\alpha(j_3)} \wedge \cdots \wedge e_{\alpha(j_t)}) \\
&\quad = -(-1)^{p-1} \operatorname{sgn}(\alpha) e_{j_2} \wedge e_{\alpha(j_3)} \wedge \cdots \wedge e_{\alpha(j_p)} \wedge e_{j_1} \wedge e_{\alpha(j_{p+1})} \wedge \cdots e_{\alpha(j_t)} \\
&\quad = (-1)^{p-2} \operatorname{sgn}(\alpha) e_{j_2} \wedge e_{\alpha(j_3)} \wedge \cdots \wedge e_{\alpha(j_p)} \wedge e_{j_1} \wedge e_{\alpha(j_{p+1})} \wedge \cdots e_{\alpha(j_t)}
\end{aligned}$$

This explains Eq. (3) in Case 2c.

Case 2d: $j_2 < \alpha(j_3) < \cdots < \alpha(j_t) < j_1$. This case is handled as in Case 2c.

Case 3: ι has the cycle-notation form $\iota = (j_1 \ \ j_z)$ where $j_z = \min\{j_2, j_3, \ldots, j_t\}$. In this case, we are trying to prove the following:

$$e_{j_z} \wedge e_{j_2} \wedge \cdots \wedge e_{j_{z-1}} \wedge e_{j_1} \wedge e_{j_{z+1}} \wedge \cdots \wedge e_{j_t} = -e_{j_1} \wedge e_{j_2} \wedge \cdots \wedge e_{j_{z-1}} \wedge e_{j_z} \wedge e_{j_{z+1}} \wedge \cdots \wedge e_{j_t}.$$

This follows from the next sequence, where the first and third equalities are by our induction hypothesis

$$\begin{aligned}
&e_{j_z} \wedge e_{j_2} \wedge \cdots \wedge e_{j_{z-1}} \wedge e_{j_1} \wedge e_{j_{z+1}} \wedge \cdots \wedge e_{j_t} \\
&\quad = (-1)^{z-2} e_{j_z} \wedge e_{j_1} \wedge e_{j_2} \wedge \cdots \wedge e_{j_{z-1}} \wedge e_{j_{z+1}} \wedge \cdots \wedge e_{j_t} \\
&\quad = -(-1)^{z-2} e_{j_1} \wedge e_{j_z} \wedge e_{j_2} \wedge \cdots \wedge e_{j_{z-1}} \wedge e_{j_{z+1}} \wedge \cdots \wedge e_{j_t} \\
&\quad = -e_{j_1} \wedge e_{j_2} \wedge \cdots \wedge e_{j_{z-1}} \wedge e_{j_z} \wedge e_{j_{z+1}} \wedge \cdots \wedge e_{j_t}
\end{aligned}$$

and the second equality is from Case 2.

Case 4: ι has the cycle-notation form $\iota = (j_1 \ \ j_q)$ for some $q \geqslant 2$. Set $j_z = \min\{j_2, j_3, \ldots, j_t\}$.

Case 4a: $j_1 < j_z$. In this case, we have $j_1 = \min\{j_2, \ldots, j_{q-1}, j_1, j_{q+1}, \ldots, j_t\}$, so Case 3 implies that

$$e_{j_q} \wedge e_{j_2} \wedge \cdots \wedge e_{j_{q-1}} \wedge e_{j_1} \wedge e_{j_{q+1}} \wedge \cdots \wedge e_{j_t}$$
$$= -e_{j_1} \wedge e_{j_2} \wedge \cdots \wedge e_{j_{q-1}} \wedge e_{j_q} \wedge e_{j_{q+1}} \wedge \cdots \wedge e_{j_t}.$$

which is the desired equality in this case.

Case 4b: $j_z < j_1$. In this case, we have

$$\min\{j_2, j_3, \ldots, j_t\} = j_z = \min\{j_2, \ldots, j_{q-1}, j_1, j_{q+1}, \ldots, j_t\}$$

so Case 3 explains the first and third equalities in the next sequence

$$\begin{aligned}
&e_{j_q} \wedge e_{j_2} \wedge \cdots \wedge e_{j_{z-1}} \wedge e_{j_z} \wedge e_{j_{z+1}} \wedge \cdots \wedge e_{j_{q-1}} \wedge e_{j_1} \wedge e_{j_{q+1}} \wedge \cdots \wedge e_{j_t}\\
&= -e_{j_z} \wedge e_{j_2} \wedge \cdots \wedge e_{j_{z-1}} \wedge e_{j_q} \wedge e_{j_{z+1}} \wedge \cdots \wedge e_{j_{q-1}} \wedge e_{j_1} \wedge e_{j_{q+1}} \wedge \cdots \wedge e_{j_t}\\
&= e_{j_z} \wedge e_{j_2} \wedge \cdots \wedge e_{j_{z-1}} \wedge e_{j_1} \wedge e_{j_{z+1}} \wedge \cdots \wedge e_{j_{q-1}} \wedge e_{j_q} \wedge e_{j_{q+1}} \wedge \cdots \wedge e_{j_t}\\
&= -e_{j_1} \wedge e_{j_2} \wedge \cdots \wedge e_{j_{z-1}} \wedge e_{j_z} \wedge e_{j_{z+1}} \wedge \cdots \wedge e_{j_{q-1}} \wedge e_{j_q} \wedge e_{j_{q+1}} \wedge \cdots \wedge e_{j_t}
\end{aligned}$$

and Case 1 explains the second equality.[12] This is the desired conclusion here.

Case 5: the general case. Cases 1–4 show that the desired result holds for any transposition that fixes $C := \{1, \ldots, n\} \smallsetminus \{j_1, \ldots, j_t\}$. In general, since ι fixes C, it is a product $\iota = \tau_1 \cdots \tau_m$ of transpositions that fix C. (For instance, this can be seen by decomposing ι, considered as an element of S_t.) Since the result holds for each τ_i, induction on m shows that the desired result holds for ι. The main point is the following: if the result holds for permutations δ and σ that fix C, then

$$\begin{aligned}
e_{\delta\sigma(j_1)} \wedge \cdots \wedge e_{\delta\sigma(j_t)} &= \operatorname{sgn}(\delta) e_{\sigma(j_1)} \wedge \cdots \wedge e_{\sigma(j_t)}\\
&= \operatorname{sgn}(\delta)\operatorname{sgn}(\sigma) e_{j_1} \wedge \cdots \wedge e_{j_t}\\
&= \operatorname{sgn}(\delta\sigma) e_{j_1} \wedge \cdots \wedge e_{j_t}
\end{aligned}$$

so the result holds for $\delta\sigma$. □

10.20 (Sketch of Solution to Exercise 4.30). Eventually, we will argue by induction on s. To remove technical issues from the induction argument, we deal with some degenerate cases first. If $i_p = i_q$ for some $p < q$, then by definition, we have

$$\begin{aligned}
(e_{i_1} \wedge \cdots \wedge e_{i_p} \wedge \cdots \wedge e_{i_q} \wedge \cdots \wedge e_{i_s}) \wedge (e_{j_1} \wedge \cdots \wedge e_{j_t})&\\
&= 0 \wedge (e_{j_1} \wedge \cdots \wedge e_{j_t})\\
&= 0\\
&= e_{i_1} \wedge \cdots \wedge e_{i_p} \wedge \cdots \wedge e_{i_q} \wedge \cdots \wedge e_{i_s} \wedge e_{j_1} \wedge \cdots \wedge e_{j_t}.
\end{aligned}$$

If $j_p = j_q$ for some $p < q$, then the same logic applies.

---

[12]This visual argument assumes that $z < q$. The case $q < z$ is handled similarly, and the case $z = q$ follows from Case 3.

For the rest of the proof, we assume that $i_p \neq i_q$ and $j_p \neq j_q$ for all $p < q$. We argue by induction on s.

Base case: $s = 1$. There are two cases to consider. If $i = j_q$ for some q, then Definitions 4.24 and 4.28 imply that

$$e_{i_1} \wedge (e_{j_1} \wedge \cdots \wedge e_{j_t}) = 0 = e_{i_1} \wedge e_{i_s} \wedge e_{j_1} \wedge \cdots \wedge e_{j_t}.$$

If $i \neq j_q$ for all q, then Definition 4.28 gives the desired equality directly

$$e_{i_1} \wedge (e_{j_1} \wedge \cdots \wedge e_{j_t}) = e_{i_1} \wedge e_{i_s} \wedge e_{j_1} \wedge \cdots \wedge e_{j_t}.$$

Induction step: assume that $s \geqslant 2$ and that

$$(e_{k_1} \wedge e_{k_2} \wedge \cdots \wedge e_{k_{s-1}}) \wedge (e_{j_1} \wedge \cdots \wedge e_{j_t}) = e_{k_1} \wedge e_{k_2} \wedge \cdots \wedge e_{k_{s-1}} \wedge e_{j_1} \wedge \cdots \wedge e_{j_t}$$

for all sequences $k_1, \ldots, k_{s-1}$ of distinct elements in $\{1, \ldots, n\}$. Let α be the permutation of $\{i_1, \ldots, i_s\}$ such that $\alpha(i_1) < \cdots < \alpha(i_s)$, and let β be the permutation of $\{j_1, \ldots, j_t\}$ such that $\beta(j_1) < \cdots < \beta(j_t)$. Exercise 4.29 explains the first step in the next sequence, and the second equality is from Definition 4.24:

$$\begin{aligned}
&(e_{i_1} \wedge \cdots \wedge e_{i_s}) \wedge (e_{j_1} \wedge \cdots \wedge e_{j_t}) \\
&\quad = \operatorname{sgn}(\alpha)\operatorname{sgn}(\beta)(e_{\alpha(i_1)} \wedge \cdots \wedge e_{\alpha(i_s)}) \wedge (e_{\beta(j_1)} \wedge \cdots \wedge e_{\beta(j_t)}) \\
&\quad = \operatorname{sgn}(\alpha)\operatorname{sgn}(\beta)e_{\alpha(i_1)} \wedge [(e_{\alpha(i_2)} \wedge \cdots \wedge e_{\alpha(i_s)}) \wedge (e_{\beta(j_1)} \wedge \cdots \wedge e_{\beta(j_t)})] \\
&\quad = \operatorname{sgn}(\alpha)\operatorname{sgn}(\beta)e_{\alpha(i_1)} \wedge (e_{\alpha(i_2)} \wedge \cdots \wedge e_{\alpha(i_s)} \wedge e_{\beta(j_1)} \wedge \cdots \wedge e_{\beta(j_t)}).
\end{aligned}$$

The third equality follows from our induction hypothesis. If $i_p = j_q$ for some p and q, then $\alpha(i_{p'}) = \beta(j_{q'})$ for some p' and q', so the first paragraph of this proof (or the base case, depending on the situation) implies that

$$\begin{aligned}
&(e_{i_1} \wedge \cdots \wedge e_{i_s}) \wedge (e_{j_1} \wedge \cdots \wedge e_{j_t}) \\
&\quad = \operatorname{sgn}(\alpha)\operatorname{sgn}(\beta)e_{\alpha(i_1)} \wedge (e_{\alpha(i_2)} \wedge \cdots \wedge e_{\alpha(i_s)} \wedge e_{\beta(j_1)} \wedge \cdots \wedge e_{\beta(j_t)}) \\
&\quad = 0 \\
&\quad = e_{i_1} \wedge \cdots \wedge e_{i_s} \wedge e_{j_1} \wedge \cdots \wedge e_{j_t}.
\end{aligned}$$

Thus, we assume that $i_p \neq j_q$ for all p and q. In particular, the permutations α and β combine to give a permutation γ of $\{i_1, \ldots, i_s, j_1, \ldots, j_t\}$ given by the rules $\gamma(i_p) = \alpha(i_p)$ and $\gamma(j_q) = \beta(j_q)$ for all p and q. Furthermore, one has $\operatorname{sgn}(\gamma) = \operatorname{sgn}(\alpha)\operatorname{sgn}(\beta)$. (To see this, write α as a product $\alpha = \tau_1 \cdots \tau_u$ of transpositions on $\{i_1, \ldots, i_s\}$, write β as a product $\beta = \pi_1 \cdots \pi_u$ of transpositions on $\{j_1, \ldots, j_t\}$, and observe that $\gamma = \tau_1 \cdots \tau_u \pi_1 \cdots \pi_u$ is a product of transpositions on $\{i_1, \ldots, i_s, j_1, \ldots, j_t\}$.) This explains the second equality in the next sequence:

$$
\begin{aligned}
(e_{i_1} \wedge \cdots \wedge e_{i_s}) \wedge (e_{j_1} \wedge \cdots \wedge e_{j_t})
&= \operatorname{sgn}(\alpha)\operatorname{sgn}(\beta) e_{\alpha(i_1)} \wedge (e_{\alpha(i_2)} \wedge \cdots \wedge e_{\alpha(i_s)} \wedge e_{\beta(j_1)} \wedge \cdots \wedge e_{\beta(j_t)}) \\
&= \operatorname{sgn}(\gamma) e_{\gamma(i_1)} \wedge (e_{\gamma(i_2)} \wedge \cdots \wedge e_{\gamma(i_s)} \wedge e_{\gamma(j_1)} \wedge \cdots \wedge e_{\gamma(j_j)}) \\
&= \operatorname{sgn}(\gamma) e_{\gamma(i_1)} \wedge e_{\gamma(i_2)} \wedge \cdots \wedge e_{\gamma(i_s)} \wedge e_{\gamma(j_1)} \wedge \cdots \wedge e_{\gamma(j_j)} \\
&= e_{i_1} \wedge \cdots \wedge e_{i_s} \wedge e_{j_1} \wedge \cdots \wedge e_{j_t}.
\end{aligned}
$$

The third equality is by our base case. The fourth equality is by Exercise 4.29. □

10.21 (Sketch of Solution to Exercise 4.31). (a) Multiplication in $\bigwedge R^n$ is distributive and unital, by definition. We check associativity first on basis elements, using Exercise 4.30:

$$
\begin{aligned}
[(e_{i_1} \wedge \cdots \wedge e_{i_s}) \wedge (e_{j_1} \wedge \cdots \wedge e_{j_t})] \wedge (e_{k_1} \wedge \cdots \wedge e_{k_u})
&= (e_{i_1} \wedge \cdots \wedge e_{i_s} \wedge e_{j_1} \wedge \cdots \wedge e_{j_t}) \wedge (e_{k_1} \wedge \cdots \wedge e_{k_u}) \\
&= e_{i_1} \wedge \cdots \wedge e_{i_s} \wedge e_{j_1} \wedge \cdots \wedge e_{j_t} \wedge e_{k_1} \wedge \cdots \wedge e_{k_u} \\
&= e_{i_1} \wedge \cdots \wedge e_{i_s} \wedge (e_{j_1} \wedge \cdots \wedge e_{j_t} \wedge e_{k_1} \wedge \cdots \wedge e_{k_u}) \\
&= e_{i_1} \wedge \cdots \wedge e_{i_s} \wedge [(e_{j_1} \wedge \cdots \wedge e_{j_t}) \wedge (e_{k_1} \wedge \cdots \wedge e_{k_u})].
\end{aligned}
$$

The general associativity holds by distributivity: if $\alpha_1, \ldots, \alpha_a, \beta_1, \ldots, \beta_b, \gamma_1, \ldots, \gamma_c$ are basis elements in $\bigwedge R^n$, then we have

$$
\begin{aligned}
\Big[\big(\textstyle\sum_i x_i\alpha_i\big) \wedge \big(\sum_j y_j\beta_j\big)\Big] \wedge \big(\sum_k z_k\gamma_k\big) &= \textstyle\sum_{i,j,k} x_i y_j z_k [(\alpha_i \wedge \beta_j) \wedge \gamma_k] \\
&= \textstyle\sum_{i,j,k} x_i y_j z_k [\alpha_i \wedge (\beta_j \wedge \gamma_k)] \\
&= \big(\textstyle\sum_i x_i\alpha_i\big) \wedge \Big[\big(\sum_j y_j\beta_j\big) \wedge \big(\sum_k z_k\gamma_k\big)\Big].
\end{aligned}
$$

(b) As in part (a), it suffices to consider basis vectors $\alpha = e_{i_1} \wedge \cdots \wedge e_{i_s} \in \bigwedge^s R^n$ and $\beta = e_{j_1} \wedge \cdots \wedge e_{j_t} \in \bigwedge^t R^n$, and prove that $\alpha \wedge \beta = (-1)^{st}\beta \wedge \alpha$. Note that we are assuming that $i_1 < \cdots < i_s$ and $j_1 < \cdots < j_t$. If $i_p = j_q$ for some p and q, then Exercise 4.30 implies that

$$\alpha \wedge \beta = 0 = (-1)^{st}\beta \wedge \alpha$$

as desired. Thus, we assume that $i_p \neq j_q$ for all p and q, that is, there are no repetitions in the list $i_1, \ldots, i_s, j_1, \ldots, j_t$.

We proceed by induction on s.

Base case: $s = 1$. In this case, the first and third equalities in the next sequence are from Exercise 4.30, using the condition $s = 1$:

$$\alpha \wedge \beta = e_{i_1} \wedge e_{j_1} \wedge \cdots \wedge e_{j_t}$$

$$
\begin{aligned}
&= (-1)^t e_{j_1} \wedge \cdots \wedge e_{j_t} \wedge e_{i_1} \\
&= (-1)^{st} \beta \wedge \alpha.
\end{aligned}
$$

The second equality is from Exercise 4.29.

Induction step: assume that $s \geqslant 2$ and that

$$(e_{i_2} \wedge \cdots \wedge e_{i_s}) \wedge (e_{j_1} \wedge \cdots \wedge e_{j_t}) = (-1)^{(s-1)t} (e_{j_1} \wedge \cdots \wedge e_{j_t}) \wedge (e_{i_2} \wedge \cdots \wedge e_{i_s}).$$

The first, third, and fifth equalities in the next sequence are by associativity:

$$
\begin{aligned}
\alpha \wedge \beta &= e_{i_1} \wedge [(e_{i_2} \wedge \cdots \wedge e_{i_s}) \wedge (e_{j_1} \wedge \cdots \wedge e_{j_t})] \\
&= e_{i_1} \wedge [(-1)^{(s-1)t} (e_{j_1} \wedge \cdots \wedge e_{j_t}) \wedge (e_{i_2} \wedge \cdots \wedge e_{i_s})] \\
&= (-1)^{(s-1)t} [e_{i_1} \wedge (e_{j_1} \wedge \cdots \wedge e_{j_t})] \wedge (e_{i_2} \wedge \cdots \wedge e_{i_s}) \\
&= (-1)^{(s-1)t} (-1)^t [(e_{j_1} \wedge \cdots \wedge e_{j_t}) \wedge e_{i_1}] \wedge (e_{i_2} \wedge \cdots \wedge e_{i_s}) \\
&= (-1)^{st} (e_{j_1} \wedge \cdots \wedge e_{j_t}) \wedge [e_{i_1} \wedge (e_{i_2} \wedge \cdots \wedge e_{i_s})] \\
&= (-1)^{st} \beta \wedge \alpha.
\end{aligned}
$$

The second equality is by our induction hypothesis, the fourth equality is from our base case, and the fifth equality is by Exercise 4.30.

(c) Let $\alpha \in \bigwedge^s R^n$ such that s is odd. If α is a basis vector, then $\alpha \wedge \alpha = 0$ by Exercise 4.30. In general, we have $\alpha = \sum_i z_i v_i$ where the v_i are basis vectors in $\bigwedge^s R^n$, and we compute:

$$
\begin{aligned}
\alpha \wedge \alpha &= \left(\textstyle\sum_i z_i v_i\right) \wedge \left(\textstyle\sum_i z_i v_i\right) \\
&= \textstyle\sum_{i,j} z_i z_j v_i \wedge v_j \\
&= \textstyle\sum_{i=j} z_i z_j v_i \wedge v_j + \sum_{i \neq j} z_i z_j v_i \wedge v_j \\
&= \textstyle\sum_i z_i^2 \underbrace{v_i \wedge v_i}_{=0} + \sum_{i<j} z_i z_j (v_i \wedge v_j + v_j \wedge v_i) \\
&= \textstyle\sum_{i<j} z_i z_j (v_i \wedge v_j + (-1)^{s^2} v_i \wedge v_j) \\
&= \textstyle\sum_{i<j} z_i z_j (v_i \wedge v_j - v_i \wedge v_j) \\
&= 0.
\end{aligned}
$$

The vanishing $v_i \wedge v_i = 0$ follows from the fact that v_i is a basis vector, so the fifth equality follows from part (b). The sixth equality follows from the fact that s is odd, and the other equalities are routine.

(d) The map $R \to \bigwedge R^n$ is a ring homomorphism since our rule for multiplication $\bigwedge^0 R^n \times \bigwedge^0 R^n \to \bigwedge^0 R^n$ is defined as the usual scalar multiplication $R \times \bigwedge^0 R^n \to \bigwedge^0 R^n$. To see that the image is in the center, let $r \in R = \bigwedge^0 R^n$ and let $z_i \in \bigwedge^i R^n$ for $i = 0, \ldots, n$. Using part (b), we have

$$r \wedge \left(\sum_i z_i\right) = \sum_i (r \wedge z_i) = \sum_i (z_i \wedge r) = \left(\sum_i z_i\right) \wedge r$$

so r in the center of $\bigwedge R^n$. □

10.22 (Sketch of Solution to Exercise 4.32). We argue by cases.

Case 1: $j_p = j_q$ for some $p < q$. In this case, we have

$$\partial_i^{\tilde{K}^R(\mathbf{x})}(e_{j_1} \wedge \cdots \wedge e_{j_t}) = \partial_i^{\tilde{K}^R(\mathbf{x})}(0) = 0$$

so we need to prove that $\sum_{s=1}^t (-1)^{s+1} x_{j_s} e_{j_1} \wedge \cdots \wedge \widehat{e_{j_s}} \wedge \cdots \wedge e_{j_t} = 0$. In this sum, if $s \notin \{p, q\}$, then we have $e_{j_1} \wedge \cdots \wedge \widehat{e_{j_s}} \wedge \cdots \wedge e_{j_t} = 0$ since the repetition $j_p = j_q$ occurs in this element. This explains the first equality in the next sequence:

$$\begin{aligned}
&\sum_{s=1}^t (-1)^{s+1} x_{j_s} e_{j_1} \wedge \cdots \wedge \widehat{e_{j_s}} \wedge \cdots \wedge e_{j_t}\\
&= (-1)^{p+1} x_{j_p} e_{j_1} \wedge \cdots \wedge \widehat{e_{j_p}} \wedge \cdots \wedge e_{j_t} + (-1)^{q+1} x_{j_p} e_{j_1} \wedge \cdots \wedge \widehat{e_{j_q}} \wedge \cdots \wedge e_{j_t}\\
&= (-1)^{p+1} x_{j_p} e_{j_1} \wedge \cdots \wedge e_{j_{p-1}} \wedge e_{j_{p+1}} \wedge \cdots \wedge e_{j_{q-1}} \wedge e_{j_q} \wedge e_{j_{q+1}} \wedge \cdots \wedge e_{j_t}\\
&\quad + (-1)^{q+1} x_{j_p} e_{j_1} \wedge \cdots \wedge e_{j_{p-1}} \wedge e_{j_p} \wedge \underbrace{e_{j_{p+1}} \wedge \cdots \wedge e_{j_{q-1}}}_{q-p-1 \text{ factors}} \wedge e_{j_{q+1}} \wedge \cdots \wedge e_{j_t}\\
&= (-1)^{p+1} x_{j_p} e_{j_1} \wedge \cdots \wedge e_{j_{p-1}} \wedge e_{j_{p+1}} \wedge \cdots \wedge e_{j_{q-1}} \wedge e_{j_q} \wedge e_{j_{q+1}} \wedge \cdots \wedge e_{j_t}\\
&\quad + (-1)^{q+1+q-p-1} x_{j_p} e_{j_1} \wedge \cdots \wedge e_{j_{p-1}} \wedge e_{j_{p+1}} \wedge \cdots \wedge e_{j_{q-1}} \wedge e_{j_p} \wedge e_{j_{q+1}} \wedge \cdots \wedge e_{j_t}\\
&= (-1)^{p+1} x_{j_p} e_{j_1} \wedge \cdots \wedge e_{j_{p-1}} \wedge e_{j_{p+1}} \wedge \cdots \wedge e_{j_{q-1}} \wedge e_{j_q} \wedge e_{j_{q+1}} \wedge \cdots \wedge e_{j_t}\\
&\quad + (-1)^{p} x_{j_p} e_{j_1} \wedge \cdots \wedge e_{j_{p-1}} \wedge e_{j_{p+1}} \wedge \cdots \wedge e_{j_{q-1}} \wedge e_{j_p} \wedge e_{j_{q+1}} \wedge \cdots \wedge e_{j_t}\\
&= 0.
\end{aligned}$$

The second equality is just a rewriting of the terms, and the third equality is from Exercise 4.29. The fourth and fifth equalities are routine simplification.

Case 2: $j_p \neq j_q$ for all $p \neq q$.

Claim. Given a permutation ι of $j_1, \ldots, j_t$, setting

$$\hat{\iota}(j_1, \ldots, j_t) := \sum_{s=1}^t (-1)^{s+1} x_{\iota(j_s)} e_{\iota(j_1)} \wedge \cdots \wedge \widehat{e_{\iota(j_s)}} \wedge \cdots \wedge e_{\iota(j_t)}$$

we have the following where id is the identity permutation.

$$\hat{\iota}(j_1, \ldots, j_t) = \operatorname{sgn}(\iota)\widehat{\mathrm{id}}(j_1, \ldots, j_t) \tag{4}$$

To prove this, write ι as a product $\iota = \tau_1 \cdots \tau_u$ of adjacent transpositions, that is, $\tau_i = (j_{p_i}\ \ j_{p_i+1})$ for some $p_i < t$. We argue by induction on u.

Base case: $u = 1$. In this case, $\iota = \tau_1$ is a transposition $\iota = (j_p\ \ j_{p+1})$ for some $p < t$. In this case, we are proving the equality $\widehat{\iota}(j_1, \ldots, j_t) = -\widehat{\mathrm{id}}(j_1, \ldots, j_t)$. We write out the terms of $\widehat{\iota}(j_1, \ldots, j_t)$, starting with the case $s < p$:

$$\begin{aligned}
&(-1)^{s+1} x_{\iota(j_s)} e_{\iota(j_1)} \wedge \cdots \wedge \widehat{e_{\iota(j_s)}} \wedge \cdots \wedge e_{\iota(j_{p-1})} \wedge e_{\iota(j_p)} \wedge e_{\iota(j_{p+1})} \wedge e_{\iota(j_{p+2})} \wedge \cdots \wedge e_{\iota(j_t)} \\
&\qquad = (-1)^{s+1} x_{j_s} e_{j_1} \wedge \cdots \wedge \widehat{e_{j_s}} \wedge \cdots \wedge e_{j_{p-1}} \wedge e_{j_{p+1}} \wedge e_{j_p} \wedge e_{j_{p+2}} \wedge \cdots \wedge e_{j_t} \\
&\qquad = -(-1)^{s+1} x_{j_s} e_{j_1} \wedge \cdots \wedge \widehat{e_{j_s}} \wedge \cdots \wedge e_{j_{p-1}} \wedge e_{j_p} \wedge e_{j_{p+1}} \wedge e_{j_{p+2}} \wedge \cdots \wedge e_{j_t}.
\end{aligned}$$

This is the sth term of $-\widehat{\mathrm{id}}(j_1, \ldots, j_t)$ in this case. Similar reasoning yields the same conclusion when $s > p + 1$. In the case $s = p$, the pth term of $\widehat{\iota}(j_1, \ldots, j_t)$ is

$$\begin{aligned}
&(-1)^{p+1} x_{\iota(j_p)} e_{\iota(j_1)} \wedge \cdots \wedge e_{\iota(j_{p-1})} \wedge \widehat{e_{\iota(j_p)}} \wedge e_{\iota(j_{p+1})} \wedge e_{\iota(j_{p+2})} \wedge \cdots \wedge e_{\iota(j_t)} \\
&\qquad\qquad = (-1)^{p+1} x_{j_{p+1}} e_{j_1} \wedge \cdots \wedge e_{j_{p-1}} \wedge \widehat{e_{j_{p+1}}} \wedge e_{j_p} \wedge e_{j_{p+2}} \wedge \cdots \wedge e_{j_t} \\
&\qquad\qquad = (-1)^{p+1} x_{j_{p+1}} e_{j_1} \wedge \cdots \wedge e_{j_{p-1}} \wedge e_{j_p} \wedge \widehat{e_{j_{p+1}}} \wedge e_{j_{p+2}} \wedge \cdots \wedge e_{j_t} \\
&\qquad\qquad = -(-1)^{p+2} x_{j_{p+1}} e_{j_1} \wedge \cdots \wedge e_{j_{p-1}} \wedge e_{j_p} \wedge \widehat{e_{j_{p+1}}} \wedge e_{j_{p+2}} \wedge \cdots \wedge e_{j_t}.
\end{aligned}$$

This is exactly the $p + 1$st term of $-\widehat{\mathrm{id}}(j_1, \ldots, j_t)$. Similar reasoning shows that the $p + 1$st term of $\widehat{\iota}(j_1, \ldots, j_t)$ is exactly the pth term of $-\widehat{\mathrm{id}}(j_1, \ldots, j_t)$. It follows that Eq. (4) holds in the base case.

Induction step: Assume that $u \geqslant 2$ and that the result holds for the permutation $\iota' := \tau_2 \cdots \tau_u$. Note that this implies that $\iota = \tau_1 \iota'$, so we have $\mathrm{sgn}(\iota) = -\mathrm{sgn}(\iota')$. This explains the first and last equalities in the next sequence:

$$\begin{aligned}
\widehat{\iota}(j_1, \ldots, j_t) &= \widehat{\tau_1 \iota'}(j_1, \ldots, j_t) \\
&= \widehat{\tau_1}(\iota'(j_1), \ldots, \iota'(j_t)) \\
&= -\widehat{\mathrm{id}}(\iota'(j_1), \ldots, \iota'(j_t)) \\
&= -\widehat{\iota'}(j_1, \ldots, j_t) \\
&= -\mathrm{sgn}(\iota')\widehat{\mathrm{id}}(j_1, \ldots, j_t) \\
&= \mathrm{sgn}(\iota)\widehat{\mathrm{id}}(j_1, \ldots, j_t).
\end{aligned}$$

The second and fourth equalities follow easily from the definition of $\widehat{*}$. The third equality is from our base case, and the fifth equality is from our induction hypothesis. This concludes the proof of the claim.

Now we complete the proof of the exercise. Let ι be a permutation of $j_1, \ldots, j_t$ such that $\iota(j_1) < \cdots < \iota(j_t)$. Exercise 4.29 explains the first equality in the next sequence, and the second, third, and fifth equalities are by definition:

$$\begin{aligned}
\partial_i^{\tilde{K}^R(\mathbf{x})}(e_{j_1}\wedge\cdots\wedge e_{j_t}) &= \operatorname{sgn}(\iota)\partial_i^{\tilde{K}^R(\mathbf{x})}(e_{\iota(j_1)}\wedge\cdots\wedge e_{\iota(j_t)})\\
&= \operatorname{sgn}(\iota)\textstyle\sum_{s=1}^{t}(-1)^{s+1}x_{\iota(j_s)}e_{\iota(j_1)}\wedge\cdots\wedge\widehat{e_{\iota(j_s)}}\wedge\cdots\wedge e_{\iota(j_t)}\\
&= \operatorname{sgn}(\iota)\hat{\iota}(j_1,\ldots,j_t)\\
&= \widehat{\mathrm{id}}(j_1,\ldots,j_t)\\
&= \textstyle\sum_{s=1}^{t}(-1)^{s+1}x_{j_s}e_{j_1}\wedge\cdots\wedge\widehat{e_{j_s}}\wedge\cdots\wedge e_{j_t}
\end{aligned}$$

The fourth equality is by the claim we proved above. □

10.23 (Sketch of Solution to Exercise 4.33). It suffices to verify the formula for basis vectors $\alpha = e_{j_1}\wedge\cdots\wedge e_{j_s}\in\bigwedge^s R^n$ and $\beta = e_{j_{s+1}}\wedge\cdots\wedge e_{j_{s+t}}\in\bigwedge^t R^n$.

$$\begin{aligned}
\partial_{s+t}^{\tilde{K}^R(\mathbf{x})}(\alpha\wedge\beta) &= \partial_{s+t}^{\tilde{K}^R(\mathbf{x})}(e_{j_1}\wedge\cdots\wedge e_{j_{s+t}})\\
&= \textstyle\sum_{u=1}^{s+t}(-1)^{u+1}x_{i_u}e_{j_1}\wedge\cdots\wedge\widehat{e_{j_u}}\wedge\cdots\wedge e_{j_{s+t}}\\
&= \textstyle\sum_{u=1}^{s}(-1)^{u+1}x_{i_u}e_{j_1}\wedge\cdots\wedge\widehat{e_{j_u}}\wedge\cdots\wedge e_{j_s}\wedge e_{j_{s+1}}\wedge\cdots\wedge e_{j_{s+t}}\\
&\quad+\textstyle\sum_{u=s+1}^{s+t}(-1)^{u+1}x_{i_u}e_{j_1}\wedge\cdots\wedge e_{j_s}\wedge e_{j_{s+1}}\wedge\cdots\wedge\widehat{e_{j_u}}\wedge\cdots\wedge e_{j_{s+t}}\\
&= \textstyle\sum_{u=1}^{s}(-1)^{u+1}x_{i_u}e_{j_1}\wedge\cdots\wedge\widehat{e_{j_u}}\wedge\cdots\wedge e_{j_s}\wedge e_{j_{s+1}}\wedge\cdots\wedge e_{j_{s+t}}\\
&\quad+\textstyle\sum_{u=1}^{t}(-1)^{u-s+1}x_{i_{s+u}}e_{j_1}\wedge\cdots\wedge e_{j_s}\wedge e_{j_{s+1}}\wedge\cdots\wedge\widehat{e_{j_{s+u}}}\wedge\cdots\wedge e_{j_{s+t}}
\end{aligned}$$

The first and second equalities are by definition, and the third equality is obtained by splitting the sum. The fourth equality is by reindexing. Thus, the first equality in the next sequence is by distributivity.

$$\begin{aligned}
&\partial_{s+t}^{\tilde{K}^R(\mathbf{x})}(\alpha\wedge\beta)\\
&= \left[\textstyle\sum_{u=1}^{s}(-1)^{u+1}x_{i_u}e_{j_1}\wedge\cdots\wedge\widehat{e_{j_u}}\wedge\cdots\wedge e_{j_s}\right]\wedge(e_{j_{s+1}}\wedge\cdots\wedge e_{j_{s+t}})\\
&\quad+(-1)^s(e_{j_1}\wedge\cdots\wedge e_{j_s})\wedge\left[\textstyle\sum_{u=1}^{t}(-1)^{u+1}x_{i_{s+u}}e_{j_{s+1}}\wedge\cdots\wedge\widehat{e_{j_{s+u}}}\wedge\cdots\wedge e_{j_{s+t}}\right]\\
&= \partial_s^{\tilde{K}^R(\mathbf{x})}(e_{j_1}\wedge\cdots\wedge e_{j_s})\wedge(e_{j_{s+1}}\wedge\cdots\wedge e_{j_{s+t}})\\
&\quad+(-1)^s(e_{j_1}\wedge\cdots\wedge e_{j_s})\wedge\partial_t^{\tilde{K}^R(\mathbf{x})}(e_{j_{s+1}}\wedge\cdots\wedge e_{j_{s+t}})\\
&= \partial_s^{\tilde{K}^R(\mathbf{x})}(\alpha)\wedge\beta+(-1)^s\alpha\wedge\partial_t^{\tilde{K}^R(\mathbf{x})}(\beta).
\end{aligned}$$

The second equality is by Exercise 4.32, and the last equality is by definition. □

10.24 (Sketch of Solution to Exercise 5.4). The fact that multiplication in A is distributive implies that the map $\mu^A\colon A\otimes_R A\to A$ given by $\mu^A(a\otimes b)=ab$ is well defined and R-linear. To see that it is a chain map, we compute:

$$\begin{aligned}
\partial_{|a|+|b|}^A(\mu_{|a|+|b|}^A(a\otimes b)) &= \partial_{|a|+|b|}^A(ab)\\
&= \partial_{|a|}^A(a)b+(-1)^{|a|}a\partial_{|b|}^A(b)
\end{aligned}$$

$$= \mu^A_{|a|+|b|-1}(\partial^A_{|a|}(a) \otimes b + (-1)^{|a|} a \otimes \partial^A_{|b|}(b))$$
$$= \mu^A_{|a|+|b|-1}(\partial^{A \otimes_R A}_{|a|+|b|}(a \otimes b)).$$

Since the multiplication on A maps $A_0 \times A_0 \to A_{0+0} = A_0$, we conclude that A_0 is closed under multiplication. Also, the fact that A_0 is an R-module implies that it is closed under addition and subtraction. Thus, the fact that A_0 is a commutative ring can be shown by restricting the axioms of A to A_0. To show that A_0 is an R-algebra, we define a map $R \to A_0$ by $r \mapsto r1_A$. Since A_0 is a ring and an R-module, it is routine to show that this is a ring homomorphism, so A_0 is an R-algebra. □

10.25 (Sketch of Solution to Exercise 5.6). (a) It is straightforward to show that the map $R \to A$ given by $r \mapsto r \cdot 1_A$ respects multiplication. The fact that it is a morphism of DG R-algebras follows from the commutativity of the next diagram

$$\begin{array}{ccccccccccc}
R = & \cdots & \longrightarrow & 0 & \longrightarrow & 0 & \longrightarrow & R & \longrightarrow & 0 \\
& & & \downarrow & & \downarrow & & \downarrow & & \downarrow \\
A = & \cdots & \longrightarrow & A_2 & \longrightarrow & A_1 & \longrightarrow & A_0 & \longrightarrow & 0
\end{array}$$

which is easily checked.

(b) Argue as in part (a).

(c) With Exercise 3.10 and Lemma 4.18, the essential point is the commutativity of the following diagram

$$\begin{array}{ccccccccccc}
K = & \cdots & \longrightarrow & R^{\binom{n}{2}} & \longrightarrow & R^n & \xrightarrow{(x_1 \ \dots \ x_n)} & R & \longrightarrow & 0 \\
& & & \downarrow & & \downarrow & & \downarrow & & \downarrow \\
R/(\mathbf{x}) = & \cdots & \longrightarrow & 0 & \longrightarrow & 0 & \longrightarrow & R/(\mathbf{x}) & \longrightarrow & 0
\end{array}$$

which is easily checked. □

10.26 (Sketch of Solution to Exercise 5.7). (a) The condition $A_{-1} = 0$ implies that $\mathrm{H}_0(A) \cong A_0/\operatorname{Im}(\partial^A_1)$, which is a homomorphic image of A_0. To show that $\mathrm{H}_0(A)$ is an A_0-algebra, it suffices to show that $\operatorname{Im}(\partial^A_1)$ is an ideal of A_0. To this end, the fact that ∂^A_1 is R-linear implies that $\operatorname{Im}(\partial^A_1)$ is nonempty and closed under subtraction. To show that it is closed under multiplication by elements of A_0, let $a_0 \in A_0$ and $\partial^A_1(a_1) \in \operatorname{Im}(\partial^A_1)$, and use the Leibniz Rule

$$\partial^A_1(a_0 a_1) = \partial^A_0(a_0) a_1 + (-1)^0 a_0 \partial^A_1(a_1) = a_0 \partial^A_1(a_1)$$

to see that $a_0 \partial^A_1(a_1) \in \operatorname{Im}(\partial^A_1)$.

(b) To see that A_i is an A_0-module, first observe that multiplication in A maps $A_0 \times A_i$ to A_i. Thus, A_i is closed under scalar multiplication by A_0. Since A_i is an R-module, it is nonempty and closed under addition and subtraction. The remaining A_0-module axioms follow from the DG algebra axioms on A.

To show that $\mathrm{H}_i(A)$ is an $\mathrm{H}_0(A)$-module, the essential point is to show that the scalar multiplication $\overline{a_0}\,\overline{a_i} := \overline{a_0 a_i}$ is well defined. (Then the axioms follow directly from the fact that A_i is an A_0-module.) The well-definedness boils down to showing that the products $\mathrm{Im}(\partial_1^A)\,\mathrm{Ker}(\partial_i^A)$ and $\mathrm{Ker}(\partial_0^A)\,\mathrm{Im}(\partial_{i+1}^A)$ are contained in $\mathrm{Im}(\partial_{i+1}^A)$. For the first of these, let $\partial_1^A(a_1) \in \mathrm{Im}(\partial_1^A)$ and $z_i \in \mathrm{Ker}(\partial_i^A)$:

$$\partial_{i+1}^A(a_1 z_i) = \partial_1^A(a_1) z_i + (-1)^1 a_1 \underbrace{\partial_i^A(z_i)}_{=0} = \partial_1^A(a_1) z_i.$$

It follows that $\partial_1^A(a_1) z_i = \partial_{i+1}^A(a_1 z_i) \in \mathrm{Im}(\partial_{i+1}^A)$, as desired. The containment $\mathrm{Ker}(\partial_0^A)\,\mathrm{Im}(\partial_{i+1}^A) \subseteq \mathrm{Im}(\partial_{i+1}^A)$ is verified similarly. □

10.27 (Sketch of Solution to Exercise 5.10). Assume that A is a DG R-algebra such that each A_i is finitely generated. Since R is noetherian, each A_i is noetherian over R. Hence, each submodule $\mathrm{Ker}(\partial_i^A) \subseteq A_i$ is finitely generated over R, so each quotient $\mathrm{H}_i(A) = \mathrm{Ker}(\partial_i^A)/\,\mathrm{Im}(\partial_{i+1}^A)$ is finitely generated over R. In particular, $\mathrm{H}_0(A)$ is finitely generated over R, so it is noetherian by the Hilbert Basis Theorem. Since $\mathrm{H}_i(A)$ is finitely generated over R and it is a module over the R-algebra $\mathrm{H}_0(A)$, it is straightforward to show that $\mathrm{H}_i(A)$ is finitely generated over $\mathrm{H}_0(A)$. □

10.28 (Sketch of Solution to Exercise 5.12). (a) By definition, every DG R-module is, in particular, an R-complex. Conversely, let X be an R-complex. We verify the DG R-module axioms. The associative, distributive, and unital axioms are automatic since X is an R-complex.

For the graded axiom, let $r \in R_i$ and $x \in X_j$. If $i \neq 0$, then $r \in R_i = 0$ implies that $rx = 0 \in X_{i+j}$. If $i = 0$, then $r \in R_i = R$ implies that $rx \in X_j = X_{0+j}$.

For the Leibniz Rule, let $r \in R_i$ and $x \in X_j$. If $i \neq 0$, then $r \in R_i = 0$ implies

$$\partial_{i+j}^X(rx) = \partial_{i+j}^X(0) = 0 = \partial_{i+j}^X(0)x + (-1)^i 0 \partial_{i+j}^X(x) = \partial_i^X(r)x + (-1)^i r \partial_j^X(x)$$

as desired. If $i = 0$, then $r \in R_i = R$, so the R-linearity of $\partial_j^X = \partial_{0+j}^X$ implies that

$$\partial_{i+j}^X(rx) = r\partial_{0+j}^X(x) = 0x + (-1)^0 r \partial_j^X(x) = \partial_i^X(r)x + (-1)^i r \partial_j^X(x)$$

as desired.

(b) Directly compare the axioms in Definitions 5.1 and 5.11.

(c) Let $f: A \to B$ a morphism of DG R-algebras, and let Y be a DG B-module. Define the DG A-module structure on Y by the formula $a_i y_j := f_i(a_i) y_j$.

We verify associativity:

$$(a_ia_j)y_k = f_{i+j}(a_ia_j)y_k = [f_i(a_i)f_j(a_j)]y_k = f_i(a_i)[f_j(a_j)y_k] = a_i(a_jy_k).$$

Distributivity, gradedness, and the Leibniz Rule follow similarly, as does unitality (using the condition $f_0(1_A) = 1_B$). □

10.29 (Sketch of Solution to Exercise 5.13). Fix an R-complex X and a DG R-algebra A. First, we show that the scalar multiplication $a(b \otimes x) := (ab) \otimes x$ is well defined. Let $a_i \in A_i$ be a fixed element. Since the map $A_j \to A_{i+j}$ given by $a_j \mapsto a_ia_j$ is well defined and R-linear, so is the induced map $A_j \otimes_R X_k \to A_{i+j} \otimes_R X_k$ which is given on generators by the formula $a_j \otimes x_k \mapsto (a_ia_j) \otimes x_k$. Assembling these maps together for all j, k provides a well-defined R-linear map

$$(A \otimes_R X)_n = \bigoplus_{j+k=n} A_j \otimes_R X_k \to \bigoplus_{j+k=n} A_{i+j} \otimes_R X_k = (A \otimes_R X)_{i+n}$$

given on generators by the formula $a_j \otimes x_k \mapsto (a_ia_j) \otimes x_k$. In other words, the given multiplication is well-defined and satisfies the graded axiom. Verification of associativity and distributivity is routine. We verify the Leibniz Rule on generators:

$$\begin{aligned}
&\partial^{A\otimes_R X}_{i+j+k}(a_i(a_j \otimes x_k)) \\
&\quad = \partial^{A\otimes_R X}_{i+j+k}((a_ia_j) \otimes x_k) \\
&\quad = \partial^A_{i+j}(a_ia_j) \otimes x_k + (-1)^{i+j}(a_ia_j) \otimes \partial^X_k(x_k) \\
&\quad = [\partial^A_i(a_i)a_j] \otimes x_k + (-1)^i[a_i\partial^A_j(a_j)] \otimes x_k + (-1)^{i+j}(a_ia_j) \otimes \partial^X_k(x_k) \\
&\quad = \partial^A_i(a_i)(a_j \otimes x_k) + (-1)^i a_i[\partial^A_j(a_j) \otimes x_k + (-1)^j a_j \otimes \partial^X_k(x_k)] \\
&\quad = \partial^A_i(a_i)(a_j \otimes x_k) + (-1)^i a_i\partial^{A\otimes_R X}_{j+k}(a_j \otimes x_k).
\end{aligned}$$

□

10.30 (Sketch of Solution to Exercise 5.14). Since scalar multiplication of A on M is well defined, so is the scalar multiplication of A on $\Sigma^i M$. To prove that $\Sigma^i M$ is a DG A-module, we check the axioms, in order, beginning with associativity:

$$\begin{aligned}
a * (b * m) &= (-1)^{i|a|}a((-1)^{i|b|}bm) = (-1)^{i(|a|+|b|)}a(bm) \\
&= (-1)^{i|ab|}(ab)m = (ab) * m.
\end{aligned}$$

Distributivity is verified similarly. For the unital axiom, recall that $|1_A| = 0$, so we have $1_A * m = (-1)^{i\cdot 0}1_A m = m$.

The graded axiom requires a bit of bookkeeping. Let $a \in A_p$ and $m \in (\Sigma^i M)_q = M_{q-i}$. Then we have $a * m = (-1)^{ip}am \in M_{p+q-i} = (\Sigma^i M)_{p+q}$, as desired.

For the Leibniz Rule, we let $a \in A_p$ and $m \in (\Sigma^i M)_q = M_{q-i}$, and we compute:

$$\begin{aligned}
\partial^{\Sigma^i M}_{p+q}(a * m) &= (-1)^i \partial^M_{p+q-i}((-1)^{ip} am) \\
&= (-1)^{ip+i}[\partial^A_p(a)m + (-1)^p a \partial^M_{q-i}(m)] \\
&= (-1)^{ip-i}[\partial^A_p(a)m + (-1)^{p+i} a \partial^{\Sigma^i M}_q(m)] \\
&= (-1)^{i(p-1)}\partial^A_p(a)m + (-1)^{p+ip} a \partial^{\Sigma^i M}_q(m) \\
&= \partial^A_p(a) * m + (-1)^p a * \partial^{\Sigma^i M}_q(m).
\end{aligned}$$

□

10.31 (Sketch of Solution to Exercise 5.15). As for Exercise 5.4; cf. 10.24. □

10.32 (Sketch of Solution to Exercise 5.22). Fix a DG R-algebra A and a chain map of R-complexes $g: X \to Y$.

(a) To prove that the chain map $A \otimes_R g: A \otimes_R X \to A \otimes_R Y$ is a morphism of DG A-modules, we check the desired formula on generators:

$$\begin{aligned}
(A \otimes_R g)_{i+j+k}(a_i(a_j \otimes x_k)) &= (A \otimes_R g)_{i+j+k}((a_i a_j) \otimes x_k)) \\
&= (a_i a_j) \otimes g_k(x_k) \\
&= a_i(a_j \otimes g_k(x_k)) \\
&= a_i(A \otimes_R g)_{j+k}(a_j \otimes x_k).
\end{aligned}$$

(b) Let k be a field, and consider the power series ring $R = k[\![X]\!]$. The residue field k is an R-algebra, hence it is a DG R-algebra. The Koszul complex $K = K^R(X)$ is a free resolution of k over R, so the natural map $K \to k$ is a quasiisomorphism. On the other hand, the induced map $k \otimes_R K \to k \otimes_R k$ is not a quasiisomorphism since $\mathrm{H}_1(k \otimes_R K) \cong k \neq 0 = \mathrm{H}_i(k \otimes_R k)$.

(c) Fix a morphism $f: A \to B$ of DG R-algebras. This explains the first equality in the next display

$$f(ab) = f(a)f(b) = af(b).$$

The second equality is from the definition of the DG A-module structure on B. □

10.33 (Sketch of Solution to Exercise 5.27). We use the following items.

Definition 10.34. Let X be an R-complex. An *R-subcomplex* of X is an R-complex Y such that each Y_i is a submodule of X_i and for all $y \in Y_i$ we have $\partial^Y_i(y) = \partial^X_i(y)$. A *DG submodule* of the DG A-module M is an R-subcomplex N such that the scalar multiplication of A on N uses the same rule as the scalar multiplication on M.

Lemma 10.35. *Let X be an R-complex, and let Y be an R-subcomplex of X.*

(a) *The quotient complex X/Y with differential induced by ∂^X is an R-complex.*
(b) *The natural sequence $0 \to Y \xrightarrow{\epsilon} X \xrightarrow{\pi} X/Y \to 0$ of chain maps is exact.*
(c) *The natural map $X \to X/Y$ is a quasiisomorphism if and only if Y is exact.*

Proof. (a) To show that the differential $\partial^{X/Y}$ given as $\partial_i^{X/Y}(\overline{x}) := \overline{\partial_i^X(x)}$ is well defined, it suffices to show that $\partial_i^X(Y_i) \subseteq Y_{i-1}$. But this condition is automatic by definition: for all $y \in Y_i$ we have $\partial_i^X(y) = \partial_i^Y(y) \in Y_{i-1}$. The remaining R-complex axioms are straightforward consequences of the axioms for X.

(b) It suffices to show that the following diagram commutes

$$\begin{array}{ccccccccc}
0 & \longrightarrow & Y_i & \xrightarrow{\epsilon_i} & X_i & \xrightarrow{\pi_i} & X_i/Y_i & \longrightarrow & 0 \\
 & & \downarrow{\scriptstyle \partial_i^Y} & & \downarrow{\scriptstyle \partial_i^X} & & \downarrow{\scriptstyle \partial_i^{X/Y}} & & \\
0 & \longrightarrow & Y_{i-1} & \xrightarrow{\epsilon_{i-1}} & X_{i-1} & \xrightarrow{\pi_{i-1}} & X_{i-1}/Y_{i-1} & \longrightarrow & 0
\end{array}$$

where ϵ_i is the inclusion and π_i is the natural surjection. The commutativity of this diagram is routine, using the assumption on ∂^Y and the definition of $\partial^{X/Y}$.

(c) Use the long exact sequence coming from part (b). □

Lemma 10.36. *Let N be a DG submodule of the DG A-module M.*

(a) *The quotient complex M/N with scalar multiplication induced by the scalar multiplication on M is a DG A-module.*
(b) *The natural maps $N \xrightarrow{\epsilon} M \xrightarrow{\pi} M/N$ are morphisms of DG A-modules.*

Proof. (a) Lemma 10.35(a) implies that M/N is an R-complex. Next, we show that the scalar multiplication $a\overline{m} := \overline{am}$ is well defined. For this, it suffices to show that $an \in N$ for all $a \in A$ and all $n \in N$. But this condition is automatic by definition, since N is closed under scalar multiplication. Now that the scalar multiplication on M/N is well defined, the DG A-module axioms on M/N follow readily from the axioms on M.

(b) As N is a DG submodule of M, the inclusion ϵ respects scalar multiplication; and π respects scalar multiplication as follows: $\pi(am) = \overline{am} = a\overline{m} = a\pi(m)$.

□

Now, we continue with the proof of the exercise. Consider the complex

$$N = \cdots \xrightarrow{\partial_{n+2}^M} M_{n+1} \xrightarrow{\partial_{n+1}^M} \operatorname{Im}(\partial_{n+1}^M) \to 0.$$

(a) Using Lemmas 10.35 and 10.36, it suffices to show that N is a DG submodule of M. By inspection, the differential on N maps $N_i \to N_{i-1}$ for all i. Since the differential and scalar multiplication on N are induced from those on M, it suffices to show that N is closed under scalar multiplication. (The other axioms are inherited from M.) For this, let $a \in A_p$ and $x \in N_q$. If $p < 0$, then $a = 0$ and we have $ax = 0 \in N_{p+q}$. Similar reasoning applies if $q < n$. Assume now that $p \geqslant 0$ and $q \geqslant n$. If $p \geqslant 1$ or $q > n$, then $ax \in M_{p+q} = N_{p+q}$, by definition. So, we are reduced to the case where $p = 0$ and $q = n$. In this case, there is an element $m \in M_{n+1}$ such that $x = \partial^M_{n+1}(m)$. Thus, the Leibniz Rule on M implies that

$$ax = 0 + ax = \partial^A_0(a)m + a\partial^M_{n+1}(m) = \partial^M_{n+1}(am) \in \operatorname{Im}(\partial^M_{n+1}) = N_n$$

as desired.

(b) Using Lemmas 10.35 and 10.36, it suffices to show that N is exact if and only if $n \geqslant \sup M$. For this, note that

$$\mathrm{H}_i(N) \cong \begin{cases} \mathrm{H}_i(M) & \text{for } i > n \\ 0 & \text{for } i \leqslant n. \end{cases}$$

It follows that the complex N is exact if and only if $\mathrm{H}_i(M) = 0$ for all $i > n$, that is, if and only if $n \geqslant \sup(M)$. □

10.37 (Sketch of Solution to Exercise 6.4). Note that $I = (x^2, xy, y^2)$ has grade two as an ideal of R since x^2, y^2 is an R-regular sequence. Also, $I = I_2(A)$, where

$$A = \begin{bmatrix} x & 0 \\ y & x \\ 0 & y \end{bmatrix}$$

Thus, I is perfect by Theorem 6.3. □

10.38 (Sketch of Solution to Exercise 6.6). Let $A = (a_{i,j})$, which is $(n+1) \times n$. Let F denote the R-complex in question, and denote the differential of F by ∂. First, we check the easy products. The fact that $F_3 = 0$ explains the next sequence.

$$\partial_2(f_j)f_k + (-1)^2 f_j\partial_2(f_k) = 0 = \partial_4(f_j f_k)$$

Next, we argue by definition and cancellation.

$$\partial_1(e_i)e_i + (-1)^1 e_i\partial_1(e_i) = (-1)^{i-1}a\det(A_i)e_i - e_i[(-1)^{i-1}a\det(A_i)] = 0 = \partial_2(e_ie_i)$$

The remaining products require some work.

Let $1 \leqslant i < j \leqslant n+1$, and consider the $n \times n$ matrices A_j and A_i. Expanding $\det(A_j)$ along the ith row, we have

$$\det(A_j) = \sum_{k=1}^{n}(-1)^{i+k}a_{i,k}\det(A^k_{i,j}). \tag{5}$$

This uses the equality $(A_j)_i^k = A_{i,j}^k$ which is a consequence of the assumption $i < j$. On the other hand, we have $(A_i)_{j-1}^k = A_{i,j}^k$ since the $(j-1)$st row of A_i is equal to the jth row of A; so when we expand $\det(A_i)$ along its $(j-1)$st row, we have

$$\det(A_i) = \sum_{k=1}^{n}(-1)^{j-1+k} a_{j,k} \det((A_i)_{j-1}^k) = \sum_{k=1}^{n}(-1)^{j-1+k} a_{j,k} \det(A_{i,j}^k). \tag{6}$$

Next, for $\ell \neq i, j$ we let $A(\ell)$ denote the matrix obtained by replacing the ith row of A with the ℓth row. It follows that we have $(A(\ell)_j)_i^k = A(\ell)_{i,j}^k = A_{i,j}^k$. Notice that the matrix $A(\ell)_j$ has two equal rows, so we have $\det(A(\ell)_j) = 0$. Expanding $\det(A(\ell)_j)$ along the ith row, we obtain the next equalities:

$$0 = \sum_{k=1}^{n}(-1)^{i+k} a_{\ell,k} \det((A(\ell)_j)_i^k) = \sum_{k=1}^{n}(-1)^{i+k} a_{\ell,k} \det(A_{i,j}^k). \tag{7}$$

Now we verify the Leibniz rule for the product $e_i e_j$, still assuming $1 \leqslant i < j \leqslant n+1$.

$$\begin{aligned}
\partial_2(e_i e_j) &= a \sum_{k=1}^{n}(-1)^{i+j+k} \det(A_{i,j}^k)\partial_2(f_k) \\
&= a \sum_{k=1}^{n}(-1)^{i+j+k} \det(A_{i,j}^k) \sum_{\ell=1}^{n+1} a_{\ell,k} e_\ell \\
&= a \sum_{\ell=1}^{n+1} \left[\sum_{k=1}^{n}(-1)^{i+j+k} a_{\ell,k} \det(A_{i,j}^k) \right] e_\ell \\
&= a \left[\sum_{k=1}^{n}(-1)^{i+j+k} a_{i,k} \det(A_{i,j}^k) \right] e_i + a \left[\sum_{k=1}^{n}(-1)^{i+j+k} a_{j,k} \det(A_{i,j}^k) \right] e_j \\
&\quad + a \sum_{\ell \neq i,j} \left[\sum_{k=1}^{n}(-1)^{i+j+k} a_{\ell,k} \det(A_{i,j}^k) \right] e_\ell \\
&= (-1)^j a \left[\sum_{k=1}^{n}(-1)^{i+k} a_{i,k} \det(A_{i,j}^k) \right] e_i \\
&\quad + (-1)^{i+1} a \left[\sum_{k=1}^{n}(-1)^{j-1+k} a_{j,k} \det(A_{i,j}^k) \right] e_j \\
&\quad + a \sum_{\ell \neq i,j} \left[\sum_{k=1}^{n}(-1)^{i+j+k} a_{\ell,k} \det(A_{i,j}^k) \right] e_\ell
\end{aligned}$$

The above equalities are by definition and simplification. In the next sequence, the second equality follows from (5) to (7), and the others are by definition and simplification.

$$\begin{aligned}\partial_2(e_i e_j) &= a\sum_{k=1}^{n}(-1)^{i+j+k}\det(A^k_{i,j})\partial_2(f_k)\\ &= (-1)^j a\det(A_j)e_i + (-1)^{i+1}a\det(A_i)e_j + (-1)^j a\sum_{\ell\neq i,j}0e_\ell\\ &= -\partial_1(e_j)e_i + \partial_1(e_i)e_j + 0\\ &= \partial_1(e_i)e_j + (-1)^{|e_i|}e_i\partial_1(e_j)\end{aligned}$$

Next, we show how the Leibniz rule for e_je_i follows from that of e_ie_j:

$$\begin{aligned}\partial_2(e_je_i) = \partial_2(-e_ie_j) &= -\partial_2(e_ie_j) = -[\partial_1(e_i)e_j - e_i\partial_1(e_j)]\\ &= e_i\partial_1(e_j) - \partial_1(e_i)e_j = \partial_1(e_j)e_i - e_j\partial_1(e_i).\end{aligned}$$

Next, we verify the Leibniz rule for products of the form $e_i f_j$ for any i and j. To begin, note that we have

$$(A_i)^j_k = \begin{cases} A^j_{k,i} & \text{if } k < i\\ A^j_{i,k+1} & \text{if } k \geqslant i\end{cases}$$

and the k,j-entry of A_i is

$$a'_{k,j} := \begin{cases} a_{k,j} & \text{if } k < i\\ a_{k+1,j} & \text{if } k \geqslant i.\end{cases}$$

Using this, we expand $\det(A_i)$ along the jth column:

$$\begin{aligned}\det(A_i) &= \sum_{k=1}^{n}(-1)^{j+k}\det((A_i)^j_k)a'_{k,j}\\ &= \sum_{k=1}^{i-1}(-1)^{j+k}\det(A^j_{k,i})a_{k,j} + \sum_{k=i}^{n}(-1)^{j+k}\det(A^j_{i,k+1})a_{k+1,j}\\ &= \sum_{k=1}^{i-1}(-1)^{j+k}\det(A^j_{k,i})a_{k,j} + \sum_{k=i+1}^{n+1}(-1)^{j+k-1}\det(A^j_{i,k})a_{k,j}\\ &= \sum_{k=1}^{i-1}(-1)^{j+k}\det(A^j_{k,i})a_{k,j} - \sum_{k=i+1}^{n+1}(-1)^{j+k}\det(A^j_{i,k})a_{k,j}.\end{aligned} \tag{8}$$

Similarly, for any $\ell \neq j$, let $A[\ell]$ denote the matrix obtained by replacing the ℓth column of A with its jth column. Thus, we have $\det(A[\ell]_i) = 0$. Expanding $\det(A[\ell]_i)$ along the ℓth column as above implies that

$$\sum_{k=1}^{i-1}(-1)^{\ell+k}\det(A^{\ell}_{k,i})a_{k,j} - \sum_{k=i+1}^{n+1}(-1)^{\ell+k}\det(A^{\ell}_{i,k})a_{k,j} = 0. \tag{9}$$

By definition, we have $e_i f_j = 0$. Thus, the Leibniz rule in this case follows from the next computation where the final equality is a consequence of (8)–(9):

$$\begin{aligned}
\partial_1(e_i)f_j - e_i\partial_2(f_j) &= (-1)^{i+1}a\det(A_i)f_j - e_i\sum_{k=1}^{n+1}a_{k,j}e_k \\
&= (-1)^{i+1}a\det(A_i)f_j - \sum_{k=1}^{n+1}a_{k,j}e_ie_k \\
&= (-1)^{i+1}a\det(A_i)f_j + \sum_{k=1}^{i-1}a_{k,j}e_ke_i - \sum_{k=i+1}^{n+1}a_{k,j}e_ie_k \\
&= (-1)^{i+1}a\det(A_i)f_j + \sum_{k=1}^{i-1}a_{k,j}a\sum_{\ell=1}^{n}(-1)^{k+i+\ell}\det(A^{\ell}_{k,i})f_\ell \\
&\qquad - \sum_{k=i+1}^{n+1}a_{k,j}a\sum_{\ell=1}^{n}(-1)^{i+k+\ell}\det(A^{\ell}_{i,k})f_\ell \\
&= (-1)^{i+1}a\det(A_i)f_j + a(-1)^i\sum_{\ell=1}^{n}\left[\sum_{k=1}^{i-1}a_{k,j}(-1)^{k+\ell}\det(A^{\ell}_{k,i})\right. \\
&\qquad \left. - \sum_{k=i+1}^{n+1}a_{k,j}(-1)^{k+\ell}\det(A^{\ell}_{i,k})\right]f_\ell \\
&= 0.
\end{aligned}$$

The final case now follows from the previous one:

$$\partial_2(f_j)e_i + f_j\partial_1(e_i) = -e_i\partial_2(f_j) + \partial_1(e_i)f_j = \partial_1(e_i)f_j - e_i\partial_2(f_j) = 0.$$

This completes the proof. □

10.39 (Sketch of Solution to Exercise 6.7). The deleted minimal R-free resolution of $R/(x^2, xy, y^2)$ is the following:

$$0 \to Rf_1 \oplus Rf_2 \xrightarrow{\left[\begin{smallmatrix} x & 0 \\ y & x \\ 0 & y \end{smallmatrix}\right]} Re_1 \oplus Re_2 \oplus Re_3 \xrightarrow{[-y^2\ xy\ -x^2]} R1 \to 0.$$

According to Theorem 6.5, the above complex has a DG R-algebra structure with

$$e_1e_2 = -e_2e_1 = yf_1$$
$$e_1e_3 = -e_3e_1 = -xf_1 + yf_2$$
$$e_2e_3 = -e_3e_2 = -xf_2$$

and $e_i^2 = 0$ for all $1 \leqslant i \leqslant 3$. □

10.40 (Sketch of Solution to Exercise 6.15). First, note that since A is a 3×3 matrix, $\mathrm{Pf}(A_{ijk}^{ijk}) = 1$ for all choices of i, j, k. Following the conditions specified by Theorem 6.14, one has the relations $e_i^2 = 0$ for all $1 \leqslant i \leqslant 3$ and

$$e_1e_2 = -e_2e_1 = f_1 - f_2 + f_3$$
$$e_1e_3 = -e_3e_1 = -f_1 - f_2 - f_3$$
$$e_2e_3 = -e_3e_2 = f_1 - f_2 + f_3$$

and $e_i f_j = f_j e_i = \delta_{ij} g$ for all $1 \leqslant i, j \leqslant 3$. □

10.41 (Sketch of Solution to Exercise 7.3).

(a) To start, we let $f = \{f_j\} \in \mathrm{Hom}_A(M, N)_q$ and prove that $\partial_q^{\mathrm{Hom}_R(M,N)}(f)$ is in $\mathrm{Hom}_A(M, N)_{q-1}$, that is, that $\partial_q^{\mathrm{Hom}_R(M,N)}(f)$ is A-linear. For this, let $a \in A_p$ and $m \in M_t$, and compute:

$$\begin{aligned}
\partial_q^{\mathrm{Hom}_R(M,N)}(f)_{p+t}(am) &= \partial_{p+t+q}^N(f_{p+t}(am)) - (-1)^q f_{p+t-1}(\partial_{p+t}^M(am)) \\
&= (-1)^{pq}\partial_{p+t+q}^N(af_t(m)) \\
&\quad - (-1)^q f_{p+t-1}\left(\partial_p^A(a)m + (-1)^p a\partial_t^M(m)\right) \\
&= (-1)^{pq}\left(\partial_p^A(a)f_t(m) + (-1)^p a\partial_{t+q}^N(f_t(m))\right) \\
&\quad - (-1)^q f_{p+t-1}\left(\partial_p^A(a)m\right) - (-1)^{q+p} f_{p+t-1}\left(a\partial_t^M(m)\right) \\
&= (-1)^{pq}\partial_p^A(a)f_t(m) + (-1)^{pq+p}a\partial_{t+q}^N(f_t(m)) \\
&\quad - (-1)^{q+(p-1)q}\partial_p^A(a)f_t(m) - (-1)^{q+p+pq}af_{t-1}\left(\partial_t^M(m)\right) \\
&= (-1)^{pq+p}a\partial_{t+q}^N(f_t(m)) - (-1)^{q+p+pq}af_{t-1}\left(\partial_t^M(m)\right) \\
&= (-1)^{p(q-1)}a\left(\partial_{t+q}^N(f_t(m)) - (-1)^q f_{t-1}(\partial_t^M(m))\right) \\
&= (-1)^{p(q-1)}a\partial_q^{\mathrm{Hom}_R(M,N)}(f)_{p+t}(m)
\end{aligned}$$

The first and last equalities are from the definition of $\partial_q^{\mathrm{Hom}_R(M,N)}(f)$. The second equality is from the A-linearity of f and the Leibniz Rule on M. The third equality is from the Leibniz Rule on N and the A-linearity of f. The fourth equality is from the A-linearity of f. The fifth equality is by cancellation since $q + (p-1)q = pq$. And the sixth equality is distributivity.

This shows that $\partial^{\mathrm{Hom}_A(M,N)}$ is well defined. Since $\mathrm{Hom}_R(M,N)$ is an R-complex, it follows readily that $\mathrm{Hom}_A(M,N)$ is also an R-complex.

With the same a and f as above, we next show that the sequence $af := \{(af)_j\}$ defined by the formula $(af)_t(m) := a(f_t(m))$ is in $\mathrm{Hom}_A(M,N)_{p+q}$. First, this rule maps $M_t \to N_{t+p+q}$ since $m \in M_t$ implies that $f_t(m) \in N_{t+q}$, which implies that $a(f_t(m)) \in N_{t+q+p}$. Next, we show that $af \in \mathrm{Hom}_R(M,N)_{p+q}$:

$$\begin{aligned}(af)_t(rm) &= a(f_t(rm)) = a(rf_t(m)) = (ar)f_t(m)\\ &= (ra)f_t(m) = r(af_t(m)) = r(af)_t(m).\end{aligned}$$

Next, we show that af is A-linear. For this, let $b \in A_s$:

$$\begin{aligned}(af)_{s+t}(bm) &= a(f_{s+t}(bm)) = (-1)^{qs}a(bf_{s+t}(m))\\ &= (-1)^{qs+ps}b(af_{s+t}(m)) = (-1)^{(q+p)s}b((af)_{s+t}(m)).\end{aligned}$$

The first and fourth equalities are by definition of af. The second equality follows because f is A-linear. The third equality is from the graded commutativity and associativity of A. Since $|b| = s$ and $|af| = q + p$, this shows that af is A-linear.

Next, we verify the DG A-module axioms for $\mathrm{Hom}_A(M,N)$. The graded axiom has already been verified. For associativity, continue with the notation from above. We need to show that $a(bf) = (ab)f$, so we compute:

$$(a(bf))_t(m) = a((bf)_t(m)) = a(b(f_t(m))) = (ab)f_t(m) = ((ab)f)_t(m).$$

The third equality is by associativity, and the other equalities are by definition. Distributivity and unitality are verified similarly. Thus, it remains to verify the Leibniz Rule. For this, we need to show that

$$\partial_{p+q}^{\mathrm{Hom}_A(M,N)}(af) = \partial_p^A(a)f + (-1)^p a\partial_q^{\mathrm{Hom}_A(M,N)}(f).$$

For this, we evaluate at m:

$$\begin{aligned}&\partial_{p+q}^{\mathrm{Hom}_A(M,N)}(af)_t(m)\\ &\quad= \partial_{p+q}^N((af)_t(m)) - (-1)^{p+q}(af)_{t-1}(\partial_t^M(m))\\ &\quad= \partial_{t+p+q}^N(a(f_t(m))) - (-1)^{p+q}a(f_{t-1}(\partial_t^M(m)))\\ &\quad= \partial_p^A(a)f_t(m) + (-1)^p a\partial_{t+q}^N(f_t(m))) - (-1)^{p+q}a(f_{t-1}(\partial_t^M(m)))\\ &\quad= \partial_p^A(a)f_t(m) + (-1)^p a[\partial_{t+q}^N(f_t(m))) - (-1)^q(f_{t-1}(\partial_t^M(m)))]\end{aligned}$$

$$= \partial_p^A(a) f_t(m) + (-1)^p a[\partial_q^{\mathrm{Hom}_A(M,N)}(f)_t(m))]$$
$$= (\partial_p^A(a) f)_t(m) + (-1)^p (a\partial_q^{\mathrm{Hom}_A(M,N)}(f))_t(m))]$$
$$= \left(\partial_p^A(a) f + (-1)^p a\partial_q^{\mathrm{Hom}_A(M,N)}(f)\right)_t (m)$$

The third equality is by the Leibniz Rule on N. The fourth step is by distributivity. The remaining equalities are by definition.

(b) Let $a \in A_p$. From the graded axiom for M, we know that the operation $m \mapsto am$ maps $M_t \to M_{p+t}$. The fact that this is R-linear follows from associativity:

$$a(rm) = (ar)m = (ra)m = r(am).$$

To show that it is A-linear, let $b \in A_s$, and compute:

$$\mu_{s+t}^{M,a}(bm) = a(bm) = (-1)^{ps} b(am) = (-1)^{ps} b\mu_t^{M,a}(m).$$

(c) Argue as in the proof of Exercise 3.4 in 10.2. □

10.42 (Sketch of Solution to Exercise 7.7). We prove that $\mathrm{Hom}_A(N, f)$ is a morphism of DG A-modules; the argument for $\mathrm{Hom}_A(f, N)$ is similar. To this end, first note that since f is a chain map of R-complexes, Exercise 3.14 shows that the induced map $\mathrm{Hom}_R(N, f)\colon \mathrm{Hom}_R(N, L) \to \mathrm{Hom}_R(N, M)$ is a chain map of R-complexes. Also, note that $\mathrm{Hom}_R(N, f)$ and $\mathrm{Hom}_A(N, f)$ are given by the same composition-with-f rule.

We need to show that $\mathrm{Hom}_A(f, N)$ is well defined. For this, let $g = \{g_j\} \in \mathrm{Hom}_A(N, L)_q$. We need to show that $\mathrm{Hom}_A(f, N)(g) \in \mathrm{Hom}_A(N, M)_q$, that is, that $\mathrm{Hom}_A(f, N)(g)$ is A-linear. For this, let $a \in A_p$ and $n \in N_t$. We need to show that $\mathrm{Hom}_A(f, N)(g)_{p+t}(an) = a \mathrm{Hom}_A(f, N)(g)_t(n)$. We compute:

$$\begin{aligned}
\mathrm{Hom}_A(f, N)(g)_{p+t}(an) &= f_{p+q+t}(g_{p+t}(an)) \\
&= (-1)^{pq} f_{p+q+t}(ag_t(n)) \\
&= (-1)^{pq} a[f_{q+t}(g_t(n))] \\
&= (-1)^{pq} a[\mathrm{Hom}_A(f, N)(g)_t(n)] \\
&= (-1)^{pq} [a \mathrm{Hom}_A(f, N)(g)]_t(n).
\end{aligned}$$

The second equality is by the A-linearity of g. The third equality is by the A-linearity of f. The remaining steps are by definition. since $|\mathrm{Hom}_A(f, N)(g)| = |g| = q$ and $|a| = p$, this is the desired equality.

Next, we need to show that $\mathrm{Hom}_A(N, f)$ respects multiplication on A. For this, we use the same letters as in the previous paragraph. We need to show that

$$\mathrm{Hom}_A(N, f)(ag) = a[\mathrm{Hom}_A(N, f)(g)]$$

so we compute:

$$\begin{aligned}
\mathrm{Hom}_A(N,f)(ag)_t(n) &= f_{t+p+q}((ag)_t(n)) \\
&= f_{t+p+q}(a(g_t(n))) \\
&= a(f_{t+q}(g_t(n))) \\
&= a(\mathrm{Hom}_A(N,f)(g)_t(n)) \\
&= a[\mathrm{Hom}_A(N,f)(g)]_t(n).
\end{aligned}$$

The third equality is by the A-linearity of f, and the others are by definition. □

10.43 (Sketch of Solution to Exercise 7.9). We begin by showing that $M \otimes_A N$ is an R-complex, using Lemma 10.35(a). Thus, we need to show that U is a subcomplex of $M \otimes_R N$. For this, it suffices to show that the differential $\partial^{M\otimes_R N}$ maps each generator of U into U. To this end, let $a \in A_p$, $m \in M_q$ and $n \in N_s$, and compute:

$$\begin{aligned}
&\partial^{M\otimes_R N}_{p+q+s}((am)\otimes n - (-1)^{pq} m \otimes (an)) \\
&\qquad = \partial^M_{p+q}(am)\otimes n + (-1)^{p+q}(am)\otimes \partial^N_s(n) \\
&\qquad\quad - (-1)^{pq}\partial^M_q(m)\otimes(an) - (-1)^{pq+q} m \otimes \partial^N_{p+s}(an) \\
&\qquad = [\partial^A_p(a)m]\otimes n + (-1)^p[a\partial^M_q(m)]\otimes n \\
&\qquad\quad + (-1)^{p+q}(am)\otimes\partial^N_s(n) - (-1)^{pq}\partial^M_q(m)\otimes(an) \\
&\qquad\quad - (-1)^{pq+q} m\otimes[\partial^A_p(a)n] - (-1)^{pq+q+p} m\otimes[a\partial^N_s(n)] \\
&\qquad = [\partial^A_p(a)m]\otimes n - (-1)^{pq+q} m\otimes[\partial^A_p(a)n] \\
&\qquad\quad + (-1)^p[a\partial^M_q(m)]\otimes n - (-1)^{pq}\partial^M_q(m)\otimes(an) \\
&\qquad\quad + (-1)^{p+q}(am)\otimes\partial^N_s(n) - (-1)^{pq+q+p} m\otimes[a\partial^N_s(n)] \\
&\qquad = [\partial^A_p(a)m]\otimes n - (-1)^{(p-1)q} m\otimes[\partial^A_p(a)n] \\
&\qquad\quad + (-1)^p\left[[a\partial^M_q(m)]\otimes n - (-1)^{p(q-1)}\partial^M_q(m)\otimes(an)\right] \\
&\qquad\quad + (-1)^{p+q}\left[(am)\otimes\partial^N_s(n) - (-1)^{pq} m\otimes[a\partial^N_s(n)]\right].
\end{aligned}$$

Since each of these terms is a multiple of a generator of U, we conclude that $\partial^{M\otimes_R N}_{p+q+s}((am)\otimes n - (-1)^{pq} m\otimes(an)) \in U$, as desired.

As in the solution to Exercise 5.13 contained in 10.29, the R-complex $M \otimes_R N$ has a well-defined DG A-module structure defined on generators by the formula $b(m\otimes n) := (bm)\otimes n$. To show that the same formula is well defined on $M \otimes_A N$, we need to show that multiplication by $b \in A_t$ maps each generator of U into U:

$$\begin{aligned} b((am)\otimes n-(-1)^{pq}m\otimes(an)) &= (b(am))\otimes n-(-1)^{pq}(bm)\otimes(an)\\ &= (-1)^{pt}(a(bm))\otimes n-(-1)^{pq}(bm)\otimes(an)\\ &= (-1)^{pt}\left[(a(bm))\otimes n-(-1)^{p(q+t)}(bm)\otimes(an)\right]. \end{aligned}$$

Since $|bm|=q+t$ and $|a|=p$, this is in U, as desired.

Now that we know that the differential and the scalar multiplication of A on $M\otimes_A N$ are well defined, the other DG A-module axioms are inherited from $M\otimes_R N$. Finally, the formula $(am)\otimes n=(-1)^{|a||m|}m\otimes(an)$ in $M\otimes_A N$ follows from the condition $(am)\otimes n-(-1)^{|a||m|}m\otimes(an)\in U$. □

10.44 (Sketch of Solution to Exercise 7.10). Let $f\colon A\to B$ be a morphism of DG R-algebras. The DG R-algebra B is a DG A-module via the scalar multiplication $a_ib:=f_i(a_i)b$, by Exercise 5.12(c). So, we know from Exercise 7.9 that $B\otimes_A M$ has a well-defined DG A-module structure. It remains to show that it has a well-defined DG A-module structure by the action $b(b'\otimes m):=(bb')\otimes m$. Notice that, once this is shown, the compatibility with the DG A-module structure is automatic:

$$a_i(b\otimes m)=(a_ib)\otimes m=(f_i(a_i)b)\otimes m=f_i(a_i)(b\otimes m).$$

Let U be the R-submodule of $B\otimes_R M$ generated by all elements of the following form: $(ab)\otimes m-(-1)^{|a||b|}b\otimes(am)$.

We show that the DG B-module structure on $B\otimes_A M$ is well defined. For this, note that $B\otimes_R M$ has a well-defined DG B-module structure via the composition $R\to A\to B$, by Exercise 5.13. Thus, it suffices to let $c\in B_p$ and show that multiplication by c maps generators of U into U:

$$\begin{aligned} &c((ab)\otimes m-(-1)^{|a||b|}b\otimes(am))\\ &\qquad= (c(ab))\otimes m-(-1)^{|a||b|}(cb)\otimes(am)\\ &\qquad= (-1)^{|c||a|}(a(cb))\otimes m-(-1)^{|a||b|}(cb)\otimes(am)\\ &\qquad= (-1)^{|c||a|}\left[(a(cb))\otimes m-(-1)^{|a|(|b|+|c|)}(cb)\otimes(am)\right]\\ &\qquad= (-1)^{|c||a|}\left[(a(cb))\otimes m-(-1)^{|a|(|cb|)}(cb)\otimes(am)\right]\in U. \end{aligned}$$

For the DG B-module axioms, the only one with any substance is the Leibniz Rule:

$$\begin{aligned} &\partial^{B\otimes_A M}_{|c|+|b|+|m|}(c(b\otimes m))\\ &\quad= \partial^{B\otimes_A M}_{|c|+|b|+|m|}((cb)\otimes m)\\ &\quad= \partial^{B}_{|c|+|b|}(cb)\otimes m+(-1)^{|c|+|b|}(cb)\otimes\partial^{M}_{|m|}m)\\ &\quad= (\partial^{B}_{|c|}(c)b)\otimes m+(-1)^{|c|}(c\partial^{B}_{|b|}(b))\otimes m+(-1)^{|c|+|b|}(cb)\otimes\partial^{M}_{|m|}m) \end{aligned}$$

$$= \partial^B_{|c|}(c)(b \otimes m) + (-1)^{|c|}c\left[(\partial^B_{|b|}(b)) \otimes m + (-1)^{|b|}(b) \otimes \partial^M_{|m|}m)\right]$$
$$= \partial^B_{|c|}(c)(b \otimes m) + (-1)^{|c|}c\partial^{B\otimes_A M}_{|c|+|b|+|m|}(b \otimes m).$$

The third equality follows from the Leibniz Rule for B, and the fourth equality is by distributivity. The remaining equalities are by definition. □

10.45 (Sketch of Solution to Exercise 7.11). Hom cancellation. For each $f = \{f_p\} \in \mathrm{Hom}_A(A, L)_i$ we have $f_0 \colon A_0 \to L_i$, hence $f_0(1_A) \in L_i$. Define $\alpha_i \colon \mathrm{Hom}_A(A, L)_i \to L_i$ by the formula $\alpha_i(f) := f_0(1_A)$. We show that $\alpha \colon \mathrm{Hom}_A(A, L) \to L$ is a morphism of DG A-modules and that it is bijective, by Remark 5.21. To show that it is a chain map over R, we compute:

$$\begin{aligned}
\alpha_{i-1}(\partial_i^{\mathrm{Hom}_A(A,L)}(f)) &= \alpha_{i-1}(\{\partial^Y_{p+i} f_p - (-1)^i f_{p-1}\partial^X_p\}) \\
&= \partial^L_i(f_0(1_A)) - (-1)^i f_{-1}\partial^A_0(1_A) \\
&= \partial^L_i(f_0(1_A)) \\
&= \partial^L_i(\alpha_i(f)).
\end{aligned}$$

To show that α is A-linear, let $a \in A_j$ and compute:

$$\alpha_{i+j}(af) = (af)_0(1_A) = af_0(1_A) = a\alpha_i(f).$$

To see that α is injective, suppose that $0 = \alpha_i(f) = f_0(1_A)$. It follows that for all $a \in A_j$ we have

$$f_j(a) = f_j(a1_A) = af_0(1_A) = a \cdot 0 = 0.$$

We conclude that α is injective. To show that α is surjective, let $x \in L_i$. As in the proof of Exercise 7.3(b) in 10.41, the map $\nu^x \colon A \to L$ given by $a \mapsto ax$ is a homomorphism of degree i. Moreover, we have

$$\alpha_i(\nu^x) = \nu^x_0(1_A) = 1_A x = x$$

so α is surjective, as desired.

Notice that the special case $L = A$ explains the isomorphism $\mathrm{Hom}_A(A, A) \cong A$.

Tensor cancellation. Define $\eta \colon A \otimes_R L \to L$ by the formula $\eta_{i+j}(a_i \otimes x_j) := a_i x_j$. This is a well-defined chain map by Exercise 5.15. Let U be the R-submodule of $A \otimes_R L$ generated by the elements of the form $(a_i b_j) \otimes x_k - (-1)^{ij} b_j \otimes (a_i x_k)$. For each such generator, we have

$$\eta_{i+j+k}((a_i b_j) \otimes x_k - (-1)^{ij} b_j \otimes (a_i x_k)) = (a_i b_j)x_k - (-1)^{ij} b_j(a_i x_k) = 0$$

since $b_j a_i = (-1)^{ij} a_i b_j$. It follows that η induces a well-defined map $\nu\colon A \otimes_A L \to L$ given by $\nu_{i+j}(a_i \otimes x_j) := a_i x_j$. Since η is a chain map, it follows readily that ν is also a chain map. Moreover, it is A-linear because

$$\nu_{i+j+k}(a_i(b_j \otimes x_k)) = \nu_{i+j+k}((a_i b_j) \otimes x_k) = (a_i b_j)x_k = a_i(b_j x_k) = a_i \nu_{j+k}(b_j \otimes x_k).$$

To show that ν is an isomorphism, we construct a two-sided inverse. Let $\beta\colon L \to A \otimes_A L$ be given by $\beta_i(x_i) = 1 \otimes x_i$. As in previous exercises, this is a well-defined morphism of DG A-modules. To see that it is a two-sided inverse for ν, we compute:

$$\beta_{i+j}(\nu_{i+j}(a_i \otimes x_j)) = 1 \otimes (a_i x_j) = a_i \otimes x_j.$$

This shows that $\beta\nu$ is the identity on $A \otimes_A L$. The fact that $\nu\beta$ is the identity on L is even easier.

Again, the special case $L = A$ explains the isomorphism $A \otimes_A A \cong A$.

Tensor commutativity. By Exercise 4.2(c), the map $L \otimes_R M \xrightarrow{\gamma} M \otimes_R L$ given by $x_i \otimes y_j \mapsto (-1)^{ij} y_j \otimes x_i$ is a well-defined isomorphism of R-complexes. Let V be the submodule of $L \otimes_R M$ generated over R by the elements of the form $(a_i x_j) \otimes y_k - (-1)^{ij} x_j \otimes (a_i y_k)$. Let W be the R-submodule of $M \otimes_R L$ generated by the elements of the form $(a_i y_j) \otimes x_k - (-1)^{ij} y_j \otimes (a_i x_k)$.

For each element $(a_i x_j) \otimes y_k - (-1)^{ij} x_j \otimes (a_i y_k) \in U$, we have

$$\begin{aligned}
\gamma_{i+j+k}((a_i x_j) \otimes y_k - (-1)^{ij} x_j \otimes (a_i y_k)) \\
&= (-1)^{(i+j)k} y_k \otimes (a_i x_j) - (-1)^{ij+(i+k)j} (a_i y_k) \otimes x_j \\
&= (-1)^{(i+j)k} y_k \otimes (a_i x_j) - (-1)^{jk} (a_i y_k) \otimes x_j \\
&= -(-1)^{jk} [(a_i y_k) \otimes x_j - (-1)^{ik} y_k \otimes (a_i x_j)] \in W.
\end{aligned}$$

It follows that γ factors through the natural epimorphisms $L \otimes_R M \to L \otimes_A M$ and $M \otimes_R L \to M \otimes_A L$, that is, the map $\overline{\gamma}\colon L \otimes_A M \to M \otimes_A L$ given by $x_i \otimes y_j \mapsto (-1)^{ij} y_j \otimes x_i$ is well defined. To show that $\overline{\gamma}$ is A-linear, we compute:

$$\begin{aligned}
\overline{\gamma}_{i+j+k}(a_i(x_j \otimes y_k)) &= \overline{\gamma}_{i+j+k}((a_i x_j) \otimes y_k) \\
&= (-1)^{(i+j)k} y_k \otimes (a_i x_j) \\
&= (-1)^{(i+j)k+ik} (a_i y_k) \otimes x_j \\
&= (-1)^{jk} a_i (y_k \otimes x_j) \\
&= a_i \overline{\gamma}_{j+k}(x_j \otimes y_k).
\end{aligned}$$

Similarly, the map $\overline{\delta}\colon M \otimes_A L \to L \otimes_A M$ given by $y_j \otimes x_i \mapsto (-1)^{ij} x_i \otimes y_j$ is well defined and A-linear. It is straightforward to show that the compositions $\overline{\gamma}\overline{\delta}$ and $\overline{\delta}\overline{\gamma}$ are the respective identities, so that $\overline{\gamma}$ is the desired isomorphism. □

10.46 (Sketch of Solution to Exercise 7.15). For the map $N \otimes_A f$, let U be the R-submodule of $N \otimes_R L$ generated by the elements $(a_i x_j) \otimes y_k - (-1)^{ij} x_j \otimes (a_i y_k)$, and let V be the R-submodule of $N \otimes_R M$ generated by the elements of the form $(a_i x_j) \otimes y_k - (-1)^{ij} x_j \otimes (a_i y_k)$. To show that $N \otimes_A f$ is well defined, it suffices to show that the map $N \otimes_R f : N \otimes_R L \to N \otimes_R M$ sends each generator of U into V:

$$\begin{aligned}(N \otimes_R f)_{i+j+k}((a_i x_j) \otimes y_k - (-1)^{ij} x_j \otimes (a_i y_k))& \\ &= (a_i x_j) \otimes f_k(y_k) - (-1)^{ij} x_j \otimes f_{i+k}(a_i y_k) \\ &= (a_i x_j) \otimes f_k(y_k) - (-1)^{ij} x_j \otimes (a_i f_k(y_k)) \in V.\end{aligned}$$

To show that $N \otimes_A f$ is A-linear, we compute similarly:

$$\begin{aligned}(N \otimes_A f)_{i+j+k}(a_i(x_j \otimes y_k)) &= (N \otimes_A f)_{i+j+k}((a_i x_j) \otimes y_k)) \\ &= (a_i x_j) \otimes f_k(y_k) \\ &= a_i(x_j \otimes f_k(y_k)) \\ &= a_i(N \otimes_A f)_{j+k}(x_j \otimes y_k)\end{aligned}$$

The map $f \otimes_R N$ is treated similarly. □

10.47 (Sketch of Solution to Exercise 7.17). We are working over R as a DG R-algebra, which has $R^\natural = R$. Since R is local, we know that a direct sum $\bigoplus_i M_i$ of R-modules is free if and only if each M_i is free. (In general, the M_i are projective; since R is local, we know that projective implies free.) If L is a semi-free DG R-module, then it is bounded below by definition, and the module $\bigoplus_i L_i$ is free over R, so each L_i is free, as desired. The converse is handled similarly.

If F is a free resolution of M, then the previous paragraph implies that F is semi-free. Exercise 3.10 implies that there is a quasiisomorphism $F \xrightarrow{\simeq} M$ over R, so this is a semi-free resolution by definition. □

10.48 (Sketch of Solution to Exercise 7.18). It is straightforward to show that M is exact (as an R-complex) if and only if the natural map $0 \to M$ is a quasiisomorphism, since the induced map on homology is the natural map $0 \to \mathrm{H}_i(M)$. Exercise 7.17 implies that 0 is semi-free, so the map $0 \to M$ is a quasiisomorphism if and only if it is a semi-free resolution.

Since A is bounded below, so are $\Sigma^n A$ and $\bigoplus_{n \geqslant n_0} \Sigma^n A^{\beta_n}$. To show that $\Sigma^n A$ is semi-free, we need to show that $1_A \in (\Sigma^n A)_n$ is a semibasis. The only subtlety here is in the signs. If n is odd, then we have

$$\begin{aligned}(\textstyle\sum_i a_i) * 1_A &= \textstyle\sum_i (-1)^i a_i \\ (\textstyle\sum_i (-1)^i a_i) * 1_A &= \textstyle\sum_i a_i.\end{aligned}$$

The first of these shows that 1_A is linearly independent: if $(\sum_i a_i) * 1_A = 0$, then $\sum_i (-1)^i a_i = 0$ so $a_i = 0$ for all i, which implies that $\sum_i a_i = 0$. The second of these shows that 1_A spans $A^\natural$ over $A^\natural$: for all $\sum_i a_i \in A^\natural$, we have $\sum_i a_i = (\sum_i (-1)^i a_i) * 1_A \in A^\natural \cdot 1_A$. If n is even, then the relevant formula is $(\sum_i a_i) * 1_A = \sum_i a_i$.

To show that $\Sigma^n A^{\beta_n}$ is semi-free, use the previous paragraph to show that the sequence of standard basis vectors $(1_A, 0, \ldots, 0), (0, 1_A, \ldots), \ldots, (0, 0, \ldots, 1_A)$ form a semibasis. To show that $\bigoplus_{n \geqslant n_0} \Sigma^n A^{\beta_n}$ is semi-free, let E_n be a semibasis for each $\Sigma^n A^{\beta_n}$ and show that $\cup_n E_n$ is a semibasis for $\bigoplus_{n \geqslant n_0} \Sigma^n A^{\beta_n}$. □

10.49 (Sketch of Solution to Exercise 7.19). (a) Exercise 7.10 shows that $K \otimes_R F$ is a DG K-module, so we only need to show that it is semi-free. For each $i \in \mathbb{Z}$, let E_i be a basis of the free R-module F_i, and set $E_i' = \{1_K \otimes e \in K \otimes_R F \mid e \in E_i\}$. We claim that $E' := \cup_i E_i'$ is a semibasis for $K \otimes_R F$. To show that E' spans $(K \otimes_R F)^\natural$, it suffices to show that for each $x \in K_i$ and each $y \in F_j$ the generator $x \otimes y \in (K \otimes_R F)_{i+j}$ is in the K-span of E'. For this, write $y = \sum_{e \in E_j} r_e e$, and compute:

$$x \otimes y = x \otimes (\textstyle\sum_{e \in E_j} r_e e) = \sum_{e \in E_j} r_e x (1_K \otimes e) \in K \cdot E'.$$

To show that E' is linearly independent takes a bit of bookkeeping. Suppose that

$$0 = \sum_{i=1}^{m} x_i (1_K \otimes e_i) \tag{10}$$

in $K \otimes_A F$ for some $x_i \in K^\natural$ and distinct elements $e_1, \ldots, e_m \in E$. Since $(K \otimes_R F)^\natural$ is a graded $K^\natural$-module, we may assume without loss of generality that each x_i is homogeneous and that the degree $|x_i \otimes e_i| = |x_i| + |e_i| = n$ is the same for all i. Moreover, $(K \otimes_R F)^\natural$ is a bi-graded $K^\natural$-module (with gradings coming from K and F) so we may assume without loss of generality that the degree $|x_i| = p$ is the same for all i and that the degree $|e_i| = q$ is the same for all i. Thus, Eq. (10) becomes $0 = \sum_{i=1}^m x_i \otimes e_i$. This sum occurs in the following submodule of $K_p \otimes_R F_q$:

$$\bigoplus_{i=1}^{m} K_p \otimes_R Re_i \cong \bigoplus_{i=1}^{m} K_p.$$

Under this isomorphism, the element $0 = \sum_{i=1}^m x_i \otimes e_i$ corresponds to the vector $0 = (x_1, \ldots, x_m)$ which implies that $x_i = 0$ for $i = 1, \ldots, m$. It follows that E' is linearly independent, as desired.

(b) Let $F \xrightarrow{\simeq} M$ be a semi-free resolution of a DG R-module M. Part (a) implies that $K \otimes_R F$ is semi-free over K. Since K is a bounded below complex of projective R-modules, Fact 4.9 implies that the induced map $K \otimes_R F \xrightarrow{\simeq} K \otimes_R M$ is a quasiisomorphism, so it is a semi-free resolution by definition. □

10.50 (Sketch of Solution to Exercise 7.26). To show that $\chi^A_M : A \to \operatorname{Hom}_A(M, M)$ has the desired properties, we first note that Exercise 7.3(b) shows that it maps A_i to $\operatorname{Hom}_A(M, M)_i$ for all i. Next, we check that χ^A_M is a chain map:

$$\partial_i^{\operatorname{Hom}_R(M,M)}((\chi^A_M)_i(a_i)) = \{\partial^M_{i+p}\mu_p^{M,a_i} - (-1)^i \mu_{p-1}^{M,a_i}\partial^M_p\}$$
$$(\chi^A_M)_{i-1}(\partial^A_i(a_i)) = \{\mu_p^{M,\partial^A_i(a_i)}\}.$$

To see that these are equal, we evaluate at $m_p \in M_p$:

$$\begin{aligned}
\partial^M_{i+p}(\mu_p^{M,a_i}(m_p)) - (-1)^i \mu_{p-1}^{M,a_i}(\partial^M_p(m_p)) \\
&= \partial^M_{i+p}(a_i m_p) - (-1)^i a_i \partial^M_p(m_p) \\
&= \partial^A_i(a_i)m_p + (-1)^i a_i \partial^M_p(m_p) - (-1)^i a_i \partial^M_p(m_p) \\
&= \partial^A_i(a_i)m_p \\
&= \mu_p^{M,\partial^A_i(a_i)}(m_p)
\end{aligned}$$

To complete the proof, we check that χ^A_M is a A-linear. For this, we need to show that $(\chi^A_M)_{i+j}(a_i b_j) = a_i(\chi^A_M)_j(b_j)$. To show this, we evaluate at $m_p \in M_p$:

$$(\chi^A_M)_{i+j}(a_i b_j)_p(m_p) = a_i b_j m_p = a_i(\chi^A_M)_j(b_j)_p(m_p)$$

as desired. □

10.51 (Sketch of Solution to Exercise 7.32). Let M and N be R-modules. Exercise 7.17 implies that each free resolution F of M gives rise to a semi-free resolution $F \xrightarrow{\simeq} M$. Thus, the module $\operatorname{Ext}^i_R(M, N)$ defined in 7.31 is $\operatorname{H}_{-i}(\operatorname{Hom}_R(F, M))$, which is the usual $\operatorname{Ext}^i_R(M, N)$. □

10.52 (Sketch of Solution to Exercise 8.4). We begin with the graded vector space $W'' = 0 \bigoplus Fw_2 \bigoplus Fw_1 \bigoplus Fw_0 \bigoplus 0$. The differential ∂'' consists of two matrices of size 1×1:

$$0 \to Fw_2 \xrightarrow{x_2} Fw_1 \xrightarrow{x_1} Fw_0 \to 0.$$

The condition $\partial''_{i-1}\partial''_i = 0$ is only nontrivial for $i = 2$, in which case it boils down to the following:

$$0 = \partial''_1(\partial''_2(w_2)) = \partial''_1(x_2 w_1) = x_1 x_2 w_0.$$

We conclude that (W'', ∂'') is an R-complex if and only if

$$x_1 x_2 = 0. \tag{11}$$

The scalar multiplication of U on W'' is completely described by specifying ew_0 and ew_1, and this requires two more elements $y_0, y_1 \in F$ so that we have $ew_0 = y_0 w_1$ and $ew_1 = y_1 w_2$. The associative law (which was not a concern for W and W') says that we must have

$$0 = 0w_0 = e^2 w_0 = e(ew_0) = e(y_0 w_1) = y_0 y_1 w_2$$

so we conclude that

$$y_0 y_1 = 0. \tag{12}$$

Note that once this is satisfied, the general associative law follows. This leaves the Leibniz Rule for the products ew_0, ew_1, and ew_2. We begin with ew_0:

$$\begin{aligned} \partial_1''(ew_0) &= \partial_1''^{U}(e)w_0 + (-1)^{|e|} e \partial_0''(w_0) \\ \partial_1''(y_0 w_1) &= 0w_0 + (-1)^{|e|} e0 \\ x_1 y_0 w_0 &= 0 \end{aligned}$$

which implies that

$$x_1 y_0 = 0. \tag{13}$$

A similar computation for ew_2 shows that

$$x_2 y_1 = 0. \tag{14}$$

A last computation for ew_1 yields $x_2 y_1 + x_1 y_0 = 0$, which is redundant because of Eqs. (13)–(14). Thus, $\mathrm{Mod}^U(W'')$ consists of all ordered quadruplets $(x_1, x_2, y_0, y_1) \in \mathbb{A}_F^4$ satisfying Eqs. (11)–(14). It is possibly worth noting that the ideal defined by Eqs. (11)–(14) has a simple primary decomposition:

$$(x_1 x_2, y_0 y_1, x_1 y_0, x_2 y_1) = (x_1 y_1) \cap (x_2, y_0).$$

Next, we repeat this process for $W''' = 0 \bigoplus F z_2 \bigoplus (F z_{1,1} \bigoplus F z_{1,2}) \bigoplus F z_0 \bigoplus 0$. The differential ∂''' in this case has the following form:

$$0 \to F z_2 \xrightarrow{\binom{a_{2,1}}{a_{2,2}}} F z_{1,1} \bigoplus F z_{1,2} \xrightarrow{(a_{1,1}\ a_{1,2})} F z_0 \to 0$$

meaning that $\partial_2'''(z_2) = a_{2,1} z_{1,1} + a_{2,2} z_{1,2}$ and $\partial_1'''(z_{1,i}) = a_{1,i} z_0$ for $i = 1, 2$. Scalar multiplication also requires more letters:

$$\begin{aligned} ez_0 &= b_{0,1}z_{1,1} + b_{0,2}z_{1,2} \\ ez_{1,1} &= b_{1,1}z_2 \\ ez_{1,2} &= b_{1,2}z_2. \end{aligned}$$

The condition $\partial'''_{i-1}\partial'''_i = 0$ is equivalent to the following equation:

$$a_{1,1}a_{2,1} + a_{1,2}a_{2,2} = 0. \tag{15}$$

The associative law is equivalent to the next equation:

$$b_{0,1}b_{1,1} + b_{0,2}b_{1,2} = 0. \tag{16}$$

For the Leibniz Rule, we need to consider the products ez_0, $ez_{1,j}$ and ez_2, so this axiom is equivalent to the following equations:

$$a_{1,1}b_{0,1} + a_{1,2}b_{0,2} = 0 \tag{17}$$

$$a_{2,i}b_{1,j} + a_{1,j}b_{0,i} = 0 \qquad \text{for all } i = 1,2 \text{ and } j = 1,2 \tag{18}$$

$$a_{2,1}b_{1,1} + a_{2,2}b_{1,2} = 0. \tag{19}$$

So, $\mathrm{Mod}^U(W'')$ consists of all $(a_{1,1}, a_{1,2}, a_{2,1}, a_{2,2}, b_{0,1}, b_{0,2}, b_{1,1}, b_{1,2}) \in \mathbb{A}^8_F$ satisfying Eqs. (15)–(19). □

10.53 (Sketch of Solution to Exercise 8.7). We recall that

$$\begin{aligned} W'' &= 0 \bigoplus Fw_2 \bigoplus Fw_1 \bigoplus Fw_0 \bigoplus 0 \\ W''' &= 0 \bigoplus Fz_2 \bigoplus (Fz_{1,1} \bigoplus Fz_{1,2}) \bigoplus Fz_0 \bigoplus 0. \end{aligned}$$

It follows that

$$\begin{aligned} \mathrm{End}_F(W'')_0 &= \bigoplus_{i=0}^{2} \mathrm{Hom}_F(Fw_i, Fw_i) \cong F^3 = \mathbb{A}^3_F \\ \mathrm{GL}_F(W'')_0 &= \bigoplus_{i=0}^{2} \mathrm{Aut}_F(Fw_i) \cong (F^\times)^3 = U_{u_2u_1u_0} \subset \mathbb{A}^3_F \\ \mathrm{End}_F(W''')_0 &\cong \mathrm{Hom}_F(F,F) \bigoplus \mathrm{Hom}_F(F^2,F^2) \bigoplus \mathrm{Hom}_F(F,F) \\ &\cong F \times F^4 \times F = \mathbb{A}^6_F \\ \mathrm{GL}_F(W''')_0 &= \mathrm{Aut}_F(F) \bigoplus \mathrm{Aut}_F(F^2) \bigoplus \mathrm{Aut}_F(F) \\ &\cong F^\times \times \mathrm{GL}_2(F) \times F^\times = U_{c_2(c_{11}c_{22}-c_{12}c_{21})c_0} \subset \mathbb{A}^6_F. \end{aligned}$$

□

10.54 (Sketch of Solution to Exercise 8.10). We continue with the notation of Example 5.17 and the solutions to Exercises 8.4 and 8.7. Under the isomorphism $\mathrm{GL}_F(W'')_0 \cong U_{u_2u_1u_0} \subseteq \mathbb{A}^3_F$, an ordered triple $(u_0, u_1, u_2) \in U_{u_2u_1u_0}$ corresponds to the isomorphism

$$\begin{array}{cccccccccc}
\partial: & & 0 & \longrightarrow & Fw_2 & \xrightarrow{x_2} & Fw_1 & \xrightarrow{x_1} & Fw_0 & \longrightarrow & 0 \\
\alpha \downarrow \cong & & & & \downarrow u_2 & & \downarrow u_1 & & \downarrow u_0 & & \\
\tilde{\partial}: & & 0 & \longrightarrow & F\tilde{w}_2 & \xrightarrow{\tilde{x}_2} & F\tilde{w}_1 & \xrightarrow{\tilde{x}_1} & F\tilde{w}_0 & \longrightarrow & 0.
\end{array}$$

Let $e \cdot_\alpha \tilde{w}_j = \tilde{y}_j \tilde{w}_{j+1}$ for $j = 0, 1$. Then direct computations as in Example 8.9 show that $\tilde{x}_i = u_{i-1} x_i u_i^{-1}$ for $i = 1, 2$ and $\tilde{y}_j = u_{j+1} y_j u_j^{-1}$ for $j = 0, 1$.

Under the isomorphism $\mathrm{GL}_F(W''')_0 \cong U_{c_2(c_{11}c_{22}-c_{12}c_{21})c_0} \subset \mathbb{A}^6_F$, an ordered sextuple $(c_2, c_{11}, c_{22}, c_{12}, c_{21}, c_0) \in U_{c_2(c_{11}c_{22}-c_{12}c_{21})c_0}$ corresponds to the isomorphism

$$\begin{array}{cccccccccc}
\partial: & & 0 & \longrightarrow & Fz_2 & \xrightarrow{\left(\begin{smallmatrix} a_{2,1} \\ a_{2,2} \end{smallmatrix}\right)} & Fz_{1,1} \oplus Fz_{1,2} & \xrightarrow{(a_{1,1}\ a_{1,2})} & Fz_0 & \longrightarrow & 0 \\
\alpha \downarrow \cong & & & & \downarrow c_2 & & \downarrow \left(\begin{smallmatrix} c_{11} & c_{12} \\ c_{21} & c_{22} \end{smallmatrix}\right) & & \downarrow c_0 & & \\
\tilde{\partial}: & & 0 & \longrightarrow & F\tilde{z}_2 & \xrightarrow[\left(\begin{smallmatrix} \tilde{a}_{2,1} \\ \tilde{a}_{2,2} \end{smallmatrix}\right)]{} & F\tilde{z}_{1,1} \oplus F\tilde{z}_{1,2} & \xrightarrow[(\tilde{a}_{1,1}\ \tilde{a}_{1,2})]{} & F\tilde{z}_0 & \longrightarrow & 0.
\end{array}$$

Set $\Delta = \det\left(\begin{smallmatrix} c_{11} & c_{12} \\ c_{21} & c_{22} \end{smallmatrix}\right) = c_{11}c_{22} - c_{12}c_{21}$. Thus, we have the following:

$$\begin{pmatrix} \tilde{a}_{2,1} \\ \tilde{a}_{2,2} \end{pmatrix} = \begin{pmatrix} c_{11} & c_{12} \\ c_{21} & c_{22} \end{pmatrix} \begin{pmatrix} \tilde{a}_{2,1} \\ \tilde{a}_{2,2} \end{pmatrix} (c_2^{-1}) = \begin{pmatrix} (c_{11}a_{2,1} + c_{12}a_{2,2})c_2^{-1} \\ (c_{21}a_{2,1} + c_{22}a_{2,2})c_2^{-1} \end{pmatrix}$$

$$\begin{aligned}
(\tilde{a}_{1,1}\ \tilde{a}_{1,2}) &= (c_0) \begin{pmatrix} a_{2,1} \\ a_{2,2} \end{pmatrix} \begin{pmatrix} c_{11} & c_{12} \\ c_{21} & c_{22} \end{pmatrix}^{-1} \\
&= \left(\Delta^{-1} c_0 (a_{1,1}c_{22} - a_{1,2}c_{21})\ \ \Delta^{-1} c_0(-a_{1,1}c_{12} + a_{1,2}c_{11})\right).
\end{aligned}$$

In other words, we have

$$\begin{aligned}
\tilde{a}_{2,i} &= (c_{i1}a_{2,1} + c_{i2}a_{2,2})c_2^{-1} \\
\tilde{a}_{1,1} &= \Delta^{-1} c_0 (a_{1,1}c_{22} - a_{1,2}c_{21}) \\
\tilde{a}_{1,2} &= \Delta^{-1} c_0 (-a_{1,1}c_{12} + a_{1,2}c_{11})
\end{aligned}$$

for $i = 1, 2$. For the scalar multiplication, we have

$$e \cdot_\alpha \tilde{z}_2 = 0$$

and using the rule $e \cdot_\alpha \tilde{z}_* = \alpha_2(e\alpha_1^{-1}(z_*))$, we find that

$$e \cdot_\alpha \tilde{z}_{1,1} = c_2\Delta^{-1}(c_{22}b_{1,1} - c_{21}b_{1,2})\tilde{z}_2$$
$$e \cdot_\alpha \tilde{z}_{1,2} = c_2\Delta^{-1}(-c_{12}b_{1,1} + c_{11}b_{1,2})\tilde{z}_2$$
$$e \cdot_\alpha \tilde{z}_0 = c_0^{-1}(c_{11}b_{0,1} + c_{12}b_{0,2})\tilde{z}_{1,1} + c_0^{-1}(c_{21}b_{0,1} + c_{22}b_{0,2})\tilde{z}_{1,2}.$$

In other words, we have

$$\tilde{b}_{1,1} = c_2\Delta^{-1}(c_{22}b_{1,1} - c_{21}b_{1,2})$$
$$\tilde{b}_{1,2} = c_2\Delta^{-1}(-c_{12}b_{1,1} + c_{11}b_{1,2})\tilde{z}_2$$
$$\tilde{b}_{0,i} = c_0^{-1}(c_{i1}b_{0,1} + c_{i2}b_{0,2})$$

for $i = 1, 2$. □

Acknowledgments We are grateful to the Department of Mathematics and Statistics at the University of Regina for its hospitality during the Workshop on Connections Between Algebra and Geometry. We are also grateful to the organizers and participants for providing such a stimulating environment, and to Saeed Nasseh for carefully reading parts of this manuscript. We are also thankful to the referee for valuable comments and corrections. Sean Sather-Wagstaff was supported in part by a grant from the NSA.

References

1. Apassov, D.: Homological dimensions over differential graded rings. In: Complexes and Differential Graded Modules. Ph.D. thesis, Lund University, pp. 25–39 (1999)
2. Artin, M., Bertin, J.E., Demazure, M., Gabriel, P., Grothendieck, A., Raynaud, M., Serre, J.-P.: Schémas en groupes. Fasc. 1: Exposés 1 à 4. In: Séminaire de Géométrie Algébrique de l'Institut des Hautes Études Scientifiques, vol. 1963. Institut des Hautes Études Scientifiques, Paris (1963/1964). MR 0207702 (34 #7517)
3. Auslander, M., Ding, S., Solberg, Ø.: Liftings and weak liftings of modules. J. Algebra **156**, 273–397 (1993). MR 94d:16007
4. Avramov, L.L.: Infinite free resolutions. In: Six Lectures on Commutative Algebra (Bellaterra, 1996). Progress in Mathematics, vol. 166, pp. 1–118. Birkhäuser, Basel (1998). MR 99m:13022
5. Avramov, L.L., Foxby, H.-B.: Ring homomorphisms and finite Gorenstein dimension. Proc. Lond. Math. Soc. **75**(2), 241–270 (1997). MR 98d:13014
6. Avramov, L.L., Halperin, S.: Through the looking glass: a dictionary between rational homotopy theory and local algebra. In: Algebra, Algebraic Topology and Their Interactions (Stockholm, 1983). Lecture Notes in Mathematics, vol. 1183, pp. 1–27. Springer, Berlin (1986). MR 846435 (87k:55015)
7. Avramov, L.L., Foxby, H.-B., Lescot, J.: Bass series of local ring homomorphisms of finite flat dimension. Trans. Am. Math. Soc. **335**(2), 497–523 (1993). MR 93d:13026
8. Avramov, L.L., Foxby, H.-B., Herzog, B.: Structure of local homomorphisms. J. Algebra **164**, 124–145 (1994). MR 95f:13029
9. Avramov, L.L., Gasharov, V.N., Peeva, I.V.: Complete intersection dimension. Inst. Hautes Études Sci. Publ. Math. **86**, 67–114 (1997). MR 1608565 (99c:13033)

10. Avramov, L.L., Foxby, H.-B., Halperin, S.: Differential graded homological algebra (in preparation)
11. Bruns, W., Herzog, J.: Cohen-Macaulay Rings, revised ed. Studies in Advanced Mathematics, vol. 39. University Press, Cambridge (1998). MR 1251956 (95h:13020)
12. Buchsbaum, D.A., Eisenbud, D.: Algebra structures for finite free resolutions, and some structure theorems for ideals of codimension 3. Am. J. Math. **99**(3), 447–485 (1977). MR 0453723 (56 #11983)
13. Burch, L.: On ideals of finite homological dimension in local rings. Proc. Camb. Philos. Soc. **64**, 941–948 (1968). MR 0229634 (37 #5208)
14. Christensen, L.W.: Semi-dualizing complexes and their Auslander categories. Trans. Am. Math. Soc. **353**(5), 1839–1883 (2001). MR 2002a:13017
15. Christensen, L.W., Sather-Wagstaff, S.: A Cohen-Macaulay algebra has only finitely many semidualizing modules. Math. Proc. Camb. Philos. Soc. **145**(3), 601–603 (2008). MR 2464778
16. Christensen, L.W., Sather-Wagstaff, S.: Descent via Koszul extensions. J. Algebra **322**(9), 3026–3046 (2009). MR 2567408
17. Cooper, S.M., Sather-Wagstaff, S.: Multiplicities of semidualizing modules. Commun. Algebra **41**(12), 4549–4558 (2013)
18. Demazure, M., Gabriel, P.: Introduction to Algebraic Geometry and Algebraic Groups. North-Holland Mathematics Studies, vol. 39. North-Holland, Amsterdam (1980). (Translated from the French by J. Bell.) MR 563524 (82e:14001)
19. Foxby, H.-B.: Gorenstein modules and related modules. Math. Scand. **31**, 267–284 (1972). MR 48 #6094
20. Foxby, H.-B., Thorup, A.: Minimal injective resolutions under flat base change. Proc. Am. Math. Soc. **67**(1), 27–31 (1977). MR 56 #11984
21. Frankild, A., Sather-Wagstaff, S.: Reflexivity and ring homomorphisms of finite flat dimension. Commun. Algebra **35**(2), 461–500 (2007). MR 2294611
22. Gabriel, P.: Finite representation type is open. In: Proceedings of the International Conference on Representations of Algebras, Carleton University, Ottawa, ON, 1974. Paper No. 10, Carleton Mathematical Lecture Notes, No. 9, p. 23, 1974. MR 0376769 (51 #12944)
23. Golod, E.S.: G-dimension and generalized perfect ideals. Trudy Mat. Inst. Steklov. **165**, 62–66 (1984). (Algebr. Geom. Appl.) MR 85m:13011
24. Grothendieck, A.: Éléments de géométrie algébrique. IV. Étude locale des schémas et des morphismes de schémas. I. Inst. Hautes Études Sci. Publ. Math. **20**, 259 (1964). MR 0173675 (30 #3885)
25. Happel, D.: Selforthogonal modules. In: Abelian Groups and Modules (Padova, 1994). Mathematics and Its Applications, vol. 343, pp. 257–276. Kluwer Academic, Dordrecht (1995). MR 1378204 (97d:16016)
26. Herzog, J.: Komplexe, auflösungen, und dualität in der lokalen algebra. Habilitationsschrift, Regensburg (1973)
27. Hilbert, D.: über die Theorie der algebraischen Formen. Math. Ann. **36**, 473–534 (1890). MR 1510634
28. Kustin, A.R.: Gorenstein algebras of codimension four and characteristic two. Commun. Algebra **15**(11), 2417–2429 (1987). MR 912779 (88j:13020)
29. Kustin, A.R., Miller, M.: Algebra structures on minimal resolutions of Gorenstein rings of embedding codimension four. Math. Z. **173**(2), 171–184 (1980). MR 583384 (81j:13013)
30. Mantese, F., Reiten, I.: Wakamatsu tilting modules. J. Algebra **278**(2), 532–552 (2004). MR 2071651 (2005d:16023)
31. Matsumura, H.: Commutative Ring Theory. Studies in Advanced Mathematics, 2nd edn., vol. 8. Cambridge University Press, Cambridge (1989). MR 90i:13001
32. Nasseh, S., Sather-Wagstaff, S.: A local ring has only finitely many semidualizing complexes up to shift-isomorphism. Preprint (2012). `arXiv:1201.0037`
33. Nasseh, S., Sather-Wagstaff, S.: Liftings and quasi-liftings of DG modules. J. Algebra **373**, 162–182 (2013). MR 2995021

34. Reiten, I.: The converse to a theorem of Sharp on Gorenstein modules. Proc. Am. Math. Soc. **32**, 417–420 (1972). MR 0296067 (45 #5128)
35. Sather-Wagstaff, S.: Semidualizing modules and the divisor class group. Ill. J. Math. **51**(1), 255–285 (2007). MR 2346197
36. Sather-Wagstaff, S.: Complete intersection dimensions and Foxby classes. J. Pure Appl. Algebra **212**(12), 2594–2611 (2008). MR 2452313 (2009h:13015)
37. Sather-Wagstaff, S.: Bass numbers and semidualizing complexes. In: Commutative Algebra and Its Applications, pp. 349–381. Walter de Gruyter, Berlin (2009). MR 2640315
38. Sather-Wagstaff, S.: Lower bounds for the number of semidualizing complexes over a local ring. Math. Scand. **110**(1), 5–17 (2012). MR 2900066
39. Sather-Wagstaff, S.: Semidualizing modules (in preparation)
40. Sharp, R.Y.: Finitely generated modules of finite injective dimension over certain Cohen-Macaulay rings. Proc. Lond. Math. Soc. **25**, 303–328 (1972). MR 0306188 (46 #5315)
41. Vasconcelos, W.V.: Divisor Theory in Module Categories. North-Holland Mathematics Studies, vol. 14. Notas de Matemática No. 53 [Notes on Mathematics, No. 53]. North-Holland, Amsterdam (1974). MR 0498530 (58 #16637)
42. Voigt, D.: Induzierte Darstellungen in der Theorie der endlichen, algebraischen Gruppen. Lecture Notes in Mathematics, vol. 592. Springer, Berlin (1977). Mit einer englischen Einführung. MR 0486168 (58 #5949)
43. Wakamatsu, T.: On modules with trivial self-extensions. J. Algebra **114**(1), 106–114 (1988). MR 931903 (89b:16020)

Four Lectures on Secant Varieties

Enrico Carlini, Nathan Grieve, and Luke Oeding

To Tony, friend and mentor

Abstract This paper is based on the first author's lectures at the 2012 University of Regina Workshop "Connections Between Algebra and Geometry." Its aim is to provide an introduction to the theory of higher secant varieties and their applications. Several references and solved exercises are also included.

1 Introduction

Secant varieties have travelled a long way from nineteenth century geometry to nowadays where they are as popular as ever before. There are different reasons for this popularity, but they can be summarized in one word: applications. These applications are both pure and applied in nature. Indeed, not only does the geometry of secant varieties play a role in the study projections of a curve, a surface or a threefold, but it also in locating a transmitting antenna [62].

E. Carlini (✉)
Department of Mathematical Sciences, Politecnico di Torino, 10129 Turin, Italy

School of Mathematical Sciences, Monash University, Melbourne, VIC 3800, Australia
e-mail: enrico.carlini@polito.it; enrico.carlini@monash.edu

N. Grieve
Department of Mathematics and Statistics, Queen's University, Kingston, ON, Canada

Department of Mathematics and Statistics, McGill University, Montreal, QC, Canada
e-mail: nathangrieve@mast.queensu.ca; ngrieve@math.mcgill.ca

L. Oeding
Department of Mathematics, University of California, Berkeley, Berkeley, CA, USA

Department of Mathematics and Statistics, Auburn University, Auburn, AL, USA
e-mail: oeding@auburn.edu

S.M. Cooper and S. Sather-Wagstaff (eds.), *Connections Between Algebra, Combinatorics, and Geometry*, Springer Proceedings in Mathematics & Statistics 76,
DOI 10.1007/978-1-4939-0626-0_2,

In these lectures we introduce the reader to the study of (higher) secant varieties by providing the very basic definition and properties, and then moving the most direct applications. In this way we introduce the tools and techniques which are central for any further study in the topic. In the last lecture we present some more advanced material and provide pointers to some relevant literature. Several exercises are included and they are meant to be a way for the reader to familiarize himself or herself with the main ideas presented in the lectures. So, have fun with secant varieties and their many applications!

The paper is structured as follows. In Sect. 2 we provide the basic definition and properties of higher secant varieties. In particular, we introduce one of the basic results in the theory, namely Terracini's Lemma, and one main source of examples and problems, namely Veronese varieties. In Sect. 3 we introduce Waring problems and we explore the connections with higher secant varieties of Veronese varieties. Specifically, we review the basic result by Alexander and Hirschowitz. In Sect. 4, Apolarity Theory makes in its appearance with the Apolarity Lemma. We see how to use Hilbert functions and sets of points to investigate Waring problems and higher secant varieties to Veronese varieties. In Sect. 5 we give pointers to the literature giving reference to the topics we treated in the paper. We also provide a brief description and references for the many relevant topics which we were not able to include in the lectures because of time constraints. Finally, we provide solutions to the exercises in Sect. 6.

2 Lecture One

In what follows, $X \subset \mathbb{P}^N$ will denote an irreducible, reduced algebraic variety; we work over an algebraically closed field of characteristic zero, which we assume to be $\mathbb{C}$.

The topic of this lecture are *higher secant varieties* of X.

Definition 2.1. The s-th higher secant variety of X is

$$\sigma_s(X) = \overline{\bigcup_{P_1,\ldots,P_s \in X} \langle P_1, \ldots, P_s \rangle},$$

where the over bar denotes the Zariski closure.

In words, $\sigma_s(X)$ is the closure of the union of s-secant spaces to X.[1]

[1]Some authors use the notation $S(X)$ for the (first) secant variety of X, which corresponds to $\sigma_2(X)$, and $S_k(X)$ to denote the kth secant variety to X, which corresponds to $\sigma_{k+1}(X)$. We prefer to reference the number of points used rather than the dimension of their span because this is often more relevant for applications because of its connection to rank.

Example 2.2. If $X \subset \mathbb{P}^2$ is a curve and not a line, then $\sigma_2(X) = \mathbb{P}^2$, the same is true for hypersurfaces which are not hyperplanes. But, if $X \subset \mathbb{P}^3$ is a non-degenerate curve (i.e. not contained in a hyperplane), then $\sigma_2(X)$ can be, in principle, either a surface or a threefold.

We note that the closure operation is in general necessary, but there are cases in which it is not.

Exercise 2.3. Show that the union of chords (secant lines) to a plane conic is closed. However, the union of the chords of the twisted cubic curve in $\mathbb{P}^3$ is not.

In general, we have a sequence of inclusions

$$X = \sigma_1(X) \subseteq \sigma_2(X) \subseteq \ldots \subseteq \sigma_r(X) \subseteq \ldots \subseteq \mathbb{P}^N.$$

If X is a linear space, then $\sigma_i(X) = X$ for all i and all of the elements of the sequence are equal.

Remark 2.4. If $X = \sigma_2(X)$ then X, is a linear space. To see this consider a point $P \in X$ and the projection map $\pi_P : \mathbb{P}^N \dashrightarrow \mathbb{P}^{N-1}$. Let $X_1 = \pi_P(X)$ and notice that $\dim X_1 = \dim X - 1$ and that $\sigma_2(X_1) = X_1$. If X_1 is a linear space also X is so and we are done. Otherwise iterate the process constructing a sequence of varieties $X_2, \ldots, X_m$ of decreasing dimension. The process will end with X_m equal to a point and then X_{m-1} a linear space. Thus X_{m-2} is a linear space and so on up to the original variety X.

Exercise 2.5. For $X \subset \mathbb{P}^N$, show that, if $\sigma_i(X) = \sigma_{i+1}(X) \neq \mathbb{P}^N$, then $\sigma_i(X)$ is a linear space and hence $\sigma_j(X) = \sigma_i(X)$ for all $j \geq i$.

Using this remark and Exercise 2.5, we can refine our chain of inclusions for X a non-degenerate variety (i.e. not contained in a hyperplane).

Exercise 2.6. If $X \subseteq \mathbb{P}^N$ is non-degenerate, then there exists an $r \geq 1$ with the property that

$$X = \sigma_1(X) \subsetneq \sigma_2(X) \subsetneq \ldots \subsetneq \sigma_r(X) = \mathbb{P}^N.$$

In particular, all inclusions are strict and there is a higher secant variety that coincides with the ambient space.

It is natural to ask: what is the smallest r such that $\sigma_r(X) = \mathbb{P}^N$? Or more generally: what is the value of $\dim \sigma_i(X)$ for all i?

As a preliminary move in this direction, we notice that there is an expected value for the dimension of any higher secant variety of X that arises just from the naive dimension count. That is to say, if the secant variety doesn't fill the ambient space, a point is obtained by choosing s points from an n-dimensional variety and one point in the $\mathbb{P}^{s-1}$ that they span.

Definition 2.7. For $X \subset \mathbb{P}^N$, set $n = \dim X$. The *expected dimension* of $\sigma_s(X)$ is

$$\operatorname{expdim}(\sigma_s(X)) = \min\{sn + s - 1, N\}.$$

Notice also that the expected dimension is also the maximum dimension of the secant variety. Moreover, if the secant line variety $\sigma_2(X)$ does not fill the ambient $\mathbb{P}^N$, then X can be isomorphically projected into a $\mathbb{P}^{N-1}$. This interest in minimal codimension embeddings is one reason secants were classically studied.

Exercise 2.8. Let $X \subseteq \mathbb{P}^n$ be a curve. Prove that $\sigma_2(X)$ has dimension 3 unless X is contained in a plane. (This is why every curve is isomorphic to a space curve but only birational to a plane curve.)

There are cases in which $\text{expdim}(\sigma_i(X)) \neq \dim(\sigma_i(X))$ and these motivate the following:

Definition 2.9. If $\text{expdim}(\sigma_i(X)) \neq \dim(\sigma_i(X))$, then X is said to be *i-defective* or simply *defective*.

Remark 2.10. Notice that $\dim(\sigma_{i+1}(X)) \leq \dim(\sigma_i(X)) + n + 1$, where $n = \dim X$. This means that if $\sigma_i(X) \neq \mathbb{P}^N$ and X is i-defective, then X is j-defective for $j \leq i$.

Let's now see the most celebrated example of a defective variety, the Veronese surface in $\mathbb{P}^5$.

Example 2.11. Consider the polynomial ring $S = \mathbb{C}[x, y, z]$ and its homogeneous pieces S_d. The Veronese map ν_2 is the morphism

$$\nu_2 \colon \mathbb{P}(S_1) \longrightarrow \mathbb{P}(S_2) \text{ defined by } [L] \mapsto [L^2].$$

In coordinates this map can be described in terms of the standard monomial basis $\langle x, y, z\rangle$ for S_1 and the standard monomial basis $\langle x^2, 2xy, 2xz, y^2, 2yz, z^2\rangle$ for S_2. Thus the Veronese map can be written as the map

$$\nu_2 \colon \mathbb{P}^2 \longrightarrow \mathbb{P}^5 \text{ defined by } [a : b : c] \mapsto [a^2 : ab : ac : b^2 : bc : c^2].$$

The Veronese surface is then defined as the image of this map, i.e. the Veronese surface is $X = \nu_2(\mathbb{P}^2) \subseteq \mathbb{P}^5$.

We now want to study the higher secant varieties of the Veronese surface X. In particular we ask: is $\dim \sigma_2(X) = \text{expdim}(\sigma_2(X)) = 5$? In other words, does $\sigma_2(X)$ equal $\mathbb{P}^5$?

To answer this question, it is useful to notice that elements in S_2 are quadratic forms and, hence, are uniquely determined by 3×3 symmetric matrices. In particular, $P \in \mathbb{P}^5$ can be seen as $P = [Q]$ where Q is a 3×3 symmetric matrix. If $P \in X$, then Q also has rank equal one. Thus we have

$$\sigma_2(X) = \overline{\bigcup_{P_1, P_2} \langle P_1, P_2\rangle}$$

$$= \overline{\{[Q_1 + Q_2] : Q_i \text{ is a } 3 \times 3 \text{ symmetric matrix and } \text{rk}(Q_i) = 1\}} \subseteq H,$$

where H is the projective variety determined by the set of 3×3 symmetric matrices of rank at most two. Clearly H is the hypersurface defined by the vanishing of the determinant of the general 3×3 symmetric matrix and hence X is 2-defective.

Exercise 2.12. Let M be an $n \times n$ symmetric matrix of rank r. Prove that M is a sum of r symmetric matrices of rank 1.

Exercise 2.13. Show that $H = \sigma_2(X)$.

Exercise 2.14. Repeat the same argument for $X = \nu_2(\mathbb{P}^3)$. Is X 2-defective?

In order to deal with the problem of studying the dimension of the higher secant varieties of X we need to introduce a celebrated tool, namely Terracini's Lemma, see [152].

Lemma 2.15 (Terracini's Lemma). *Let $P_1, \ldots, P_s \in X$ be general points and $P \in \langle P_1, \ldots, P_s \rangle \subset \sigma_s(X)$ be a general point. Then the tangent space to $\sigma_s(X)$ in P is*

$$T_P(\sigma_s(X)) = \langle T_{P_1}(X), \ldots, T_{P_s}(X) \rangle.$$

Remark 2.16. To get a (affine) geometric idea of why Terracini's Lemma holds, we consider an affine curve $\gamma(t)$. A general point on $P \in \sigma_2(\gamma)$ is described as $\gamma(s_0) + \lambda_0[\gamma(t_0) - \gamma(s_0)]$. A neighborhood of P is then described as

$$\gamma(s) + \lambda[\gamma(t) - \gamma(s)].$$

Hence the tangent space $T_P(\sigma_s(\gamma))$ is spanned by

$$\gamma'(s_0) - \lambda_0\gamma'(s_0), \lambda_0\gamma'(t_0), \gamma(t_0) - \gamma(s_0),$$

and this is the affine span of the affine tangent spaces $\{\gamma(s_0) + \alpha\gamma'(s_0) : \alpha \in \mathbb{R}\}$ and $\{\gamma(t_0) + \beta\gamma'(t_0) : \beta \in \mathbb{R}\}$.

As a first application of Terracini's Lemma, we consider the twisted cubic curve.

Example 2.17. Let X be the twisted cubic curve in $\mathbb{P}^3$, i.e. $X = \nu_3(\mathbb{P}^1)$ where ν_3 is the map

$$\nu_3 \colon \mathbb{P}^1 \longrightarrow \mathbb{P}^3 \text{ defined by } [s : t] \mapsto [s^3 : s^2t : st^2 : t^3].$$

We want to compute $\dim \sigma_2(X) = \dim T_P(\sigma_2(X))$ at a generic point P. Using Terracini's Lemma it is enough to choose generic points $P_1, P_2 \in X$ and to study the linear span

$$\langle T_{P_1}(X), T_{P_2}(X) \rangle.$$

In particular, $\sigma_2(X) = \mathbb{P}^3$ if and only if the lines $T_{P_1}(X)$ and $T_{P_2}(X)$ do not intersect, that is, if and only if there does not exist a hyperplane containing both lines.

If $H \subset \mathbb{P}^3$ is a hyperplane, then the points of $H \cap X$ are determined by finding the roots of the degree three homogeneous polynomial $g(s,t)$ defining $\nu_3^{-1}(H) \subset \mathbb{P}^1$.

If $H \supset T_{P_1}(X)$ then g has a double root. However, the homogeneous polynomial is smooth and thus, in the general case, no hyperplane exists containing both tangent lines.

In conclusion, $\sigma_2(X) = \mathbb{P}^3$.

Exercise 2.18. Prove that if $H \supset T_P(X)$, then the polynomial defining $\nu_3^{-1}(H)$ has a double root.

We now introduce the Veronese variety in general.

Definition 2.19. Consider the polynomial ring $S = \mathbb{C}[x_0, \dots, x_n]$ and its homogeneous pieces S_d. The *d-th Veronese map* ν_d is the morphism

$$\nu_d : \mathbb{P}(S_1) \longrightarrow \mathbb{P}(S_d) \text{ defined by } [L] \mapsto [L^d].$$

In coordinates, using suitable monomial bases for S_1 and S_d, ν_d is the morphism

$$\nu_d : \mathbb{P}^n \longrightarrow \mathbb{P}^N \text{ defined by}$$

$$[a_0 : \dots : a_n] \mapsto [M_0(a_0, \dots, a_n) : \dots : M_N(a_0, \dots, a_n)]$$

where $N = \binom{n+d}{d} - 1$ and where $M_0, \dots, M_N$ are monomials which form a basis for S_d.

We call $\nu_d(\mathbb{P}^n)$ a *Veronese variety*.

Example 2.20. A relevant family of Veronese varieties are the *rational normal curves* which are Veronese varieties of dimension one, i.e. $n = 1$. In this situation $S = \mathbb{C}[x_0, x_1]$ and S_d is the vector space of degree d binary forms. The rational normal curve $\nu_d(\mathbb{P}(S_1)) \subseteq \mathbb{P}(S_d)$ is represented by d-th powers of binary linear forms.

Example 2.21. The rational normal curve $X = \nu_2(\mathbb{P}^1) \subset \mathbb{P}^2$ is an irreducible conic. It is easy to see that $\sigma_2(X) = \mathbb{P}^2 = \mathbb{P}(S_2)$. This equality can also be explained by saying that any binary quadratic form Q is the sum of two squares of linear forms, i.e. $Q = L^2 + M^2$.

Exercise 2.22. Consider the rational normal curve in $\mathbb{P}^3$, i.e. the twisted cubic curve $X = \nu_3(\mathbb{P}(S_1)) \subset \mathbb{P}(S_3)$. We know that $\sigma_2(X)$ fills up all the space. Can we write any binary cubic as the sum of two cubes of linear forms? Try $x_0 x_1^2$.

Exercise 2.23. We described the Veronese variety $X = \nu_d(\mathbb{P}^n)$ in parametric form by means of the relation: $[F] \in X$ if and only if $F = L^d$. Use this description and standard differential geometry to compute $T_{[L^d]}(X)$ (describe this as a vector space of homogeneous polynomials). This can be used to apply Terracini's Lemma, for example, to the twisted cubic curve.

3 Lecture Two

In the last lecture we spoke about higher secant varieties in general. Now we focus on the special case of Veronese varieties. Throughout this lecture we will consider the polynomial ring $S = \mathbb{C}[x_0, \dots, x_n]$.

An explicit description of the tangent space to a Veronese variety will be useful, so we give it here.

Remark 3.1. Let $X = \nu_d(\mathbb{P}^n)$ and consider $P = [L^d] \in X$ where $L \in S_1$ is a linear form. Then

$$T_P(X) = \langle [L^{d-1}M] : M \in S_1 \rangle.$$

We can use this to revisit the Veronese surface example.

Example 3.2. Consider the Veronese surface $X = \nu_2(\mathbb{P}^2) \subset \mathbb{P}^5$. To compute $\dim(\sigma_2(X))$ we use Terracini's Lemma. Hence we choose two general points $P = [L^2], Q = [N^2] \in X$ and we consider the linear span of their tangent spaces

$$T = \langle T_P(X), T_Q(X) \rangle.$$

By applying Grassmann's formula, and noticing that $T_P(X) \cap T_Q(X) = [LN]$ we get $\dim T = 3 + 3 - 1 - 1 = 4$ and hence $\sigma_2(X)$ is a hypersurface.

The study of higher secant varieties of Veronese varieties is strongly connected with a problem in polynomial algebra: the *Waring problem for forms*, i.e. for homogeneous polynomials, see [99]. We begin by introducing the notion of Waring rank.

Definition 3.3. Let $F \in S$ be a degree d form. The *Waring rank of* F is denoted $\mathrm{rk}(F)$ and is defined to be the *minimum* s such that we have

$$F = L_1^d + \dots + L_s^d$$

for some linear forms $L_i \in S_1$.

Remark 3.4. It is clear that $\mathrm{rk}(L^d) = 1$ if L is a linear form. However in general, if L and N are linear forms, $\mathrm{rk}(L^d + N^d) \leq 2$. It is 1 if L and N are proportional and 2 otherwise. For more than two factors the computation of the Waring rank for a sum of powers of linear form is not trivial.

We can now state the *Waring problem for forms*, which actually comes in two fashions. The *big* Waring problem asks for the computation of

$$g(n, d)$$

the minimal integer such that

$$\mathrm{rk}(F) \leq g(n, d)$$

for a *generic* element $F \in S_d$, i.e. for a generic degree d form in $n+1$ variables. The *little* Waring problem is more ambitious and asks us to determine the smallest integer

$$G(n, d)$$

such that

$$\mathrm{rk}(F) \leq G(n, d)$$

for any $F \in S_d$.

Remark 3.5. To understand the difference between the big and the little Waring problem we can refer to a probabilistic description. Pick a random element $F \in S_d$, then $\mathrm{rk}(F) \leq G(n, d)$ and with *probability one* $\mathrm{rk}(F) = g(n, d)$ (actually equality holds). However, if the choice of F is unlucky, it could be that $\mathrm{rk}(F) > g(n, d)$. Note also that these notions are field dependent, see [65] for example.

Remark 3.6. To make the notion of a generic element precise we use topology. Specifically, the big Waring problem asks us to bound the Waring rank for all elements belonging to a nonempty Zariski open subset of $\mathbb{P}S_d$; since non-empty Zariski open subsets are dense this also explains the probabilistic interpretation.

The big Waring problem has a nice geometric interpretation using Veronese varieties—this interpretation allows for a complete solution to the problem. Also the little Waring problem has a geometric aspect but this problem, in its full generality, is still unsolved.

Remark 3.7. As the Veronese variety $X = \nu_d(\mathbb{P}^n) \subset \mathbb{P}^N$ parameterizes pure powers in S_d, it is clear that $g(n, d)$ is the smallest s such that $\sigma_s(X) = \mathbb{P}^N$. Thus solving the big Waring problem is equivalent to finding the smallest s such that secant variety $\sigma_s(X)$ fills up $\mathbb{P}^N$. On the other hand, as taking the Zariski closure of the set $\bigcup_{P_1,\ldots,P_s \in X} \langle P_1, \ldots, P_s \rangle$ is involved in defining $\sigma_s(X)$, this is not equivalent to solving the little Waring problem.

Remark 3.8. To solve the little Waring problem one has to find the smallest s such that every element $[F] \in \mathbb{P}S_d$ lies on the span of some collection of s points of X.

Let's consider two examples to better understand the difference between the two problems.

Example 3.9. Let $X = \nu_2(\mathbb{P}^1) \subset \mathbb{P}^2$ be the rational normal curve in $\mathbb{P}^2$, i.e. a non-degenerate conic. We know that $\sigma_2(X) = \mathbb{P}^2$ and hence $g(n = 1, d = 2) = 2$. But we also know that each point of $\mathbb{P}^2$ lies on the span of two distinct points of X—every 2×2 symmetric matrix is the sum of two rank-one symmetric matrices—thus $G(n = 1, d = 2) = 2$. In particular this means that the Waring rank of a binary quadratic form is always at most two.

Example 3.10. Let $X = \nu_3(\mathbb{P}^1) \subset \mathbb{P}^3$ be the rational normal curve. Again, we know that $\sigma_2(X) = \mathbb{P}^3$ and hence $g(n = 1, d = 3) = 2$. However, there are degree three binary forms F such that $\mathrm{rk}(F) = 3$, and actually $G(n = 1, d = 3) = 3$.

To understand which the bad forms are, consider the projection map π_P from any point $P = [F] \in \mathbb{P}^3$. Clearly, if $P \notin X$, $\pi_p(X)$ is a degree 3 rational plane curve. Hence, it is singular, and being irreducible, only two possibilities arise. If the singularity is a node, then $P = [F]$ lies on a chord of X, and thus $F = L^3 + N^3$. But, if the singularity is a cusp, this is no longer true as P lies on a tangent line to X and not on a chord. Thus, the bad binary cubics lie on tangent lines to the twisted cubic curve. In other words, the bad binary cubics are of the form L^2N.

Exercise 3.11. For binary forms, we can stratify $\mathbb{P}S_2$ using the Waring rank: rank 1 elements correspond to points of the rational normal curve, while all the points outside the curve have rank 2. Do the same for binary cubics and stratify $\mathbb{P}S_3 = \mathbb{P}^3$.

We can produce a useful interpretation of Terracini's Lemma in the case of Veronese varieties. We consider the Veronese variety $X = \nu_d(\mathbb{P}^n) \subset \mathbb{P}^N$.

Remark 3.12. If $H \subset \mathbb{P}^N$ is a hyperplane, then $\nu_d^{-1}(H)$ is a degree d hypersurface. To see this, notice that H has an equation of the form $a_0z_0 + \ldots + a_Nz_N$ where z_i are the coordinates of $\mathbb{P}^N$. To determine an equation for $\nu_d^{-1}(H)$ it is enough to substitute each z_i with the corresponding degree d monomial in the $x_0, \ldots, x_n$.

Remark 3.13. If $H \subset \mathbb{P}^N$ is a hyperplane and $[L^d] \in H$, then $\nu_d^{-1}(H)$ is a degree d hypersurface passing through the point $[L] \in \mathbb{P}^n$. This is clearly true since $\nu_d^{-1}([L^d]) = [L]$.

Remark 3.14. If $H \subset \mathbb{P}^N$ is a hyperplane such that $T_{[L^d]}(X) \subset H$, then $\nu_d^{-1}(H)$ is a degree d hypersurface singular at the point $[L] \in \mathbb{P}^n$. This can be seen using *apolarity* or by direct computation choosing $L^d = x_0^d$.

We illustrate the last remark in an example.

Example 3.15. Consider the Veronese surface $X \subseteq \mathbb{P}^5$, let

$$P = [1 : 0 : 0 : 0 : 0 : 0] = [x^2] \in X,$$

and let $\mathbb{C}[z_0, z_1, \ldots, z_5]$ be the coordinate ring of $\mathbb{P}^5$. If H is a hyperplane containing P, then H has equation

$$a_1z_1 + a_2z_2 + a_3z_3 + a_4z_4 + a_5z_5 = 0$$

and hence $\nu_2^{-1}(H)$ is the plane conic determined by the equation

$$a_1xy + a_2xz + a_3y^2 + a_4yz + a_5z^2 = 0,$$

which passes through the point $\nu^{-1}(P) = [1 : 0 : 0]$. The tangent space $T_P(X)$ is the projective space associated with the linear span of the forms

$$x^2, xy, xz,$$

and hence it is the linear span of the points

$$[1 : 0 : 0 : 0 : 0 : 0], [0 : 1 : 0 : 0 : 0 : 0], [0 : 0 : 1 : 0 : 0 : 0].$$

Thus, if $H \supset T_P(X)$, then $a_1 = a_2 = 0$ and the corresponding conic has equation

$$a_3 y^2 + a_4 yz + a_5 z^2 = 0,$$

which is singular at the point $[1 : 0 : 0]$.

Exercise 3.16. Repeat the argument above to prove the general statement: if one has $T_{[L^d]}(\nu_d(\mathbb{P}^n)) \subset H$, then $\nu_d^{-1}(H)$ is a degree d hypersurface singular at the point $[L] \in \mathbb{P}^n$.

We will now elaborate on the connection between double point schemes and higher secant varieties to Veronese varieties.

Definition 3.17. Let $P_1, \dots, P_s \in \mathbb{P}^n$ be points with defining ideals $\wp_1, \dots, \wp_s$, respectively. The scheme defined by the ideal $\wp_1^2 \cap \dots \cap \wp_s^2$ is called a 2-*fat point* scheme or a *double point* scheme.

Remark 3.18. Let $X = \nu_d(\mathbb{P}^n) \subset \mathbb{P}^N$. There is a bijection between

$$\{H \subset \mathbb{P}^N \text{ a hyperplane } : H \supset \langle T_{P_1}(X), \dots, T_{P_s}(X)\rangle\},$$

and

$$\{\text{degree } d \text{ hypersurfaces of } \mathbb{P}^n \text{ singular at } P_1, \dots, P_s\} = (\wp_1^2 \cap \dots \cap \wp_s^2)_d.$$

Using the double point interpretation of Terracini's Lemma we get the following criterion to study the dimension of higher secant varieties to Veronese varieties.

Lemma 3.19. *Let $X = \nu_d(\mathbb{P}^n) \subset \mathbb{P}^N$ and choose generic points $P_1, \dots, P_s \in \mathbb{P}^n$ with defining ideals $\wp_1, \dots, \wp_s$ respectively. Then*

$$\dim \sigma_s(X) = N - \dim(\wp_1^2 \cap \dots \cap \wp_s^2)_d.$$

Example 3.20. We consider, again, the Veronese surface X in $\mathbb{P}^5$. To determine $\dim \sigma_2(X)$ we choose generic points $P_1, P_2 \in \mathbb{P}^2$ and look for conics singular at both points, i.e. elements in $(\wp_1^2 \cap \wp_s^2)_2$. Exactly one such conic exists (the line through P_1 and P_2 doubled) and hence $\sigma_2(X)$ is a hypersurface.

Exercise 3.21. Solve the big Waring problem for $n = 1$ using the double points interpretation.

We now return to the big Waring problem. Notice that the secant variety interpretation and a straightforward dimension count yield an expected value for $g(n, d)$ which is

$$\left\lceil \frac{\binom{d+n}{n}}{n+1} \right\rceil.$$

This expectation turns out to be true except for a short list of exceptions. A complete solution for the big Waring problem is given by a celebrated result by Alexander and Hirschowitz, see [8].

Theorem 3.22 ([8]). *Let F be a generic degree d form in $n + 1$ variables. Then*

$$\mathrm{rk}(F) = \left\lceil \frac{\binom{d+n}{n}}{n+1} \right\rceil,$$

unless

- $d = 2$, *any n where* $\mathrm{rk}(F) = n + 1$.
- $d = 4, n = 2$ *where* $\mathrm{rk}(F) = 6$ *and not* 5 *as expected.*
- $d = 4, n = 3$ *where* $\mathrm{rk}(F) = 10$ *and not* 9 *as expected.*
- $d = 3, n = 4$ *where* $\mathrm{rk}(F) = 8$ *and not* 7 *as expected.*
- $d = 4, n = 4$ *where* $\mathrm{rk}(F) = 15$ *and not* 14 *as expected.*

Remark 3.23. A straightforward interpretation of the Alexander and Hirschowitz result in terms of higher secants is as follows. The number $g(n, d)$ is the smallest s such that $\sigma_s(\nu_d(\mathbb{P}^n)) = \mathbb{P}^N$, unless n and d fall into one of the exceptional cases above.

Remark 3.24. Actually the Alexander and Hirschowitz result gives more for higher secant varieties of the Veronese varieties, namely that $\nu_d(\mathbb{P}^n)$ is not defective, for all s, except for the exceptional cases.

Let's now try to explain some of the defective cases of the Alexander–Hirschowitz result.

Example 3.25. For $n = 2, d = 4$ we consider $X = \nu_4(\mathbb{P}^2) \subset \mathbb{P}^{14}$. In particular, we are looking for the smallest s such that $\sigma_s(X) = \mathbb{P}^{14}$. We expect $s = 5$ to work and we want to check whether this is the case or not. To use the double point interpretation, we choose 5 generic points $P_1, \ldots, P_5 \in \mathbb{P}^2$ and we want to determine

$$\dim(\wp_1^2 \cap \ldots \cap \wp_5^2)_4.$$

To achieve this, we want to know the dimension of the space of quartic curves that are singular at each P_i. Counting conditions we have $15 - 5 \times 3 = 0$ and expect that

$$\dim(\wp_1^2 \cap \ldots \cap \wp_5^2)_4 = 0.$$

In fact, there exists a conic passing through the points P_i and this conic doubled is a quartic with the required properties. Thus,

$$\dim(\wp_1^2 \cap \ldots \cap \wp_5^2)_4 \geq 1,$$

and $\dim \sigma_5(X) \leq 14 - 1 = 13$.

Exercise 3.26. Show that $\sigma_5(\nu_4(\mathbb{P}^2))$ is a hypersurface, i.e. that it has dimension equal to 13.

Exercise 3.27. Explain the exceptional cases $d = 2$ any n.

Exercise 3.28. Explain the exceptional cases $d = 4$ and $n = 3, 4$.

Exercise 3.29. Explain the exceptional case $d = 3$ and $n = 4$. (Hint: use Castelnuovo's Theorem which asserts that there exists a (unique) rational normal curve passing through $n + 3$ generic points in $\mathbb{P}^n$.)

4 Lecture Three

In the last lecture we explained the solution to the big Waring problem and showed how to determine the Waring rank $\mathrm{rk}(F)$ for F a generic form. We now focus on a more general question: given any form F what can we say about $\mathrm{rk}(F)$?

The main tool we will use is *Apolarity* and, in order to do this, we need the following setting. Let $S = \mathbb{C}[x_0, \ldots, x_n]$ and $T = \mathbb{C}[y_0, \ldots, y_n]$. We make T act on S via differentiation, i.e. we define

$$y_i \circ x_j = \frac{\partial}{\partial x_i} x_j,$$

i.e. $y_i \circ x_j = 1$ if $i = j$ and it is zero otherwise. We then extend the action to all T so that $\partial \in T$ is seen as a differential operator on elements of S; from now on we will omit $\circ$. If A is a subset of a graded ring, we let A_d denote the degree d graded piece of A.

Definition 4.1. Given $F \in S_d$ we define the *annihilator*, or *perp ideal*, of F as follows:

$$F^\perp = \{\partial \in T : \partial F = 0\}.$$

Exercise 4.2. Show that $F^\perp \subset T$ is an ideal and that it also is Artinian, i.e. $(T/F^\perp)_i$ is zero for $i > d$.

Exercise 4.3. Let S_i and T_i denote the degree i homogenous pieces of S and T, respectively. Show that the map

$$S_i \times T_i \longrightarrow \mathbb{C}$$

$$(F, \partial) \mapsto \partial F$$

is a perfect pairing, i.e.

$$(F, \partial_0) \mapsto 0, \forall F \in S_i \Longrightarrow \partial_0 = 0,$$

and

$$(F_0, \partial) \mapsto 0, \forall \partial \in T_i \Longrightarrow F_0 = 0.$$

Remark 4.4. Recall that Artinian Gorenstein rings are characterized by the property that they are all of the form $A = T/F^{\perp}$. Moreover, a property of such an A is that it is finite dimensional, and the Hilbert function is symmetric.

Remark 4.5. Actually even more is true, and $A = T/F^{\perp}$ is Artinian and Gorenstein with socle degree d. Using the perfect pairing $S_i \times T_i \longrightarrow \mathbb{C}$ we see that $\dim A_d = \dim A_0 = 1$ and that A_d is the socle of A.

In what follows we will make use of Hilbert functions, thus we define them here.

Definition 4.6. For an ideal $I \subset T$ we define the *Hilbert function* of T/I as

$$HF(T/I, t) = \dim(T/I)_t.$$

Example 4.7. Let $F \in S_d$. We see that $HF(T/F^{\perp}, t) = 0$ for all $t > d$, in fact all partial differential operators of degree $t > d$ will annihilate the degree d form F and hence $(T/F^{\perp})_t = 0$, for $t > d$. From the remark above we also see that $HF(T/F^{\perp}, d) = 1$.

Exercise 4.8. Given $F \in S_d$ show that $HF(T/F^{\perp}, t)$ is a symmetric function with respect to $\frac{d+1}{2}$ of t.

An interesting property of the ideal $F^{\perp}$ is described by Macaulay's Theorem (see [128]).

Theorem 4.9. *If $F \in S_d$, then $T/F^{\perp}$ is an Artinian Gorenstein ring with socle degree d. Conversely, if T/I is an Artinian Gorenstein ring with socle degree d, then $I = F^{\perp}$ for some $F \in S_d$.*

Let's now see how apolarity relates to the Waring rank. Recall that $s = \mathrm{rk}(F)$ if and only if $F = \sum_1^s L_i^d$ and no shorter presentation exists.

Example 4.10. We now compute the possible Waring ranks for a binary cubic, i.e. for $F \in S_3$ where $S = \mathbb{C}[x_0, x_1]$. We begin by describing the Hilbert function of $F^{\perp}$. There are only two possibilities:

case 1.

t	0	1	2	3	4	
$HF(T/F^{\perp},t)$	1	1	1	1	0	$\rightarrow$

case 2.

t	0	1	2	3	4	
$HF(T/F^{\perp},t)$	1	2	2	1	0	$\rightarrow$

We want to show that in **case 1** we have $F = L^3$. From the Hilbert function we see that $(F^{\perp})_1 = \langle \partial_1 \rangle$. From the perfect pairing property we see that

$$\{L \in S_1 : \partial_1 L = 0\} = \langle L_1 \rangle.$$

Thus we can find $L_0 \in S_1$ such that $\partial_1 L_0 = 1$ and

$$S_1 = \langle x_0, x_1 \rangle = \langle L_0, L_1 \rangle.$$

We now perform a linear change of variables and we obtain a polynomial

$$G(L_0, L_1) = aL_0^3 + bL_0^2 L_1 + cL_0 L_1^2 + dL_1^3$$

such that

$$G(L_0, L_1) = F(x_0, x_1).$$

As $\partial_1 L_0 \neq 0$ and $\partial_1 L_1 = 0$ we get

$$0 = \partial_1 G = 2bL_0 L_1 + cL_1^2 + 3dL_1^2,$$

and hence $G = F = aL_0^3$ thus $\mathrm{rk}(F) = 1$.

We want now to show that in **case 2** we have $\mathrm{rk}(F) = 2$ or $\mathrm{rk}(F) = 3$. We note that $\mathrm{rk}(F) \neq 1$, otherwise $(F^{\perp})_1 \neq 0$. As in this case $(F^{\perp})_1 = 0$, we consider the degree two piece, $(F^{\perp})_2 = \langle Q \rangle$. We have to possibilities

$$Q = \partial\partial', \text{ where } \partial \text{ and } \partial' \text{ are not proportional, or } Q = \partial^2.$$

If $Q = \partial\partial'$, where ∂ and ∂' are not proportional, we can construct a basis for $S_1 = \langle L, L' \rangle$ in such a way that

$$\partial L = \partial' L' = 1,$$

and

$$\partial' L = \partial L' = 0.$$

Then we perform a change of variables and obtain

$$F(x_0, x_1) = G(L_0, L_1) = aL_0^3 + bL_0^2L_1 + cL_0L_1^2 + dL_1^3.$$

We want to show that $F(x_0, x_1) = aL_0^3 + dL_1^3$. To do this we define

$$H(x_0, x_1) = G(L_0, L_1) - aL_0^3 - dL_1^3,$$

and show that the degree 3 polynomial H is the zero polynomial. To do this, it is enough to show that $(H^\perp)_3 = T_3$. We now compute that

$$\partial^3 H = 6aL - 6aL = 0 \text{ and },$$

$$\partial'^3 H = 6dL' - 6dL' = 0.$$

We then notice that $\partial^2\partial' = \partial Q \in F^\perp$ and $\partial^2\partial' H = 0$; similarly for $\partial\partial'^2$. Thus $H = 0$ and $F(x_0, x_1) = aL_0^3 + dL_1^3$. As $(F^\perp)_1 = 0$ this means that $\text{rk}(F) = 2$.

Finally, if $Q = \partial^2$ we assume by contradiction that $\text{rk}(F) = 2$, thus $F = N^3 + M^3$ for some linear forms N and M. There exist linearly independent differential operators $\partial_N, \partial_M \in S_1$ such that

$$\partial_N N = \partial_M M = 1,$$

and

$$\partial_N M = \partial_M N = 0.$$

And then $\partial_N \partial_M \in F^\perp$ and this is a contradiction as Q is the only element in $(F^\perp)_2$ and it is a square.

Remark 4.11. We consider again the case of binary cubic forms. We want to make a connection between the Waring rank of F and certain ideals contained in $F^\perp$. If $\text{rk}(F) = 1$, then we saw that $F^\perp \supset (\partial_1)$ and this is the ideal of one point in $\mathbb{P}^1$. If $\text{rk}(F) = 2$, then $F^\perp \supset (\partial\partial')$ and this is the ideal of two distinct points in $\mathbb{P}^1$; as $(F^\perp)_1 = 0$ there is no ideal of one point contained in the annihilator. Finally, if $\text{rk}(F) = 3$, then $F^\perp \supset (\partial^2)$ and there is no ideal of two points, or one point, contained in the annihilator. However, $(F^\perp)_3 = T_3$ and we can find many ideals of three points.

There is a connection between $\text{rk}(F)$ and set of points whose ideal I is such that $I \subset F^\perp$. This connection is the content of the *Apolarity Lemma*, see [108].

Lemma 4.12. *Let $F \in S_d$ be a degree d form in $n+1$ variables. Then the following facts are equivalent:*

- $F = L_1^d + \ldots + L_s^d$;
- $F^\perp \supset I$ *such that I is the ideal of a set of s distinct points in* $\mathbb{P}^n$.

Example 4.13. We use the Apolarity Lemma to explain the Alexander–Hirschowitz defective case $n = 2$ and $d = 4$. Given a generic $F \in S_4$ we want to show that $\mathrm{rk}(F) = 6$ and not 5 as expected. To do this we use Hilbert functions. Clearly, if $I \subset F^\perp$, then $HF(T/I, t) \geq HF(T/F^\perp, t)$ for all t. Thus by computing $HF(T/F^\perp, t)$ we get information on the Hilbert function of any ideal contained in the annihilator, and in particular for ideal of sets of points.

t	0	1	2	3	4	5	6
$HF(T/F^\perp, t)$	1	3	6	3	1	0	$\rightarrow$

In particular, $HF(T/F^\perp, 2) = 6$ means that for no set of five points its defining ideal I could be such that $I \subset F^\perp$.

Exercise 4.14. Use the Apolarity Lemma to compute $\mathrm{rk}(x_0 x_1^2)$. Then try the binary forms $x_0 x_1^d$.

Exercise 4.15. Use the Apolarity Lemma to explain Alexander–Hirschowitz exceptional cases.

It is in general very difficult to compute the Waring rank of a given form and (aside from brute force) no algorithm exists which can compute it in all cases. Lim and Hillar show that this problem is an instance of the fact that, as their title states, "Most tensor problems are NP-Hard," [105]. However, we know $\mathrm{rk}(F)$ when F is a quadratic form, and we do have an efficient algorithm when F is a binary form.

Remark 4.16. There is an algorithm, attributed to Sylvester, to compute $\mathrm{rk}(F)$ *for a binary form and it uses the Apolarity Lemma.* The idea is to notice that $F^\perp = (\partial_1, \partial_2)$, i.e. the annihilator is a complete intersection ideal, say, with generators in degree $d_1 = \deg \partial_1 \leq d_2 = \deg \partial_2$. If ∂_1 is square free, then we are done and $\mathrm{rk}(F) = d_1$. If not, as ∂_1 and ∂_2 do not have common factors, there is a square free degree d_2 element in $F^\perp$. Hence, $\mathrm{rk}(F) = d_2$.

Exercise 4.17. Compute $\mathrm{rk}(F)$ when F is a quadratic form.

Remark 4.18. The Waring rank for monomials was determined in 2011 in a paper of Carlini, Catalisano, and Geramita, see [41], and independently by Buczyńska, Buczyński, and Teitler, see [35]. In particular, it was shown that

$$\mathrm{rk}(x_0^{a_0} \ldots x_n^{a_n}) = \frac{1}{(a_0 + 1)} \Pi_{i=0}^{n} (a_i + 1),$$

where $1 \leq a_0 \leq a_1 \leq \ldots \leq a_n$.

We conclude this lecture by studying the Waring rank of degree d forms of the kind $L_1^d + \ldots + L_s^d$. Clearly, $\mathrm{rk}(L_1^d) = 1$ and $\mathrm{rk}(L_1^d + L_2^d) = 2$, if L_1 and L_2 are linearly independent. If the linear forms L_i are not linearly independent, then the situation is more interesting.

Example 4.19. Consider the binary cubic form $F = ax_0^3 + bx_1^3 + (x_0 + x_1)^3$. We want to know $\mathrm{rk}(F)$. For a generic choice of a and b, we have $\mathrm{rk}(F) = 2$, but for special values of a and b $\mathrm{rk}(F) = 3$. The idea is that the rank 3 elements of $\mathbb{P}S_3$ lay on the tangent developable of the twisted cubic curve, which is an irreducible surface. Hence, the general element of the plane

$$\langle [x_0^3], [x_1^3], [(x_0 + x_1)^3] \rangle$$

has rank 2, but there are rank 3 elements.

Exercise 4.20. Prove that $\mathrm{rk}(L^d + M^d + N^d) = 3$ whenever L, M, and N are linearly independent linear forms.

5 Lecture Four

In the last lecture we introduced the Apolarity Lemma and used it to study the Waring rank of a given specific form. In this lecture we will go back to the study of higher secant varieties of varieties that are not Veronese varieties.

The study of higher secant varieties of Veronese varieties is connected to Waring's problems, and hence with sum of powers decompositions of forms. We now want to consider tensors in general and not only homogenous polynomials, which correspond to symmetric tensors.

Consider $\mathbb{C}$-vector spaces $V_1, \dots, V_t$ and the tensor product

$$V = V_1 \otimes \dots \otimes V_t.$$

Definition 5.1. A tensor $v_1 \otimes \dots \otimes v_t \in V$ is called *elementary*, or *indecomposable* or *rank-1* tensor.

Elementary tensors are the building blocks of V. More specifically, there is a basis of V consisting of elementary tensors, so any tensor $T \in V$ can be written as a linear combination of elementary tensors; in this sense, elementary tensors are analogous to monomials.

A natural question is: given a tensor T what is the minimum s such that $T = \sum_1^s T_i$ where each T_i is an elementary tensor? The value s is called the *tensor rank* of T and is the analogue of Waring rank for forms. Of course we could state tensor versions of the Waring's problems and try to solve them as well.

In order to study these problems geometrically, we need to introduce a new family of varieties.

Definition 5.2. Given vector spaces $V_1, \dots, V_t$ the *Segre map* is the map

$$\mathbb{P}V_1 \times \dots \times \mathbb{P}V_t \longrightarrow \mathbb{P}(V_1 \otimes \dots \otimes V_t)$$
$$([v_1], \dots, [v_t]) \mapsto [v_1 \otimes \dots \otimes v_t]$$

and the image variety is called a *Segre variety* or *the Segre product* of $\mathbb{P}V_1, \dots, \mathbb{P}V_t$.

If the vector spaces are such that $\dim V_i = n_i + 1$, then we will often denote by

$$X = \mathbb{P}^{n_1} \times \ldots \times \mathbb{P}^{n_t}$$

the image of the Segre map. In particular, $X \subset \mathbb{P}(V_1 \otimes \ldots \otimes V_t) = \mathbb{P}^N$ where $N + 1 = \Pi(n_i + 1)$. Note that $\dim X = n_1 + \ldots + n_t$.

By choosing bases of the vector spaces V_i we can write the Segre map in coordinates

$$\mathbb{P}^{n_1} \times \mathbb{P}^{n_2} \times \ldots \times \mathbb{P}^{n_t} \longrightarrow \mathbb{P}^N$$

$$[a_{0,1} : \ldots : a_{n_1,1}] \times [a_{0,2} : \ldots : a_{n_2,2}] \times \ldots \times [a_{0,t} : \ldots : a_{n_t,t}] \mapsto$$
$$[a_{0,1}a_{0,2} \ldots a_{0,t} : a_{0,1}a_{0,2} \ldots a_{1,t} : \ldots : a_{n_1,1}a_{n_2,2} \ldots a_{n_t,t}].$$

Example 5.3. Consider $X = \mathbb{P}^1 \times \mathbb{P}^1$, then $X \subset \mathbb{P}^3$ is a surface. The Segre map is

$$\mathbb{P}^1 \times \mathbb{P}^1 \longrightarrow \mathbb{P}^3$$

$$[a_0 : a_1] \times [b_0 : b_1] \mapsto [a_0b_0 : a_0b_1 : a_1b_0 : a_1b_1].$$

If z_0, z_1, z_2, and z_3 are the coordinates of $\mathbb{P}^3$, then it is easy to check that X has equation

$$z_0z_3 - z_1z_2 = 0.$$

Thus X is a smooth quadric in $\mathbb{P}^3$.

Example 5.4. Consider again $X = \mathbb{P}^1 \times \mathbb{P}^1 \subset \mathbb{P}^3$. We identify $\mathbb{P}^3$ with the projectivization of the vector space of 2×2 matrices. Using this identification we can write the Segre map as

$$\mathbb{P}^1 \times \mathbb{P}^1 \longrightarrow \mathbb{P}^3$$

$$[a_0 : a_1] \times [b_0 : b_1] \mapsto \begin{pmatrix} a_0b_0 & a_0b_1 \\ a_1b_0 & a_1b_1 \end{pmatrix}.$$

Thus X represents the set of 2×2 matrices of rank at most one and the ideal of X is generated by the vanishing of the determinant of the generic matrix $\begin{pmatrix} z_0 & z_1 \\ z_2 & z_3 \end{pmatrix}$.

Exercise 5.5. Work out a matrix representation for the Segre varieties with two factors $\mathbb{P}^{n_1} \times \mathbb{P}^{n_2}$.

Before entering into the study of the higher secant varieties of Segre varieties, we provide some motivation coming from Algebraic Complexity Theory.

Example 5.6. The multiplication of two 2×2 matrices can be seen as bilinear map

$$T : \mathbb{C}^4 \times \mathbb{C}^4 \to \mathbb{C}^4$$

or, equivalently, as a tensor

$$T \in \mathbb{C}^{4^*} \otimes \mathbb{C}^{4^*} \otimes \mathbb{C}^4.$$

It is interesting to try to understand how many multiplications over the ground field are required to compute the map T.

If we think of T as a tensor, then we can write it as a linear combination of elementary tensors and each elementary tensor represents a multiplication. The naive algorithm for matrix multiplication, which in general uses n^3 scalar multiplications to compute the product of two $n \times n$ matrices, implies that T can be written as $T = \sum_1^8 \alpha_i \otimes \beta_i \otimes c_i$. However, Strassen in [150] proved that

$$[T] \in \sigma_7(\mathbb{P}^3 \times \mathbb{P}^3 \times \mathbb{P}^3),$$

and, even more, that T is the sum of 7 elementary tensors

$$T = \sum_1^7 \alpha_i \otimes \beta_i \otimes c_i.$$

Strassen's algorithm actually holds for multiplying matrices over any algebra. Thus by viewing a given $n \times n$ matrix in one with size a power of 2, one can use Strassen's algorithm iteratively. So $2^m \times 2^m$ matrices can be multiplied using 7^m multiplications. In particular, this method lowers the upper bound for the complexity of matrix multiplication from n^3 to $n^{\log_2 7} \simeq n^{2.81}$.

After Strassen's result, it was shown that the rank of T is not smaller than 7. On the other hand, much later, Landsberg proved that the border rank of T is 7, that is to say that $T \notin \sigma_6(\mathbb{P}^3 \times \mathbb{P}^3 \times \mathbb{P}^3)$ [112].

The question of the complexity of matrix multiplication has recently been called one of the most important questions in Numerical Analysis [153]. The reason for this is that the complexity of matrix multiplication also determines the complexity of matrix inversion. Matrix inversion is one of the main tools for solving a square system of linear ODE's.

Williams, in 2012, improved the Coppersmith–Winograd algorithm to obtain the current best upper bound for the complexity of matrix multiplication [161], but it would lead us to far afield to discuss this here. On the other hand, the current best lower bounds come from the algebra and geometry of secant varieties. These bounds arise by showing non-membership of the matrix multiplication tensor on certain secant varieties. To do this, one looks for nontrivial equations that vanish on certain secant varieties, but do not vanish on the matrix multiplication tensor. Indeed, the best results in this direction make use of representation theoretic descriptions of the ideals of secant varieties. For more, see [115, 119].

5.1 Dimension of Secant Varieties of Segre Varieties

In the 2×2 matrix multiplication example, we should have pointed out the fact that $\sigma_7(\mathbb{P}^3 \times \mathbb{P}^3 \times \mathbb{P}^3)$ actually fills the ambient space, so almost all tensors in $\mathbb{P}(\mathbb{C}^4 \otimes \mathbb{C}^4 \otimes \mathbb{C}^4)$ have rank 7. For this and many other reasons we would like to know the dimensions of secant varieties of Segre varieties.

Like in the polynomial case, there is an expected dimension, which is obtained by the naive dimension count. When X is the Segre product $\mathbb{P}^{n_1} \times \ldots \times \mathbb{P}^{n_t}$, the expected dimension of $\sigma_s(X)$ is

$$\min\left\{ s\sum_{i=1}^{t}(n_i+1)+s-1, \prod_{i=1}^{t}(n_i+1)-1 \right\}.$$

We would like to have an analogue of the Alexander–Hirschowitz theorem for the Segre case; however, this is a very difficult problem. See [4] for more details. There are some partial results, however. For example, Catalisano et al. in [53] show that $\sigma_s(\mathbb{P}^1 \times \cdots \times \mathbb{P}^1)$ always has the expected dimension, except for the case of four factors.

Again, the first tool one uses to study the dimensions of secant varieties is the computation of tangent spaces together with Terracini's lemma.

Exercise 5.7. Let $X = \mathbb{P}V_1 \times \cdots \times \mathbb{P}V_t$ and let $[v] = [v_1 \otimes \cdots \otimes v_t]$ be a point of X. Show that the cone over the tangent space to X at v is the span of the following vector spaces:

$$\begin{gathered} V_1 \otimes v_2 \otimes v_3 \otimes \cdots \otimes v_t, \\ v_1 \otimes V_2 \otimes v_3 \otimes \cdots \otimes v_t, \\ \vdots \\ v_1 \otimes v_2 \otimes \cdots \otimes v_{t-1} \otimes V_t. \end{gathered}$$

Exercise 5.8. Show that $\sigma_2(\mathbb{P}^1 \times \mathbb{P}^1 \times \mathbb{P}^1) = \mathbb{P}^7$.

Exercise 5.9. Use the above description of the tangent space of the Segre product and Terracini's lemma to show that $\sigma_3(\mathbb{P}^1 \times \mathbb{P}^1 \times \mathbb{P}^1 \times \mathbb{P}^1)$ is a hypersurface in $\mathbb{P}^{15}$ and not the entire ambient space as expected. This shows that the four-factor Segre product of $\mathbb{P}^1$s is defective.

There are two main approaches to the study of the dimensions of secant varieties of Segre products: [49] and [4]. In [49] the authors introduce and use what they call the affine-projective method. In this way, the study of the dimension of higher secant varieties of Segre, and Segre–Veronese, varieties reduces to the study of the postulation of non-reduced schemes supported on linear spaces. In [4], the authors show that the "divide and conquer" method of Alexander and Hirschowitz can be used to set up a multi-step induction proof for the non-defectivity of Segre products. They are able to obtain partial results on non-defectivity by then checking many initial cases, often using the computer. On the other hand, for the remaining cases there are many more difficult computations to do in order to get the full result.

5.2 *Flattenings*

Often, the first tool used to understand properties of tensors is to reduce to Linear Algebra (when possible). For this, the notion of flattenings is essential. Consider for the moment the three-factor case. We may view the vector space $\mathbb{C}^a \otimes \mathbb{C}^b \otimes \mathbb{C}^c$ as a space of matrices in three essentially different ways as the following spaces of linear maps:

$$\begin{aligned}(\mathbb{C}^a)^* &\to \mathbb{C}^b \otimes \mathbb{C}^c \\ (\mathbb{C}^b)^* &\to \mathbb{C}^a \otimes \mathbb{C}^c. \\ (\mathbb{C}^c)^* &\to \mathbb{C}^a \otimes \mathbb{C}^b\end{aligned}$$

(A priori there are many more choices of flattenings, however, in the three-factor case the others are obtained by transposing the above maps.) When there are more than three factors the situation is similar, with many more flattenings to consider.

For $V_1 \otimes \cdots \otimes V_t$, a *p-flattening* is the interpretation as a space of matrices with p factors on the left:

$$(V_{i_1} \otimes \cdots \otimes V_{i_p})^* \to V_{j_1} \otimes \cdots \otimes V_{j_{t-p}}.$$

For a given tensor $T \in V_1 \otimes \cdots \otimes V_t$ we call a p-flattening of T a realization of T in one of the above flattenings. This naturally gives rise to the notion of *multi-linear rank*, which is the vector the ranks of the 1-flattenings of T, see [40].

Exercise 5.10. Show that T has rank 1 if and only if its multilinear rank is $(1, \ldots, 1)$.

Recall that a linear mapping $T\colon (\mathbb{C}^a)^* \to \mathbb{C}^b$ has rank r if the image of the map has dimension r and the kernel has dimension $a - r$. Moreover, since the rank of the transpose is also r, after re-choosing bases in $\mathbb{C}^a$ and $\mathbb{C}^b$ one can find r-dimensional subspaces in $\mathbb{C}^a$ and $\mathbb{C}^b$ so that $T \in (\mathbb{C}^r)^* \otimes \mathbb{C}^r$. This notion generalizes to tensors of higher order. In particular, it is well known that $T \in V_1 \otimes \cdots \otimes V_t$ has multilinear rank $\le (r_1, \dots, r_t)$ if and only if there exist subspaces $\mathbb{C}^{r_i} \subset V_i$ such that $T \in \mathbb{C}^{r_1} \otimes \cdots \otimes \mathbb{C}^{r_t}$. The Zariski closure of all tensors of multilinear rank $(r_1, \dots, r_t)$ is known as the subspace variety, denoted $\mathrm{Sub}_{r_1,\dots,r_t}$. See [123] for more details.

The connection between subspace varieties and secant varieties is the content of the following exercise.

Exercise 5.11. Let $X = \mathbb{P}V_1 \times \cdots \times \mathbb{P}V_t$. Show that if $r \le r_i$ for $1 \le i \le t$, then

$$\sigma_r(X) \subset \mathrm{Sub}_{r_1,\dots,r_t} .$$

Notice that for the two-factor case, $\sigma_r(\mathbb{P}^{a-1} \times \mathbb{P}^{b-1}) = \mathrm{Sub}_{r,r}$.

Aside from the case of binary tensors (tensor products of $\mathbb{C}^2$s), another case that is well understood is the case of very unbalanced tensors. (For the more refined notion of "unbalanced," see [4, Sect. 4].) Again consider $\mathbb{C}^a \otimes \mathbb{C}^b \otimes \mathbb{C}^c$ and suppose that $a \ge bc$. Then for all $r \le bc$ we have

$$\sigma_r(\mathbb{P}^{a-1} \times \mathbb{P}^{b-1} \times \mathbb{P}^{c-1}) = \sigma_r(\mathbb{P}^{a-1} \times \mathbb{P}^{bc-1}) = \mathrm{Sub}_{r,r} .$$

Therefore we can always reduce the case of very unbalanced threefold tensors to the case of matrices. More generally a tensor in $V_1 \otimes \cdots \otimes V_t$ is called very unbalanced if $(n_i + 1) \ge \prod_{j \ne i}(n_j + 1)$. In the very unbalanced case we can reduce to the case of matrices and use results and techniques from linear algebra.

5.3 *Equations of Secant Varieties of Segre Products*

Now we turn to the question of defining equations. Recall that a matrix has rank $\le r$ if and only if all of its $(r+1) \times (r+1)$ minors vanish. Similarly, a tensor has multilinear rank $\le (r_1, \dots, r_t)$ if all of its $(r_i + 1) \times (r_i + 1)$ minors vanish. It is easy to see that a tensor has rank 1 if and only if all of the 2×2 minors of flattenings vanish. Since the Segre product is closed, these equations actually define its ideal. For $\sigma_2(X)$ it was conjectured by Garcia, Stillman, and Sturmfels (GSS) that the 3×3 minors suffice to define the ideal. The first partial results were by GSS themselves as well as by Landsberg and Manivel [117, 118], Landsberg and Weyman [123] and by Allman and Rhodes [11, 12], while the full conjecture was resolved by Raicu [144].

While the result of Exercise 5.11 implies that minors of flattenings give some equations of secant varieties, often they provide no information at all. For example, consider the case of $3 \times 3 \times 3$ tensors. One expects that $\sigma_4(\mathbb{P}^2 \times \mathbb{P}^2 \times \mathbb{P}^2)$ fills the entire ambient $\mathbb{P}^{26}$, however this is not the case. On the other hand, one cannot detect this from flattenings since all of the flattenings are 3×9 matrices, which have maximum rank 3, and there are no 5×5 minors to consider.

Strassen noticed that a certain equation actually vanishes on $\sigma_4(\mathbb{P}^2\times\mathbb{P}^2\times\mathbb{P}^2)$, and moreover, he shows that it is a hypersurface. Strassen's equation was studied in more generality by Landsberg and Manivel [118], put into a broader context by Ottaviani [137] generalized by Landsberg and Ottaviani [120]. Without explaining the full generality of the construction, we can describe Ottaviani's version of Strassen's equation as follows.

Suppose $T \in V_1 \otimes V_2 \otimes V_3$ with $V_i \cong \mathbb{C}^3$ and consider the flattening

$$(V_1)^* \to V_2 \otimes V_3.$$

Choose a basis $\{v_1, v_2, v_3\}$ for V_1 and write T as a linear combination of matrices $T = v_1 \otimes T^1 + v_2 \otimes T^2 + v_3 \otimes T^3$. The 3×3 matrices T^i are called the slices of T in the V_1-direction with respect to the chosen basis. Now consider the following matrix:

$$\varphi_T = \begin{pmatrix} 0 & T^1 & -T^2 \\ -T^1 & 0 & T^3 \\ T^2 & -T^3 & 0 \end{pmatrix},$$

where all of the blocks are 3×3.

Exercise 5.12.

1. Show that if T has rank 1 then φ_T has rank 2.
2. Show that φ is additive in its argument, i.e. show that $\varphi_{T+T'} = \varphi_T + \varphi_{T'}$.

The previous exercise together with the subadditivity of matrix rank implies that if T has tensor rank r then φ_T has matrix rank $\leq 2r$. In particular, if T has tensor rank 4, the determinant of φ_T must vanish. Indeed $\det(\varphi_T)$ is Strassen's equation, and it is the equation of the degree 9 hypersurface $\sigma_4(\mathbb{P}^2 \times \mathbb{P}^2 \times \mathbb{P}^2)$.

Remark 5.13. This presentation of Strassen's equation $\det(\varphi_T)$ is very compact yet its expansion in monomials is very large, having 9,216 terms.

This basic idea of taking a tensor and constructing a large matrix whose rank depends on the rank of T is at the heart of almost all known equations of secant varieties of Segre products—see [120]. One exception is that of the degree 6 equations in the ideal of $\sigma_4(\mathbb{P}^2 \times \mathbb{P}^2 \times \mathbb{P}^3)$. The only known construction of these equations comes from representation theoretic considerations. For more details see [19, 117].

Despite this nice picture, we actually know surprisingly little about the defining equations of secant varieties of Segre products in general, and this is an ongoing area of current research.

6 Solution of the Exercises

In what follows if S is a subset of $\mathbb{P}^n$, then $\overline{S}$ denotes the smallest closed subset of $\mathbb{P}^n$ containing S with the reduced subscheme structure. On the other hand, $\langle S\rangle$ denotes the smallest linear subspace of $\mathbb{P}^n$ containing S.

Exercise 2.3

Show that the union of chords (secant lines) to a plane conic is closed. However, the union of the chords of the twisted cubic curve in $\mathbb{P}^3$ is not.

Proof (Solution). Let $X \subseteq \mathbb{P}^2$ be a plane conic. It suffices to show that $\cup_{p,q\in X}\langle p,q\rangle = \mathbb{P}^2$. If $y \in \mathbb{P}^2\backslash X$, then there exists a line containing y and intersecting X in two distinct points p and q. Thus, $y \in \langle p,q\rangle$ so $\cup_{p,q\in X}\langle p,q\rangle = \mathbb{P}^2$.

Let $X \subseteq \mathbb{P}^3$ be the twisted cubic. Exercise 2.8 implies that $\sigma_2(X) = \mathbb{P}^3$. On the other hand, direct calculation shows that the point $[0 : 1 : 0 : 0] \in \mathbb{P}^3$ lies on no secant line of X. Hence, $\cup_{p,q\in X}\langle p,q\rangle$ is not equal to its closure and hence is not closed.

Exercise 2.5

For $X \subset \mathbb{P}^N$, show that, if $\sigma_i(X) = \sigma_{i+1}(X) \neq \mathbb{P}^N$, then $\sigma_i(X)$ is a linear space and hence $\sigma_j(X) = \sigma_i(X)$ for all $j \geq i$.

Proof (Solution). It suffices to prove that, for $k \geq 1$, if $\sigma_k(X) = \sigma_{k+1}(X)$ then $\sigma_{k'}(X) = \langle X\rangle$ for $k' \geq k$.

Note

$$\sigma_{k+1}(X) = \overline{\cup_{p_i,q\in X}\langle\langle p_1,\ldots,p_k\rangle,q\rangle} \tag{1}$$

while $\sigma_k(X) = \overline{\cup_{p_i\in X}\langle p_1,\ldots,p_k\rangle}$. Since the singular locus of $\sigma_k(X)$ is a proper closed subset of $\sigma_k(X)$ there exists a nonsingular point $z \in \sigma_k(X)$ such that $z \in \cup_{p_i\in X}\langle p_1,\ldots,p_k\rangle$.

Now for all $y \in X$ the line $\langle y,z\rangle$ is contained in $\overline{\cup_{x\in X}\langle x,z\rangle}$ and passes through z. Hence $\overline{\cup_{x\in X}\langle x,z\rangle} \subseteq T_z\overline{\cup_{x\in X}\langle x,z\rangle}$. Using (1), we deduce

$$X \subseteq \overline{\cup_{x\in X}\langle x,z\rangle} \subseteq T_z\overline{\cup_{x\in X}\langle x,z\rangle} \subseteq T_z\sigma_{k+1}(X).$$

Hence $\langle X\rangle \subseteq T_z\sigma_{k+1}(X)$. In addition, since $\sigma_{k+1}(X) = \sigma_k(X)$ we deduce

$$\sigma_k(X) \subseteq \langle X\rangle \subseteq T_z\sigma_k(X).$$

Since z is nonsingular, $\dim T_z\sigma_k(X) = \dim \sigma_k(X)$. Finally, since $T_z\sigma_k(X)$ is irreducible and $\sigma_k(X)$ is reduced we conclude $\sigma_k(X) = \langle X\rangle = T_z\sigma_k(X)$. If $k' \geq k$, then $\langle X\rangle \subseteq \sigma_k(X) \subseteq \sigma_{k'}(X)$. Since $\sigma_{k'}(X) \subseteq \langle X\rangle$ we deduce $\sigma_{k'}(X) = \sigma_k(X)$.

Exercise 2.6

If $X \subseteq \mathbb{P}^N$ is non-degenerate, then there exists an $r \geq 1$ with the property that

$$X = \sigma_1(X) \subsetneq \sigma_2(X) \subsetneq \ldots \subsetneq \sigma_r(X) = \mathbb{P}^N.$$

In particular, all inclusions are strict and there is a higher secant variety that coincides with the ambient space.

Proof (Solution). In the notation of Exercise 2.5, set $k_0 := \min\{k \mid \sigma_k(X) = \langle X\rangle\}$. It suffices to prove that there exists the following chain of strict inclusions

$$X = \sigma_1(X) \subsetneq \sigma_2(X) \subsetneq \cdots \subsetneq \sigma_{k_0}(X) = \langle X\rangle.$$

If $\sigma_k(X) = \sigma_{k+1}(X)$ then, by Exercise 2.5, $\sigma_k(X) = \langle X\rangle$ and hence $k \geq k_0$.

Exercise 2.8

Let $X \subseteq \mathbb{P}^n$ be a curve. Prove that $\sigma_2(X)$ has dimension 3 unless X is contained in a plane.

Proof (Solution). Let $\mathfrak{X} := \{(\langle p,q\rangle, y) : y \in \langle p,q\rangle\} \subseteq \mathbb{G}(1,3) \times \mathbb{P}^n$ be the incident correspondence corresponding to the (closure) of the secant line map $(X \times X) - \Delta_X \to \mathbb{G}(1,3) \times \mathbb{P}^n$. Then $\mathfrak{X}$ is an irreducible closed subset of $\mathbb{G}(1,3) \times \mathbb{P}^n$ and $\sigma_2(X) = p_2(\mathfrak{X})$. Hence, $\dim \sigma_2(X) \leq 3$. Suppose now that $\dim \sigma_2(X) = 2$. Let us prove that X is contained in a plane. First note that for a fixed $p \in X$, $\overline{\bigcup_{q\in X}\langle p,q\rangle}$ is reduced, irreducible, has dimension $\dim X + 1$ and is contained in $\sigma_2(X)$. Hence, if $\sigma_2(X)$ has dimension $\dim X + 1$, $\sigma_2(X) = \overline{\bigcup_{q\in X}\langle p,q\rangle}$. Hence, there exists a non-singular $x \in \sigma_2(X)$ such that $x \in \bigcup_{q\in X}\langle p,q\rangle$ and such that $X \subseteq \overline{\bigcup_{q\in X}\langle x,q\rangle} = \overline{\bigcup_{q\in X}\langle p,q\rangle} \subseteq T_x\overline{\bigcup_{q\in X}\langle x,q\rangle} = T_x\sigma_2(X)$. Hence, $\langle X\rangle \subseteq T_x\sigma_2(X)$. Since $\sigma_2(X)$ has dimension 2 then $T_x\sigma_2(X)$ is a plane. Finally, note that if $\sigma_2(X)$ has dimension 1, then $\sigma_2(X) = \sigma_1(X)$. Hence $\sigma_2(X) = \langle X\rangle$ so $\sigma_2(X)$ and hence X is a line.

Exercise 2.12

Let M be an $n \times n$ symmetric matrix of rank r. Prove that M is a sum of r symmetric matrices of rank 1.

Proof (Solution). By performing elementary row and column operations it is possible to find an invertible $n \times n$ matrix P and complex numbers $\lambda_1, \ldots, \lambda_n$ such that

$$M = P \operatorname{diag}(\lambda_1, \ldots, \lambda_n) P^T.$$

Moreover, if M has rank r, then there exists $\{i_1, \ldots, i_r\} \subseteq \{1, \ldots, n\}$ such that $\lambda_i \neq 0$ for $i \in \{i_1, \ldots, i_r\}$ and $\lambda_i = 0$ for $i \notin \{i_1, \ldots, i_r\}$. For $k = 1, \ldots, n$, set $m_{ij}^k := \lambda_k p_{ik} p_{jk}$ and let M_k be the matrix with i, j entry m_{ij}^k. Since $m_{ij}^k = m_{ji}^k$, M_k is symmetric. Moreover, $M = \sum_{i=1}^n M_k$. It remains to show that M_k has rank at most 1. Let P_k be the matrix with i, j entry $p_{ik} p_{jk}$. Since M_k is a scalar multiple of P_k it suffices to show that P_k has rank at most 1. For this, we show that all 2×2 minors of P_k are zero. Indeed an arbitrary 2×2 minor of P_k is the determinant of the 2×2 matrix formed by omitting all rows except row α, β and columns γ, δ, that is, the determinant

$$\begin{vmatrix} p_{\alpha k} p_{\gamma k} & p_{\alpha k} p_{\delta k} \\ p_{\beta k} p_{\gamma k} & p_{\beta k} p_{\delta k} \end{vmatrix}$$

which is zero.

Exercise 2.13

Let $X = \nu_2(\mathbb{P}^2)$. Let H be the locus of 3×3 symmetric matrices of rank at most 2. Prove that $H = \sigma_2(X)$.

Proof (Solution). Using Exercise 2.12 we deduce that X is the locus of 3×3 symmetric matrices of rank at most 1. On the other hand we know that $X \subseteq \sigma_2(X) \subseteq H$. Moreover, if M is a symmetric matrix of rank 2 then, by Exercise 2.12, M is a sum of two symmetric matrices of rank 1. Hence $M \in \sigma_2(X)$.

Exercise 2.14

Let H be the locus of 4×4 symmetric matrices of rank 2. Let $X = \nu_2(\mathbb{P}^3)$. Prove that $H = \sigma_2(X)$. Is X 2-defective?

Proof (Solution). Using Exercise 2.12 we deduce that X is the locus of 4×4 symmetric matrices of rank at most 1. On the other hand we know that $X \subseteq \sigma_2(X) \subseteq H$. Moreover, if M is a symmetric matrix of rank 2 then, by Exercise 2.12, M is a sum of two symmetric matrices of rank 1. Hence $M \in \sigma_2(X)$. To see that X is 2-defective note that $\operatorname{expdim}(\sigma_2(X)) = 7$ while the locus of 4×4 symmetric matrices of rank at most 2 have dimension 6.

Exercise 2.18

Prove that if $H \supset T_P(X)$, then the polynomial defining $\nu_3^{-1}(H)$ has a double root.

Proof (Solution). This is a special case of the following more general claim which is also Exercise 3.16.

Claim. Let $L \in S_1$ be a liner form. If Λ is a hyperplane in $\mathbb{P}^N$ containing $T_{[L^d]}(\nu_d(\mathbb{P}^n))$, then $\nu_d^{-1}(\Lambda)$ is a degree d hypersurface singular at the point $[L] \in \mathbb{P}^n$.

Proof (Proof of Claim). We can find a basis for S_1 of the form $\{L, L_1, \dots, L_n\}$ where L_i are linear forms. With respect to this basis, in coordinates, and using the notation Lecture 1, the image of $[L]$ in $\mathbb{P}^N$ is the point $p = [1 : 0 : \cdots : 0]$. For $i = 0, \dots, n$ let p_i be the point of $\mathbb{P}^N$ with homogeneous coordinates $[Z_0 : \cdots : Z_N]$ given by $Z_j = \delta_{ij}$. Translating the intrinsic description of $T_{[L^d]}\nu_d(\mathbb{P}^n)$ as the linear space $\langle L^{d-1}M \mid M \in S_1\rangle$ into our coordinates for $\mathbb{P}^n$ and $\mathbb{P}^N$ with respect to the basis $\{L, L_1, \dots, L_n\}$ for S_1 we conclude that $T_P(\nu(\mathbb{P}^n)) = \langle p_0, \dots, p_n\rangle$. Let F_Λ be a linear form defining Λ. Then $F_\Lambda = a_0 Z_0 + \dots a_N Z_N$. If $T_p(\nu_d(\mathbb{P}^n)) \subseteq \Lambda$ then $F_\Lambda(p_i) = 0$, for $i = 0, \dots, n$. Hence $0 = a_0 = \cdots = a_n$. Let f be the pullback of F_Λ for $\mathbb{P}^n$. Then $f = a_{n+1}x_0^{d-2}x_1 + \cdots + a_N x_n^d$. Clearly the partial derivatives of f vanish at $q = [1 : 0 : \cdots : 0]$ which is identified with $[L]$. Hence the zero locus of f, a degree d hypersurface, is singular at q. On the other hand, the zero locus of f is $\nu_d^{-1}(\Lambda)$.

Apply the claim in the special case to complete the exercise.

Exercise 2.22

Consider the rational normal curve in $\mathbb{P}^3$, i.e. the twisted cubic curve $X = \nu_3(\mathbb{P}(S_1)) \subset \mathbb{P}(S_3)$. We know that $\sigma_2(X)$ fills up all the space. Can we write any binary cubic as the sum of two cubes of linear forms? Try $x_0 x_1^2$.

Proof (Solution). Direct calculation shows that we cannot write $x_0 x_1^2$ as a linear combination of cubes of two linear forms.

Exercise 2.23

Let $X = \nu_d(\mathbb{P}^n)$. Recall that $[F] \in X$ if and only if $[F] = L^d$ for some linear form L on $\mathbb{P}^n$. Use this description and standard differential geometry to compute $T_{[L^d]}(X)$.

Proof (Solution). Let L be a linear form. Consider an affine curve passing through L. It will have the form $(L + tM)$ where M is allowed to be any linear form. Considering the image in $\mathbb{P}^N$ we have

$$(L + tM)^d = L^d + \binom{d}{d-1} L^{d-1}tM + \text{terms containing higher powers of } t.$$

Taking the derivative with respect to t and setting $t = 0$ we deduce that

$$T_{[F]} = \langle [L^{d-1}M] \mid M \in S_1 \rangle.$$

Exercise 3.11

For binary forms, we can stratify $\mathbb{P}S_2$ using the Waring rank: rank 1 elements correspond to points of the rational normal curve, while all the points outside the curve have rank 2. Do the same for binary cubics and stratify $\mathbb{P}S_3 = \mathbb{P}^3$.

Proof (Solution). Let $X \subseteq \mathbb{P}^3$ be the rational normal curve. If p is a point of $\mathbb{P}^3$ which does not lie on X, then considering the image of X in $\mathbb{P}^2$ by projecting from p we see either that p lies on a tangent line to X or that p lies on a secant line to X. Suppose that q is not a point of X but that q lies on a tangent line. If X is the twisted cubic and $p = [L^3]$, then

$$T_p(X) = \langle [L^3], [L^2M] \rangle$$

where M is any linear form which is not a scalar multiple of L. Thus, without loss of generality, to show that p can be written as a sum of 3 cubes it suffices to show that x^2y is a sum of three cubes. For this, observe that

$$x^2y = \frac{1}{6}((x+y)^3 + (y-x)^3 - 2y^3).$$

Thus, $\mathbb{P}^3$ is stratified by Waring rank. Those points of rank 1 correspond to points of X. Those points of Waring rank 2 correspond to points which lie on no tangent line to X. Those points of Waring rank 3 correspond to points which lie on a tangent line and are not on X. All three of these sets are locally closed.

Exercise 3.16

Prove the general statement. If Λ is a hyperplane containing $T_{[L^d]}(\nu_d(\mathbb{P}^n))$, then $\nu_d^{-1}(\Lambda)$ is a degree d hypersurface singular at the point $[L] \in \mathbb{P}^n$.

Proof (Solution). See the solution to Exercise 2.18.

Exercise 3.21

Solve the big Waring problem for $n = 1$ using the double points interpretation.

Proof (Solution). We have $N = \binom{d+1}{d} - 1 = d$. On the other hand if $p_1, \ldots, p_s$ are general points of $\mathbb{P}^1$, then $\dim(\mathfrak{p}_1^2 \cap \cdots \cap \mathfrak{p}_s^2)_d = -2s + d + 1$ for $1 \leq s \leq d + 1$. So $N - 2s + d + 1 \geq N$ implies that $s \geq \lceil \frac{d+1}{2} \rceil$. In other words $g(1, d) = \lceil \frac{d+1}{2} \rceil$.

Exercise 3.26

Show that $\sigma_5(\nu_4(\mathbb{P}^2))$ is a hypersurface, i.e. that it has dimension equal 13.

Proof (Solution). Since $\sigma_5(\nu_4(\mathbb{P}^2)) \neq \mathbb{P}^{13}$ we conclude that $\dim \sigma_5(\nu_4(\mathbb{P}^2)) \geq \dim \sigma_4(\nu_4(\mathbb{P}^2)) + 2$. On the other hand we know $\dim \sigma_5(\nu_4(\mathbb{P}^2)) \leq 13$. Hence it suffices to show $\dim \sigma_4(\nu_4(\mathbb{P}^2)) \geq 11$. To see this we note that it costs at most 3 linear conditions for a plane curve to be singular at a point, hence $\dim(\mathfrak{p}_1^2 \cap \ldots \mathfrak{p}_4^2)_4 \geq 15 - 12 = 3$ for all collections of four points. Applying the double point lemma for a general collection of points, we deduce that $\dim \sigma_4(\nu_4(\mathbb{P}^2)) \geq 11$.

Exercise 3.27

Explain the exceptional cases $d = 2$ and any n.

Proof (Solution). Let us prove that $g(n, 2) = n + 1$. A general $n + 1 \times n + 1$ symmetric matrix has rank $n + 1$ and is hence a linear combination of $n + 1$ symmetric matrices of rank 1 and is not a linear combination of any smaller number of rank 1 symmetric matrices. On the other hand, symmetric matrices of rank 1 are exactly those in the image of the quadratic Veronese map. This explains the exceptional case $d = 2$ for all n.

Exercise 3.28

Explain the exceptional cases $d = 4$ and $n = 3, 4$.

Proof (Solution). We want to show that $g(3, 4) = 10$ and not 9 and that $g(4, 4) = 15$ and not 14. In both cases a parameter count shows that we can find quadrics through 9 points and 14 points in $\mathbb{P}^3$ and $\mathbb{P}^4$, respectively. Squaring these forms produces a quartic singular at these points. Applying the double point lemma we conclude that $\sigma_9(\nu_4(\mathbb{P}^3))$ and $\sigma_{14}(\nu_4(\mathbb{P}^4))$ fail to fill up the space. On the other hand a parameter count also shows that $\sigma_{10}(\nu_4(\mathbb{P}^3))$ and $\sigma_{15}(\nu_4(\mathbb{P}^4))$ fill up the space.

Exercise 3.29

Explain the exceptional case $d = 3$ and $n = 4$. (Hint: use Castelnuovo's Theorem which asserts that there exists a (unique) rational normal curve passing through $n+3$ generic points in $\mathbb{P}^n$.)

Proof (Solution). By the double point lemma it suffices to pick seven general points $p_1, \ldots, p_7 \in \mathbb{P}^4$ and prove that there exists a degree 3 hypersurface in $\mathbb{P}^4$ singular at $p_1, \ldots, p_7$. So choose $p_1, \ldots, p_7$ general points. By Castelnuovo's Theorem, there exists a rational quartic curve X in $\mathbb{P}^4$ passing through the points $p_1, \ldots, p_7$. It suffices to prove that $\sigma_2(X)$ is a degree 3 hypersurface singular along X.

To show that $\sigma_2(X)$ is singular along X if $x \in X$, then we can show that $X \subseteq T_x\sigma_2(X)$ so that $\langle X \rangle \subseteq T_x\sigma_2(X)$. Since X is non-degenerate we conclude that $T_x\sigma_2(X) = \mathbb{P}^4$, for all $x \in X$. Since $\sigma_2(X)$ has dimension 3 we conclude that $\sigma_2(X)$ is singular along X.

To compute the degree we project to $\mathbb{P}^2$ from a general secant line and count the number of nodes. Since the resulting curve is rational the number of nodes equals the arithmetic genus of a plane curve of degree 4 which is 3.

Exercise 4.2

Let $0 \neq F \in S_d$. Show that $F^\perp \subset T$ is an ideal and that it is also Artinian. i.e., $(T/F^\perp)_i = 0$ for all $i > d$.

Proof (Solution). If $\partial \in F^\perp$ and $\tilde{\partial} \in T$, then by definition of the T-action $\tilde{\partial}\partial F = (\tilde{\partial})\partial F = 0$ so $\tilde{\partial}\partial \in F^\perp$. On the other hand since F has degree d, $F^\perp$ contains all differential operators of degree greater than d. Hence $T/F^\perp$ is a finite dimensional vector space and hence Artinian.

Exercise 4.3

Show that the map $S_i \times T_i \to \mathbb{C}$, $(F, \partial) \mapsto \partial F$ is a perfect paring and that $A = T/F^{\perp}$ is Artinian and Gorenstein with socle degree d.

Proof (Solution). We use the standard monomial basis for S_i and T_i. By definition of the action we have

$$y_0^{a_0} \dots y_n^{a_n} \circ x_0^{b_0} \dots x_n^{b_n} = \begin{cases} a_0! \dots a_n! & \text{iff } a_j = b_j, \text{ for } j = 0 \dots, n \\ 0 & \text{otherwise} \end{cases}$$

from which the first assertion is clear.

It remains to show that A is Gorenstein with socle degree d. Let $\mathfrak{m}$ denote the homogeneous maximal ideal of A. Then $Soc(A) = \{x \in A \mid x\mathfrak{m} = 0\}$. Using this description we can check that

$$\dim Soc(A)_i = \begin{cases} 0 & \text{if } i < d \\ 1 & \text{if } i = d. \end{cases}$$

Hence $Soc(A)$ is 1 dimensional and nonzero only in degree d which implies that A is Gorenstein with socle degree d.

Exercise 4.8

Given $F \in S_d$ show that $HF(T/F^{\perp}, t)$ is a symmetric function of t.

Proof (Solution). Let $A = T/F^{\perp}$ and let d be the socle degree of A. It suffices to show that for $0 \le l \le d$, multiplication in A defines a perfect pairing $A_{d-l} \times A_l \to A_d$. If $l = d$ or $l = 0$, the assertion is clear. Let $0 < l < d$. Let $y \in A_l$. Suppose $yx = 0$ for all $x \in A_{d-l}$. Let's prove that $y \in Soc(A)$. Since $Soc(A)$ is zero in degrees less than d we will arrive at a contradiction.

To prove that $y \in Soc(A)$ it suffices to prove that y annihilates every homogeneous element of A which has positive degree. If $M \in A_n$ and $n > d - l$, then $yM \in A_{l+n} = 0$ since $l + n > d$. On the other hand we have by assumption that $yA_{d-l} = 0$. Descending induction with base case $d - l$ proves that $yA_l = 0$ for $0 < l < d - l$. Indeed suppose $0 < n < d - l$ and let $M \in A_n$. Suppose that $yM \neq 0$. We have $Mx_i \in A_{n+1}$ for $i = 0, \dots, n$. By induction $yMx_i = 0$. Hence $yM\mathfrak{m} = 0$ so $yM \in Soc(A)$. Since $\deg yM < d$ this is a contradiction. Hence $y \in Soc(A)$ which is also a contradiction.

Exercise 4.14

Use the Apolarity Lemma to compute $\mathrm{rk}(x_0 x_1^2)$. Then try the binary forms $x_0 x_1^d$.

Proof (Solution). $F^\perp = \langle \partial_{x_0}^2, \partial_{x_1}^{d+1} \rangle$ so rank $F = d + 1$.

Exercise 4.15

Use the Apolarity Lemma to explain the Alexander–Hirschowitz exceptional cases.

Proof (Solution). We explain the exceptional cases $d = 4$ and $n = 2, 3$, or 4. Then exceptional case $d = 3$ and $n = 4$ can be treated via syzygies.

Since $\binom{2+2}{2} = 6 > 5$ if I is the ideal of five points in $\mathbb{P}^2$, then I contains a quadric. Since $\binom{2+3}{2} = 10 > 9$ if I is the ideal of nine points in $\mathbb{P}^3$, then I contains a quadric. Since $\binom{2+4}{2} = 15 > 14$ if I is the ideal of 14 points in $\mathbb{P}^4$, then I contains a quadric.

On the other hand, if F is a general form of degree 4, in $\mathbb{C}[x_0, .., x_n]$, for $n = 2, 3$ or 4, then $F^\perp$ contains no quadrics. We work out the case $n = 2$ explicitly. The case $n = 3$ or 4 is similar.

Let $S = \mathbb{C}[x, y, z]$. Every form $F \in S_4$ determines a linear map from the vector space of differential operators of degree 2 to the space of degree 2 polynomials in S. This map is determined by applying the differential operators to F. Explicitly, if

$$F = ax^4 + bx^3y + cx^3z + dx^2y^2 + ex^2yz + fx^2z^2 + gxy^3 + hxy^2z + ixyz^2 + jxz^3 + ky^4 + ly^3z + my^2z^2 + oyz^3 + pz^4$$

then, using the basis

$$\partial_{xx}, \partial_{xy}, \partial_{xz}, \partial_{y^2}, \partial_{yz}, \partial_{z^2}$$

for the source, and the basis

$$x^2, xy, xz, y^2, yz, z^2$$

for the target space, the matrix for this map is given by

$$\begin{bmatrix} 12a & 3b & 3c & 2d & e & 2f \\ 6b & 4d & 2e & 6g & 2h & 2i \\ 6c & 2e & 4f & 2h & 2i & 6j \\ 2d & 3g & h & 12k & 3l & 2m \\ 2e & 2h & 2i & 6l & 4m & 6o \\ 2f & i & 3j & 2m & 3o & 12p \end{bmatrix}.$$

Elements in the kernel of this map correspond to elements of $(F^{\perp})_2$. The collection of forms for which this map is injective is given by the nonvanishing of the determinant of this matrix. We conclude that $(F^{\perp})_2$ is zero for a general quartic form in S_4.

Note that we have shown that if F is a general form of degree 4 in $\mathbb{C}[x_0, .., x_n]$, for $n = 2, 3$ or 4, then

t	0	1	2	3	4	5
$HF(T/F^{\perp}, t)$	1	$n+1$	$\binom{n+2}{2}$	$n+1$	1	$0 \rightarrow$

On the other hand, for every ideal I of, respectively, 5 points in $\mathbb{P}^2$, 9 points in $\mathbb{P}^3$, or 14 points in $\mathbb{P}^4$, we have shown that

$$HF(T/I, 2) < \binom{n+2}{2}.$$

Thus, using the Apolarity Lemma, a general quartic form in $\mathbb{C}[x_0, .., x_n]$, for $n = 2, 3$, or 4, cannot have rank, respectively, 5, 9, or 14.

To see an issue which arises in the case exceptional case $d = 3$ and $n = 4$, note that if F is any cubic form in $\mathbb{C}[x_0, x_1, x_2, x_3, x_4]$ then

t	0	1	2	3	4
$HF(T/F^{\perp}, t)$	1	5	5	1	$0 \rightarrow$

Thus the Hilbert function $HF(T/F^{\perp}, t)$ is the same for all

$$F \in \mathbb{C}[x_0, x_1, x_2, x_3, x_4]_3.$$

Nevertheless, the exceptional case $d = 3$ and $n = 4$ can still be explained using the Apolarity Lemma, although a more detailed study is needed to conclude that $F^{\perp}$ for a general $F \in \mathbb{C}[x_0, x_1, x_2, x_3, x_4]_3$ contains no ideal of seven points in $\mathbb{P}^3$. See the paper [145] for a more detailed discussion.

Exercise 4.17

Compute $\text{rk}(F)$ when F is a quadratic form.

Proof (Solution). $\text{rk}(F) = \text{rank}\, M_F$ where M_F is the symmetric matrix associated with F. (See Exercise 2.12.)

Exercise 4.20

Prove that $\mathrm{rk}(L^d + M^d + N^d) = 3$ whenever L, M, and N are linearly independent linear forms.

Proof (Solution). It is clear that $\mathrm{rk}(L^d + M^d + N^d) \leq 3$. To show that $\mathrm{rk}(L^d + M^d + N^d)$ is not less than 3, without loss of generality it suffices to consider the case that $F = x_0^d + x_1^d + x_2^d$. In this case $F^\perp = \langle \partial_{x_0}^{d+1}, \partial_{x_1}^{d+1}, \partial_{x_2}^{d+1}, \partial_{x_3}, \ldots, \partial_{x_n} \rangle$. Now if $F^\perp$ contained the ideal of one or two distinct points in $\mathbb{P}^n$, then $\langle \partial_{x_0}^{d+1}, \partial_{x_1}^{d+1}, \partial_{x_2}^{d+1} \rangle$ would contain a linear form which it does not.

Exercise 5.5

Workout a matrix representation for the Segre varieties with two factors $\mathbb{P}^{n_1} \times \mathbb{P}^{n_2}$.

Proof (Solution). Let $A \cong \mathbb{C}^{n_1+1}$ and $B \cong \mathbb{C}^{n_2+1}$. By definition, points in $\mathrm{Seg}(\mathbb{P}^{n_1} \times \mathbb{P}^{n_2})$ are of the form

$$[a \otimes b] \in \mathbb{P}(A \otimes B).$$

Now consider $A \otimes B$ as a space of matrices $A^* \to B$. Then, by choosing bases of A and B, we may represent a as a column vector $(a_1, \ldots, a_{n_1+1})^t$ (an element of A^*) and b as a row vector $(b_1, \ldots, b_{n_2+1})$ so that the tensor product $a \otimes b$ becomes the product of a column and a row:

$$a \otimes b = (a_1, \ldots, a_{n_1+1})^t \times (b_1, \ldots, b_{n_2+1}) = \begin{pmatrix} a_1 b_1 & a_1 b_2 & \ldots & a_1 b_{n_2+1} \\ a_2 b_1 & a_2 b_2 & \ldots & a_2 b_{n_2+1} \\ \vdots & & \ddots & \vdots \\ a_{n_1+1} b_1 & a_{n_1+1} b_2 & \ldots & a_{n_1+1} b_{n_2+1} \end{pmatrix}.$$

So we see that (up to scale) elements of $\mathrm{Seg}(\mathbb{P}^{n_1} \times \mathbb{P}^{n_2})$ correspond to rank-1 $(n_1 + 1) \times (n_2 + 1)$ matrices.

Exercise 5.7

Let $X = \mathbb{P}V_1 \times \cdots \times \mathbb{P}V_t$ and let $[v] = [v_1 \otimes \cdots \otimes v_t]$ be a point of X. Show that the cone over the tangent space to X at v is the span of the following vector spaces:

$$V_1 \otimes v_2 \otimes v_3 \otimes \cdots \otimes v_t,$$
$$v_1 \otimes V_2 \otimes v_3 \otimes \cdots \otimes v_t,$$
$$\vdots$$
$$v_1 \otimes v_2 \otimes \cdots \otimes v_{t-1} \otimes V_t.$$

Proof (Solution). The cone over the tangent space to a variety X at a point v may be computed by considering all curves $\gamma : [0,1] \to \hat{X}$ such that $\gamma(0) = v$, and taking the linear span of all derivatives at the origin:

$$\widehat{T_x X} = \left\{ \gamma'(0) \mid \gamma : [0,1] \to \hat{X}, \gamma(0) = v \right\}.$$

Now take $X = \mathbb{P}V_1 \times \cdots \times \mathbb{P}V_t$ and let $[v] = [v_1 \otimes \cdots \otimes v_t]$ be a point of X.

All curves $\gamma(t)$ on $\hat{X}$ through v are of the form $\gamma(t) = v_1(t) \otimes \cdots \otimes v_t(t)$, where $v_i(t)$ are curves in V_i such that $v_i(0) = v_i$.

Now apply the product rule, and for notational convenience, set $v_i' = v_i'(0) \in V_i$. We have

$$\gamma'(0) = v_1' \otimes v_2 \otimes \cdots \otimes v_t + v_1 \otimes v_2' \otimes v_3 \otimes \cdots \otimes v_t + \cdots + v_1 \otimes \cdots \otimes v_t'.$$

Since v_i' can be anything in V_i we get the result.

Exercise 5.8

Show that $\sigma_2(\mathbb{P}^1 \times \mathbb{P}^1 \times \mathbb{P}^1) = \mathbb{P}^7$.

Proof (Solution). Let $a \otimes b \otimes c + \tilde{a} \otimes \tilde{b} \otimes \tilde{c}$ be a general point on $\sigma_2(\mathbb{P}A \times \mathbb{P}B \times \mathbb{P}C)$, with $A \cong B \cong C \cong \mathbb{C}^2$.

By the previous exercise, we have

$$\widehat{T_{a\otimes b\otimes c}}(\mathbb{P}^1 \times \mathbb{P}^1 \times \mathbb{P}^1) = A \otimes b \otimes c + a \otimes B \otimes c + a \otimes b \otimes C,$$

and similarly

$$\widehat{T_{\tilde{a}\otimes \tilde{b}\otimes \tilde{c}}}(\mathbb{P}^1 \times \mathbb{P}^1 \times \mathbb{P}^1) = A \otimes \tilde{b} \otimes \tilde{c} + \tilde{a} \otimes B \otimes \tilde{c} + \tilde{a} \otimes \tilde{b} \otimes C.$$

Now by Terracini's lemma we have

$$\begin{aligned} &\widehat{T_{a\otimes b\otimes c + \tilde{a}\otimes \tilde{b}\otimes \tilde{c}}}\sigma_2(\mathbb{P}^1 \times \mathbb{P}^1 \times \mathbb{P}^1) \\ &= A \otimes b \otimes c + a \otimes B \otimes c + a \otimes b \otimes C + A \otimes \tilde{b} \otimes \tilde{c} + \tilde{a} \otimes B \otimes \tilde{c} + \tilde{a} \otimes \tilde{b} \otimes C. \end{aligned} \tag{2}$$

Because we chose a general point, we have $\{a, \tilde{b}\} = A$ and similarly for B and C. Now consider the linear space in (2). We see that we can get every tensor monomial in $A \otimes B \otimes C$—all monomials with 0 or 1 ˜occur in the first 3 summands, while all monomials with 2 or 3 ˜occur in the second 3 summands.

So the cone over the tangent space at a general point is eight dimensional, so the secant variety (being irreducible) fills the whole ambient space.

Exercise 5.9

Use the above description of the tangent space of the Segre product and Terracini's lemma to show that $\sigma_3(\mathbb{P}^1 \times \mathbb{P}^1 \times \mathbb{P}^1 \times \mathbb{P}^1)$ is a hypersurface in $\mathbb{P}^{15}$ and not the entire ambient space as expected. This shows that the four-factor Segre product of $\mathbb{P}^1$s is defective.

Proof (Solution). This should be done on the computer. Take the following to be a general point:

$$a \otimes b \otimes c \otimes d + \tilde{a} \otimes \tilde{b} \otimes \tilde{c} \otimes \tilde{d} +$$
$$(\alpha_1 a + \alpha_2 \tilde{a}) \otimes (\beta_1 b + \beta_2 \tilde{b}) \otimes (\gamma_1 c + \gamma_2 \tilde{c}) \otimes (\delta_1 d + \delta_2 \tilde{d}),$$

where $[\alpha_1, \alpha_2], [\beta_1, \beta_2], [\gamma_1, \gamma_2], [\delta_1, \delta_2] \in \mathbb{P}^1$.

Then compute the derivatives at the origin. If we let the parameters $a(0)'$ $\tilde{a}(0)'$ etc., vary this will produce 16 vectors that span the tangent space. Now compute the rank of the matrix with these vectors as columns. It gets hard to do by hand, but the next section produces an easier way to do the problem.

The "easier way" is to consider all three essentially different two-flattenings and show that two of them are algebraically independent.

Exercise 5.10

Show that T has rank 1 if and only if its multilinear rank is $(1, \ldots, 1)$.

Proof (Solution). It is equivalent (by taking transposes) to consider the $n-1$ flattenings. Let $\varphi_{i,T} : (V_1 \otimes \cdots \otimes \widehat{V_i} \otimes \cdots \otimes V_n)^* \to V_i$ denote the $(n-1)$-flattening to the ith factor. Denote by A_i the image of $\varphi_{i,T}$.

If $\varphi_{i,T}$ has rank 1, then $\dim A_i = 1$. So, we must have $T \in A_1 \otimes \ldots A_n$. But every tensor in $A_1 \otimes A_2 \otimes \ldots A_n$ has rank 1 if all the factors have dimension 1.

Conversely, if $T = a_1 \otimes \ldots a_n \in V_1 \otimes \cdots \otimes V_n$ the image of $\phi_{i,T}$ is the line through a_i, so the multilinear rank is $(1, 1, \ldots, 1)$.

Exercise 5.11

Let $X = \mathbb{P}V_1 \times \cdots \times \mathbb{P}V_t$. Show that if $r \leq r_i$ for $1 \leq i \leq t$,

$$\sigma_r(X) \subset \mathrm{Sub}_{r_1,\dots,r_t}.$$

Proof (Solution). A general point on $\sigma_r(X)$ is $p = \sum_{s=1}^{r} \bigotimes_{i=1}^{t} a_{i,s}$, where for fixed i the $a_{i,s} \in V_i$ are linearly independent. Set $A_i = \{a_{i,1}, \dots, a_{i,r}\}$. Then $p \in A_1 \otimes \cdots \otimes A_t$, so $p \in \mathrm{Sub}_{r_1,\dots,r_t}$. Now take the orbit closure of p to obtain the result.

Exercise 5.12

1. Show that if T has rank 1 then φ_T has rank 2.
2. Show that φ is additive in its argument, i.e. show that $\varphi_{T+T'} = \varphi_T + \varphi_{T'}$.

Proof (Solution). If T has rank 1, after change of coordinates we may assume that $T = v_1 \otimes w_1 \otimes x_1$. Then the matrix φ_T has precisely two ones in different rows and columns, so clearly has rank 2.

For the second part, notice

$$\varphi_T + \varphi_{T'} = \begin{pmatrix} 0 & T^1 & -T^2 \\ -T^1 & 0 & T^3 \\ T^2 & -T^3 & 0 \end{pmatrix} + \begin{pmatrix} 0 & T'^1 & -T'^2 \\ -T'^1 & 0 & T'^3 \\ T'^2 & -T'^3 & 0 \end{pmatrix}$$

$$= \begin{pmatrix} 0 & T^1 + T'^1 & -T^2 + T'^2 \\ -T^1 + T'^1 & 0 & T^3 + T'^3 \\ T^2 + T'^2 & -T^3 + T'^3 & 0 \end{pmatrix} = \varphi_{T+T'}. \qquad \square$$

7 Looking Forward, Further Readings

This series of lectures draws from many sources from a large group of authors. We do our best to collect representative works here so that the reader may have some starting points for further study. A general introductory reference for the material we treated is the booklet is [99]. Here we collect a few guiding questions and an extensive list of references.

7.1 Guiding Questions

In our opinion, there are a few leading topics that are still driving current research in the area of secant varieties. These topics are dimension, identifiability, decomposition, and equations. More specifically here are four leading questions:

1. For a variety X what are the dimensions of the higher secant varieties to X? There is much interest when X is the variety of elementary tensors of a given format (partially symmetric, skew symmetric, general).
2. Suppose $p \in \sigma_s(X)$ is general, when does p have a *unique* representation as a sum of s points from X? When uniqueness occurs we say that $\sigma_s(X)$ is *generically identifiable*.
3. Suppose $X \subset \mathbb{C}^N$ and $p \in \mathbb{C}^N$. If X is not degenerate, then we know that there is some s so that p has a representation as the linear combination of s points from X. For special X (Segre, Veronese, etc.,) find (a) determine the minimal s explicitly, and (b)find efficient algorithms when s is relatively small to find such a decomposition of p.
4. How do we find equations $\sigma_s(X)$ in general? What is the degree of $\sigma_s(X)$? Is $\sigma_s(X)$ a Cohen–Macaulay variety? Again, there is much interest when X is the variety of elementary tensors of a given format.

7.2 Background Material

Textbooks: [33,38,59,69,83–85,88,93,94,98,100,102,103,116,135,139,157,158, 162].

Classical: [152]. See also the nice overview in [99] and the introduction of [145].

7.3 Dimensions of Secant Varieties

For Veronese varieties, the capstone result is that of Alexander and Hirschowitz [9]. Brambilla and Ottaviani [30] provided a very nice exposition, and Postinghel [140] provided a new proof.

Related work on polynomial interpolation: [8, 31].

Waring's problem for binary forms: [61, 108, 121] Waring's problem for polynomials: [58, 64, 92, 129, 130, 137, 145]. Waring's problem for monomials was solved in [41] and an alternative proof can be found also in [35] where the apolar sets of points to monomials are described. See also [27].

The polynomial Waring problem over the reals was investigated by Comon and Ottaviani [65], with some solutions to their questions provided by Blekherman [25], Causa and Re [55], Ballico [16]. The case of real monomials in two variables is discussed in [27]. For typical ranks of tensors see [68, 89, 149].

Recent algorithms for Waring decomposition: [24, 29, 67, 108, 134].

The notion of Weak Defectivity: [15, 56, 57].

Dimensions of Secant varieties of Segre–Veronese varieties: [1–4, 6, 18, 23, 46–50, 52, 53, 111]. Nice Summary of results on dimensions of secant varieties: [45].

7.4 Equations of Secant Varieties

See [22, 43, 109, 117, 118, 120, 122–124, 132, 136, 142–144, 144, 146, 147, 150].

The Salmon Problem: [10, 19, 90, 91].

7.5 Applications

The question of best low-rank approximation of tensors is often ill posed [78], and most tensor problems are NP-Hard, [105].

Algebraic Statistics: [81, 82, 95, 151].

Phylogenetics: [11, 12, 44, 86].

Signal Processing: see [7, 62, 63, 66, 70, 72–77].

Matrix Multiplication: [112–115, 119].

7.6 Related Varieties

Grassmannians: [5, 28, 51].

Discriminants and Hyperdeterminants are intimately related to secant varieties of Segre–Veronese varieties, see [26, 54, 96, 97, 107, 133, 159, 160].

Chow varieties and monomials[13, 32, 39].

7.7 Related Concepts

Eigenvectors of Tensors: [14, 42, 106, 110, 127, 131, 138, 141].

Veronese reembeddings: [34]. Hilbert Schemes: [87].

Orbits: [79, 156].

Ranks and Decompositions: [17, 36, 37].

Asymptotic questions: [80].

7.8 Software

Symbolic computation: [60, 71, 101, 154].

Numerical Algebraic Geometry: [20, 21, 104, 125, 126, 148, 155].

Acknowledgments The second and third authors would like to thank the first author for his lectures at the workshop "Connections Between Algebra and Geometry" held at the University of Regina in 2012. The second author would like to thank the third author for tutoring these lectures and for helping to write solutions to the exercises. All three authors would like to thank the organizers S. Cooper, S. Sather-Wagstaff and D. Stanley for their efforts in organizing the workshop and for securing funding to cover the costs of the participants. It is also our pleasure to acknowledge the lecture notes of Tony Geramita [99] which have influenced our exposition here. The three authors received partial support by different sources: Enrico Carlini by GNSAGA of INDAM, Nathan Grieve by an Ontario Graduate Fellowship, and Luke Oeding by NSF RTG Award # DMS-0943745. Finally, all the authors thank the anonymous referee for the improvement to the paper produced by the referee's suggestions and remarks.

References

1. Abo, H., Brambilla, M.C.: Secant varieties of Segre-Veronese varieties $\mathbb{P}^m \times \mathbb{P}^n$ embedded by $\mathcal{O}(1,2)$. Exp. Math. **18**(3), 369–384 (2009)
2. Abo, H., Brambilla, M.C.: New examples of defective secant varieties of Segre-Veronese varieties. Collect. Math. **63**(3), 287–297 (2012)
3. Abo, H., Brambilla, M.C.: On the dimensions of secant varieties of Segre-Veronese varieties. Ann. Mat. Pura Appl. **192**(1), 61–92 (2013)
4. Abo, H., Ottaviani, G., Peterson, C.: Induction for secant varieties of Segre varieties. Trans. Am. Math. Soc. **361**(2), 767–792 (2009)
5. Abo, H., Ottaviani, G., Peterson, C.: Non-defectivity of Grassmannians of planes. J. Algebr. Geom. **21**(1), 1–20 (2012)
6. Abrescia, S.: About the defectivity of certain Segre-Veronese varieties. Can. J. Math. **60**(5), 961–974 (2008)
7. Albera, L., Ferreol, A., Comon, P., Chevalier, P.: Blind identification of overcomplete mixtures of sources (BIOME). Linear Algebra Appl. **391**, 3–30 (2004)
8. Alexander, J., Hirschowitz, A.: La méthode d'Horace éclatée: application à l'interpolation en degré quatre. Invent. Math. **107**(3), 585–602 (1992)
9. Alexander, J., Hirschowitz, A.: Polynomial interpolation in several variables. J. Algebr. Geom. **4**(2), 201–222 (1995)
10. Allman, E.: Open problem: determine the ideal defining $Sec_4(\mathbb{P}^3 \times \mathbb{P}^3 \times \mathbb{P}^3)$. http://www.dms.uaf.edu/~eallman/salmonPrize.pdf (2010)
11. Allman, E., Rhodes, J.: Phylogenetic invariants for the general Markov model of sequence mutation. Math. Biosci. **186**(2), 113–144 (2003)
12. Allman, E., Rhodes, J.: Phylogenetic ideals and varieties for the general Markov model. Adv. Appl. Math. **40**(2), 127–148 (2008)
13. Arrondo, E., Bernardi, A.: On the variety parameterizing completely decomposable polynomials. J. Pure Appl. Algebra **215**(3), 201–220 (2011)
14. Ballard, G., Kolda, T., Plantenga, T.: Efficiently computing tensor eigenvalues on a GPU. In: CSRI Summer Proceedings (2011)
15. Ballico, E.: On the weak non-defectivity of Veronese embeddings of projective spaces. Cent. Eur. J. Math. **3**(2), 183–187 (2005)

16. Ballico, E.: On the typical rank of real bivariate polynomials. ArXiv e-prints (April 2012)
17. Ballico, E., Bernardi, A.: Decomposition of homogeneous polynomials with low rank. Math. Z. **271**(3–4), 1141–1149 (2011)
18. Ballico, E., Bernardi, A., Catalisano, M.V.: Higher secant varieties of $\mathbb{P}^n \times \mathbb{P}^1$ embedded in bi-degree (a, b). Commun. Algebra **40**(10), 3822–3840 (2012)
19. Bates, D.J., Oeding, L.: Toward a salmon conjecture. Exp. Math. **20**(3), 358–370 (2011)
20. Bates, D.J., Hauenstein, J.D., Sommese, A.J., Wampler, C.W.: Software for numerical algebraic geometry: a paradigm and progress towards its implementation. In: Software for Algebraic Geometry. IMA Volumes in Mathematics and Its Applications, vol. 148, pp. 1–14. Springer, New York (2008)
21. Bates, D.J., Hauenstein, J.D., Sommese, A.J., Wampler, C.W.: Bertini: software for numerical algebraic geometry. http://www.nd.edu/~sommese/bertini (2010)
22. Bernardi, A.: Ideals of varieties parameterized by certain symmetric tensors. J. Pure Appl. Algebra **212**(6), 1542–1559 (2008)
23. Bernardi, A., Carlini, E., Catalisano, M.V.: Higher secant varieties of $\mathbb{P}^n \times \mathbb{P}^m$ embedded in bi-degree $(1, d)$. J. Pure Appl. Algebra **215**(12), 2853–2858 (2011)
24. Bernardi, A., Gimigliano, A., Idà, M.: Computing symmetric rank for symmetric tensors. J. Symb. Comput. **46**(1), 34–53 (2011)
25. Blekherman, G.: Typical real ranks of binary forms. ArXiv e-prints (May 2012)
26. Boffi, G., Weyman, J.: Koszul complexes and hyperdeterminants. J. Algebra **230**(1), 68–88 (2000)
27. Boij, M., Carlini, E., Geramita, A.V.: Monomials as sums of powers: the real binary case. Proc. Am. Math. Soc. **139**(9), 3039–3043 (2011)
28. Boralevi, A., Buczyński, J.: Secants of Lagrangian Grassmannians. Ann. Math. Pura Appl. **190**(4), 725–739 (2011)
29. Brachat, J., Comon, P., Mourrain, B., Tsigaridas, E.: Symmetric tensor decomposition. Linear Algebra Appl. **433**(11–12), 1851–1872 (2010)
30. Brambilla, M.C., Ottaviani, G.: On the Alexander-Hirschowitz theorem. J. Pure Appl. Algebra **212**(5), 1229–1251 (2008)
31. Brambilla, M.C., Ottaviani, G.: On partial polynomial interpolation. Linear Algebra Appl. **435**(6), 1415–1445 (2011)
32. Briand, E.: Covariants vanishing on totally decomposable forms. In: Liaison, Schottky Problem and Invariant Theory. Progress in Mathematics, vol. 280, pp. 237–256. Birkhäuser Verlag, Basel (2010)
33. Bruns, W., Vetter, U.: Determinantal Rings. Lecture Notes in Mathematics, vol. 1327. Springer, Berlin (1988)
34. Buczyńska, W., Buczyński, J.: Secant varieties to high degree Veronese reembeddings, catalecticant matrices and smoothable Gorenstein schemes. J. Algebr. Geom. (2010, to appear). ArXiv:1012.3563
35. Buczyńska, W., Buczyński, J., Teitler, Z.: Waring decompositions of monomials. J. Algebra **378**, 45–57 (2013)
36. Buczyński, J., Landsberg, J.M.: On the third secant variety. J. Algebr. Comb. (2011, to appear). ArXiv e-prints. http://arxiv.org/abs/1111.7005
37. Buczyński, J., Landsberg, J.M.: Ranks of tensors and a generalization of secant varieties. Linear Algebra Appl. **438**(2), 668–689 (2013)
38. Bürgisser, P., Clausen, M., Shokrollahi, M.: Algebraic Complexity Theory. Grundlehren der Mathematischen Wissenschaften [Fundamental Principles of Mathematical Sciences], vol. 315. Springer, Berlin (1997)
39. Carlini, E.: Codimension one decompositions and Chow varieties. In: Projective Varieties with Unexpected Properties, pp. 67–79. Walter de Gruyter GmbH & Co. KG, Berlin (2005)
40. Carlini, E., Kleppe, J.: Ranks derived from multilinear maps. J. Pure Appl. Algebra **215**(8), 1999–2004 (2011)
41. Carlini, E., Catalisano, M.V., Geramita, A.V.: The solution to the Waring problem for monomials and the sum of coprime monomials. J. Algebra **370**, 5–14 (2012)

42. Cartwright, D., Sturmfels, B.: The number of eigenvalues of a tensor. Linear Algebra Appl. **438**(2), 942–952 (2013)
43. Cartwright, D., Erman, D., Oeding, L.: Secant varieties of $\mathbf{P}^2 \times \mathbf{P}^n$ embedded by $\mathcal{O}(1,2)$. J. Lond. Math. Soc. **85**(1), 121–141 (2012)
44. Casanellas, M., Fernández-Sánchez, J.: Relevant phylogenetic invariants of evolutionary models. J. Math. Pures Appl. **96**(3), 207–229 (2011)
45. Catalisano, M.V.: Higher secant varieties of Segre, Segre-Veronese and Grassmann varieties. In: Interactions Between Commutative Algebra and Algebraic Geometry the 22nd Annual Route 81 Conference — in honour of Tony Geramita [Conference], Kingston, 20 October 2012. http://www.mast.queensu.ca/~ggsmith/Route81/catalisano.pdf
46. Catalisano, M.V., Geramita, A.V., Gimigliano, A.: On the rank of tensors, via secant varieties and fat points. In: Zero-dimensional Schemes and Applications - Naples, 2000, vol. 123, pp. 133–147 (2002)
47. Catalisano, M.V., Geramita, A.V., Gimigliano, A.: Ranks of tensors, secant varieties of Segre varieties and fat points. Linear Algebra Appl. **355**, 263–285 (2002)
48. Catalisano, M.V., Geramita, A.V., Gimigliano, A.: Erratum to: "ranks of tensors, secant varieties of Segre varieties and fat points" [Linear Algebra Appl. **355**, 263–285 (2002); MR1930149 (2003g:14070)]. Linear Algebra Appl. **367**, 347–348 (2003)
49. Catalisano, M.V., Geramita, A.V., Gimigliano, A.: Higher secant varieties of Segre-Veronese varieties. In: Projective Varieties with Unexpected Properties, pp. 81–107. Walter de Gruyter GmbH & Co. KG, Berlin (2005)
50. Catalisano, M.V., Geramita, A.V., Gimigliano, A.: Higher secant varieties of the Segre varieties $\mathbb{P}^1 \times \cdots \times \mathbb{P}^1$. J. Pure Appl. Algebra **201**(1–3), 367–380 (2005)
51. Catalisano, M.V., Geramita, A.V., Gimigliano, A.: Secant varieties of Grassmann varieties. Proc. Am. Math. Soc. **133**(3), 633–642 (2005)
52. Catalisano, M.V., Geramita, A.V., Gimigliano, A.: On the ideals of secant varieties to certain rational varieties. J. Algebra **319**(5), 1913–1931 (2008)
53. Catalisano, M.V., Geramita, A.V., Gimigliano, A.: Secant varieties of $\mathbb{P}^1 \times \cdots \times \mathbb{P}^1$ (n-times) are not defective for $n \geq 5$. J. Algebr. Geom. **20**(2), 295–327 (2011)
54. Cattani, E., Cueto, M.A., Dickenstein, A., Di Rocco, S., Sturmfels, B.: Mixed discriminants. Math. Z. **274**(3–4), 761–778 (2013)
55. Causa, A., Re, R.: On the maximum rank of a real binary form. Ann. Math. Pura Appl. **190**, 55–59 (2011)
56. Chiantini, L., Ciliberto, C.: Weakly defective varieties. Trans. Am. Math. Soc. **354**(1), 151–178 (2002)
57. Chiantini, L., Ciliberto, C.: On the concept of k-secant order of a variety. J. Lond. Math. Soc. **73**(2), 436–454 (2006)
58. Chipalkatti, J.: The Waring loci of ternary quartics. Exp. Math. **13**(1), 93–101 (2004)
59. Ciliberto, C., Geramita, A.V., Harbourne, B., Miró-Roig, R.M., Ranestad, K. (eds.): Projective Varieties with Unexpected Properties. Walter de Gruyter GmbH & Co. KG, Berlin (2005) (A volume in memory of Giuseppe Veronese)
60. CoCoATeam: CoCoA: a system for doing Computations in Commutative Algebra. http://cocoa.dima.unige.it
61. Comas, G., Seiguer, M.: On the rank of a binary form. Found. Comput. Math. **11**, 65–78 (2011)
62. Comon, P.: Tensors decompositions. State of the art and applications. In: IMA Conference of Mathematics in Signal Processing, 18–20 December 2000
63. Comon, P.: Canonical tensor decompositions. Research Report RR-2004-17, I3S, 17 June 2004
64. Comon, P., Mourrain, B.: Decomposition of quantics in sums of powers of linear forms. Signal Process. **53**(2), 93–107 (1996) (Special issue on High-Order Statistics)
65. Comon, P., Ottaviani, G.: On the typical rank of real binary forms. Linear Multilinear Algebra **60**(6), 657–667 (2012)

66. Comon, P., Rajih, M.: Blind identification of under-determined mixtures based on the characteristic function. Signal Process. **86**(9), 2271–2281 (2006)
67. Comon, P., Golub, G., Lim, L.-H., Mourrain, B.: Symmetric tensors and symmetric tensor rank. SIAM J. Matrix Anal. Appl. **30**(3), 1254–1279 (2008)
68. Comon, P., ten Berge, J.M.F., De Lathauwer, L., Castaing, J.: Generic and typical ranks of multi-way arrays. Linear Algebra Appl. **430**(11–12), 2997–3007 (2009)
69. Cox, D., Little, J., O'Shea, D.: Ideals, Varieties, and Algorithms: An Introduction to Computational Algebraic Geometry and Commutative Algebra. Undergraduate Texts in Mathematics, 3rd edn. Springer, New York (2007)
70. 22nd Annual IEEE Conference on Computational Complexity (CCC 2007), San Diego, CA, 13–16 June 2007. IEEE Computer Society, Los Alamitos (2007)
71. Decker, W., Greuel, G.-M., Pfister, G., Schönemann, H.: SINGULAR 3-1-1 — a computer algebra system for polynomial computations. http://www.singular.uni-kl.de
72. De Lathauwer, L.: A link between the canonical decomposition in multilinear algebra and simultaneous matrix diagonalization. SIAM J. Matrix Anal. Appl. **28**(3), 642–666 (2006)
73. De Lathauwer, L., Castaing, J.: Fourth-order cumulant-based blind identification of underdetermined mixtures. Trans. Signal Process. **55**(6), 2965–2973 (2007)
74. De Lathauwer, L., Castaing, J.: Blind identification of underdetermined mixtures by simultaneous matrix diagonalization. Trans. Signal Process. **56**(3), 1096–1105 (2008)
75. De Lathauwer, L., de Baynast, A.: Blind deconvolution of DS-CDMA signals by means of decomposition in rank-$(1, L, L)$ terms. IEEE Trans. Signal Process. **56**(4), 1562–1571 (2008)
76. De Lathauwer, L., Févotte, C., De Moor, B., Vandewalle, J.: Jacobi algorithm for joint block diagonalization in blind identification. In: Proceedings of 23rd Symposium on Information Theory in the Benelux, Louvain-la-Neuve, pp. 155–162, May 2002
77. De Lathauwer, L., De Moor, B., Vandewalle, J.: Computation of the canonical decomposition by means of a simultaneous generalized Schur decomposition. SIAM J. Matrix Anal. Appl. **26**(2), 295–327 (2004)
78. de Silva, V., Lim, L.H.: Tensor rank and the ill-posedness of the best low-rank approximation problem. SIAM J. Matrix Anal. Appl. **30**, 1084–1127 (2008)
79. Djoković, D.Ž.: Closures of equivalence classes of trivectors of an eight-dimensional complex vector space. Can. Math. Bull. **26**(1), 92–100 (1983)
80. Draisma, J., Kuttler, J.: Bounded-rank tensors are defined in bounded degree. Math. Sci. Net. ArXiv e-prints (March 2011)
81. Drton, M., Sturmfels, B., Sullivant, S.: Algebraic factor analysis: tetrads, pentads and beyond. Probab. Theory Relat. Fields **138**(3–4), 463–493 (2007)
82. Drton, M., Sturmfels, B., Sullivant, S.: Lectures on algebraic statistics. In: Oberwolfach Seminars 39, viii, 271 pp. Birkhäuser, Basel (2009)
83. Eisenbud, D.: Commutative Algebra with a View Toward Algebraic Geometry. Graduate Texts in Mathematics, vol. 150. Springer, New York (1995)
84. Eisenbud, D.: The Geometry of Syzygies, Second Course in Commutative Algebra and Algebraic Geometry. Graduate Texts in Mathematics, vol. 229. Springer, New York (2005)
85. Enriques, F., Chisini, O.: Lezioni sulla teoria geometrica delle equazioni e delle funzioni algebriche. Zanichelli, Bologna (1918)
86. Eriksson, N., Ranestad, K., Sturmfels, B., Sullivant, S.: Phylogenetic algebraic geometry. In: Projective Varieties with Unexpected Properties, pp. 237–255. Walter de Gruyter GmbH & Co. KG, Berlin (2005)
87. Erman, D., Velasco, M.: A syzygetic approach to the smoothability of zero-dimensional schemes. Adv. Math. **224**, 1143–1166 (2010)
88. Flenner, H., O'Carroll, L., Vogel, W.: Joins and Intersections. Springer Monographs in Mathematics. Springer, Berlin (1999)
89. Friedland, S.: On the generic and typical ranks of 3-tensors. Linear Algebra Appl. **436**(3), 478–497 (2012)
90. Friedland, S.: On tensors of border rank l in $\mathbb{C}^{m \times n \times l}$. Linear Algebra Appl. **438**(2), 713–737 (2013)

91. Friedland, S., Gross, E.: A proof of the set-theoretic version of the salmon conjecture. J. Algebra **356**, 374–379 (2012)
92. Fröberg, R., Ottaviani, G., Shapiro,B.: On the waring problem for polynomial rings. Proc. Natl. Acad. Sci. USA **109**(15), 5600–5602 (2012)
93. Fulton, W., Harris, J.: Representation Theory, a First Course. Graduate Texts in Mathematics, vol. 129. Springer, New York (1991)
94. Gantmacher, F.R.: Applications of the Theory of Matrices (Translated by J.L. Brenner, with the assistance of D.W. Bushaw and S. Evanusa). Interscience, New York (1959)
95. Garcia, L., Stillman, M., Sturmfels, B.: Algebraic geometry of Bayesian networks. J. Symb. Comput. **39**(3–4), 331–355 (2005)
96. Gel'fand, I.M., Zelevinskiĭ, A.V., Kapranov, M.M.: Projective-dual varieties and hyperdeterminants. Dokl. Akad. Nauk SSSR **305**(6), 1294–1298 (1989)
97. Gel'fand, I.M., Kapranov, M.M., Zelevinsky, A.V.: Hyperdeterminants. Adv. Math. **96**(2), 226–263 (1992)
98. Gel'fand, I.M., Kapranov, M.M., Zelevinsky, A.V.: Discriminants, Resultants, and Multidimensional Determinants. Mathematics: Theory & Applications. Birkhäuser, Boston (1994)
99. Geramita, A.V.: Inverse systems of fat points: Waring's problem, secant varieties of Veronese varieties and parameter spaces for Gorenstein ideals. In: The Curves Seminar at Queen's, Vol. X (Kingston, ON, 1995). Queen's Papers in Pure and Applied Mathematics, vol. 102, pp. 2–114. Queen's University, Kingston (1996)
100. Goodman, R., Wallach, N.: Representations and Invariants of the Classical Groups. Encyclopedia of Mathematics and its Applications, vol. 68. Cambridge University Press, Cambridge (1998)
101. Grayson, D.R., Stillman, M.E.: Macaulay2, a software system for research in algebraic geometry. http://www.math.uiuc.edu/Macaulay2/
102. Harris, J.: Algebraic Geometry: A First Course. Graduate Texts in Mathematics, vol. 133. Springer, New York (1992)
103. Hartshorne, R.: Algebraic Geometry. Graduate Texts in Mathematics, vol. 52. Springer, New York (1977)
104. Hauenstein, J.D., Sommese, A.J., Wampler, C.W.: Regeneration homotopies for solving systems of polynomials. Math. Comput. **80**, 345–377 (2011)
105. Hillar, C.J., Lim, L.H.: Most tensor problems are NP hard. Math. Sci. Net. ArXiv e-prints (November 2009)
106. Hu, S., Huang, Z.H., Ling, C., Qi, L.: On determinants and eigenvalue theory of tensors. J. Symb. Comput. **50**, 508–531 (2013)
107. Huggins, P., Sturmfels, B., Yu, J., Yuster, D.: The hyperdeterminant and triangulations of the 4-cube. Math. Comput. **77**(263), 1653–1679 (2008)
108. Iarrobino, A., Kanev, V.: Power Sums, Gorenstein Algebras, and Determinantal Loci. Lecture Notes in Mathematics, vol. 1721. Springer, Berlin (1999) (Appendix C by Iarrobino and Steven L. Kleiman)
109. Kanev, V.: Chordal varieties of Veronese varieties and catalecticant matrices. J. Math. Sci. (N.Y.) **94**(1), 1114–1125 (1999) (Algebraic geometry, 9)
110. Kolda, T.G., Mayo, J.R.: Shifted power method for computing tensor eigenpairs. SIAM J. Matrix Anal. Appl. **32**(4), 1095–1124 (2011)
111. Laface, A., Postinghel, E.: Secant varieties of Segre-Veronese embeddings of $(\mathbb{P}^1)^r$. Math. Ann. **356**(4), 1455–1470 (2013)
112. Landsberg, J.M.: The border rank of the multiplication of 2×2 matrices is seven. J. Am. Math. Soc. **19**(2), 447–459 (2006)
113. Landsberg, J.M.: Geometry and the complexity of matrix multiplication. Bull. Am. Math. Soc. (N.S.) **45**(2), 247–284 (2008)
114. Landsberg, J.M.: Explicit tensors of border rank at least 2n-1. ArXiv e-prints (September 2012)
115. Landsberg, J.M.: New lower bounds for the rank of matrix multiplication. Math. Sci. Net. ArXiv e-prints (June 2012)

116. Landsberg, J.M.: Tensors: Geometry and Applications. Graduate Studies in Mathematics, vol. 128. American Mathematical Society, Providence (2012)
117. Landsberg, J.M., Manivel, L.: On the ideals of secant varieties of Segre varieties. Found. Comput. Math. **4**(4), 397–422 (2004)
118. Landsberg, J.M., Manivel, L.: Generalizations of Strassen's equations for secant varieties of Segre varieties. Commun. Algebra **36**(2), 405–422 (2008)
119. Landsberg, J.M., Ottaviani, G.: New lower bounds for the border rank of matrix multiplication. ArXiv e-prints (December 2011)
120. Landsberg, J.M., Ottaviani, G.: Equations for secant varieties of veronese and other varieties. Ann. Math. Pura Appl. **4**, 1–38 (2011)
121. Landsberg, J.M., Teitler, Z.: On the ranks and border ranks of symmetric tensors. Found. Comput. Math. **10**(3), 339–366 (2010)
122. Landsberg, J.M., Weyman, J.: On tangential varieties of rational homogeneous varieties. J. Lond. Math. Soc. **76**(2), 513–530 (2007)
123. Landsberg, J.M., Weyman, J.: On the ideals and singularities of secant varieties of Segre varieties. Bull. Lond. Math. Soc. **39**(4), 685–697 (2007)
124. Landsberg, J.M., Weyman, J.: On secant varieties of compact Hermitian symmetric spaces. J. Pure Appl. Algebra **213**(11), 2075–2086 (2009)
125. Lee, T.L., Li, T.Y., Tsai, C.H.: HOM4PS-2.0: a software package for solving polynomials systems by the polyhedral homotopy continuation method. http://www.mth.msu.edu/~li/ (2010)
126. Li, T.Y.: Numerical solution of polynomial systems by homotopy continuation methods. In: Handbook of Numerical Analysis, Special Volume: Foundations of Computational Mathematics, vol. XI, pp. 209–304. North-Holland, Amsterdam (2003)
127. Lim, L.H.: Singular values and eigenvalues of tensors: a variational approach. In: 1st IEEE International Workshop on Computational Advances in Multi-Sensor Adaptive Processing, 2005, pp. 129–132, December 2005
128. Macaulay, F.S.: Some properties of enumeration in the theory of modular systems. Proc. Lond. Math. Soc. **S2-26**(1), 531–555 (1927)
129. Mella, M.: Singularities of linear systems and the Waring problem. Trans. Am. Math. Soc. **358**(12), 5523–5538 (2006)
130. Mella, M.: Base loci of linear systems and the Waring problem. Proc. Am. Math. Soc. **137**(1), 91–98 (2009)
131. Ni, G., Qi, L., Wang, F., Wang, Y.: The degree of the E-characteristic polynomial of an even order tensor. J. Math. Anal. Appl. **329**(2), 1218–1229 (2007)
132. Oeding, L.: Set-theoretic defining equations of the tangential variety of the Segre variety. J. Pure Appl. Algebra **215**(6), 1516–1527 (2011)
133. Oeding, L.: Hyperdeterminants of polynomials. Adv. Math. **231**(3–4), 1308–1326 (2012)
134. Oeding, L., Ottaviani, G.: Eigenvectors of tensors and algorithms for waring decomposition. J. Symb. Comput. **54**, 9–35 (2013)
135. Okonek, C., Schneider, M., Spindler, H.: Vector Bundles on Complex Projective Spaces. Progress in Mathematics, vol. 3. Birkhäuser, Boston (1980)
136. Ottaviani, G.: Symplectic bundles on the plane, secant varieties and Lüroth quartics revisited. In: Vector Bundles and Low Codimensional Subvarieties: State of the Art and Recent Developments. Quaderni di Matematica, vol. 21, pp. 315–352. Department of Mathematics, Seconda University of Napoli, Caserta (2007)
137. Ottaviani, G.: An invariant regarding Waring's problem for cubic polynomials. Nagoya Math. J. **193**, 95–110 (2009)
138. Ottaviani, G., Sturmfels, B.: Matrices with Eigenvectors in a Given Subspace. Proc. Am. Math. Soc. **141**, 1219–1232 (2013)
139. Pachter, L., Sturmfels, B. (eds.): Algebraic Statistics for Computational Biology. Cambridge University Press, New York (2005)
140. Postinghel, E.: A new proof of the Alexander-Hirschowitz interpolation theorem. Ann. Math. Pura Appl. **191**(1), 77–94 (2012)

141. Qi, L.: Eigenvalues of a real supersymmetric tensor. J. Symb. Comput. **40**(6), 1302–1324 (2005)
142. Raicu, C.: 3×3 Minors of Catalecticants. ArXiv e-prints (November 2010)
143. Raicu, C.: Secant varieties of Segre-Veronese varieties. ProQuest LLC, Ann Arbor, MI. Ph.D. thesis, University of California, Berkeley (2011)
144. Raicu, C.: Secant varieties of Segre–Veronese varieties. Algebra Number Theory **6–8**, 1817–1868 (2012)
145. Ranestad, K., Schreyer, F.O.: Varieties of sums of powers. J. Reine Angew. Math. **525**, 147–181 (2000)
146. Sidman, J., Sullivant, S.: Prolongations and computational algebra. Can. J. Math. **61**(4), 930–949 (2009)
147. Sidman, J., Vermeire, P.: Equations defining secant varieties: geometry and computation. In: Combinatorial Aspects of Commutative Algebra and Algebraic Geometry. Abel Symposium, vol. 6, pp. 155–174. Springer, Berlin (2011)
148. Sommese, A.J., Wampler, C.W.: Numerical Solution of Polynomial Systems Arising in Engineering and Science. World Scientific, Singapore (2005)
149. Sumi, T., Sakata, T., Miyazaki, M.: Typical ranks for $m \times n \times (m-1)n$ tensors with $m \leq n$. Linear Algebra Appl. **438**(2), 953–958 (2013)
150. Strassen, V.: Rank and optimal computation of generic tensors. Linear Algebra Appl. **52–53**, 645–685 (1983)
151. Sturmfels, B.: Open problems in algebraic statistics. In: Sullivant, S., Sturmfels, B. (eds.) Emerging Applications of Algebraic Geometry. The IMA Volumes in Mathematics and Its Applications, vol. 149, pp. 1–13. Springer, New York (2009)
152. Terracini, A.: Sulle V^k per cui la varietá degli sh (h+1)-secanti ha dimensione minore dell'ordinario. Rend. Circ. Mat. Palermo **I**, 392–396 (1911)
153. Trefethen, N.: The smart money is on numerical analysts. SIAM News **45**(9) (2012). see https://www.siam.org/news/news.php?id=2024
154. van Leeuwen, M.A.A., Coehn, A.M., Lisser, B.: LiE, A Package for Lie Group Computations. Computer Algebra Nederland, Amsterdam (1992)
155. Verschelde, J.: PHCpack: a general-purpose solver for polynomial systems by homotopy continuation. http://www.math.uic.edu/~jan/PHCpack/phcpack.html (2010)
156. Vinberg, È.B., Èlašvili, A.G.: A classification of the three-vectors of nine-dimensional space. Trudy Sem. Vektor. Tenzor. Anal. **18**, 197–233 (1978)
157. Weyl, H.: The Classical Groups. Princeton Landmarks in Mathematics. Princeton University Press, Princeton (1997) (Their invariants and representations, Fifteenth printing, Princeton Paperbacks)
158. Weyman, J.: Cohomology of Vector Bundles and Syzygies. Cambridge Tracts in Mathematics, vol. 149. Cambridge University Press, Cambridge (2003)
159. Weyman, J., Zelevinsky, A.: Multiplicative properties of projectively dual varieties. Manuscr. Math. **82**(2), 139–148 (1994)
160. Weyman, J., Zelevinsky, A.: Singularities of hyperdeterminants. Ann. Inst. Fourier (Grenoble) **46**(3), 591–644 (1996)
161. Williams, V.V.: Multiplying matrices faster than Coppersmith-Winograd. In: Karloff, H.J., Pitassi, T. (eds.) Symposium on Theory of Computing Conference, pp. 887–898. ACM, New York (2012)
162. Zak, F.L.: Tangents and Secants of Algebraic Varieties. Translations of Mathematical Monographs, vol. 127. American Mathematical Society, Providence (1993) (Translated from the Russian manuscript by the author)

Regina Lectures on Fat Points

Susan M. Cooper and Brian Harbourne

We dedicate these notes to Tony Geramita, a wonderful mentor, colleague and friend: the contagious joy he takes in life and in mathematics has been an inspiration for both of us.

Abstract These notes are a record of lectures given in the Workshop on Connections Between Algebra and Geometry at the University of Regina, May 29–June 1, 2012. The lectures were meant as an introduction to current research problems related to fat points for an audience that was not expected to have much background in commutative algebra or algebraic geometry (although Sects. 8 and 9 of these notes demand somewhat more background than earlier sections).

1 Motivation

Fat points are relevant to many areas of research. For example, one reason fat points are of interest in algebraic geometry is because of their connection to linear systems: one can identify the homogeneous components of ideals of fat points in $\mathbb{P}^n$ with the spaces of global sections of line bundles on blowings-up of $\mathbb{P}^n$ at given finite sets of points. Fat points also arise indirectly in other topics of study in algebraic geometry, such as the study of secant varieties [8]. In commutative algebra ideals of fat points give a useful class of test cases and suggest interesting questions that can be true more generally (see, for example, [43], where the authors give a

S.M. Cooper
Department of Mathematics, Central Michigan University, Mount Pleasant, MI 48859, USA
e-mail: s.cooper@cmich.edu

B. Harbourne (✉)
Department of Mathematics, University of Nebraska–Lincoln, Lincoln, NE 68588-0130, USA
e-mail: bharbour@math.unl.edu

S.M. Cooper and S. Sather-Wagstaff (eds.), *Connections Between Algebra, Combinatorics, and Geometry*, Springer Proceedings in Mathematics & Statistics 76,
DOI 10.1007/978-1-4939-0626-0_3,

conjecture for all nonreduced zero-dimensional schemes, and as evidence prove it for fat points). Fat points also arise in more applied situations, such as combinatorics and in interpolation problems [40, 46]. Regarding the latter, consider the following question.

Question 1.1. What can we say about a function $f \in \mathbf{C}[x_1, \dots, x_n]$ if all we know are values of f and certain of its partial derivatives at some finite set of points $p_1, \dots, p_s \in \mathbf{C}^n$? In particular:

1. What is the least degree among all f satisfying the given data?
2. How many such f are there up to some given degree t?
3. What is the smallest degree t guaranteed to have such an f, regardless of the choice of the points p_i? (For example, there is a linear f vanishing at three collinear points of the plane, but not at three noncollinear points, so the least degree t guaranteeing vanishing at three points in the plane without knowing the disposition of the points is $t = 2$.)

These are open problems when $n \geq 2$ even in the simplest case, where we specify points $p_1, \dots, p_s$, and an order of vanishing m_i at each point p_i, and ask to find all $f \in \mathbf{C}[x_1, \dots, x_n]$ such that $\mathrm{ord}_{p_i}(f) \geq m_i$ for all i, where, given a point p, $\mathrm{ord}_p(f) > 0$ just means $f(p) = 0$, $\mathrm{ord}_p(f) > 1$ means $f(p) = 0$ and $\frac{\partial f}{\partial x_i}(p) = 0$ for all i, $\mathrm{ord}_p(f) > 2$ means $f(p) = 0$, $\frac{\partial f}{\partial x_i}(p) = 0$ for all i, and $\frac{\partial^2 f}{\partial x_i \partial x_j}(p)(p) = 0$ for all i and j, and $\mathrm{ord}_p(f) > m$ just means $f(p) = 0$ and $\frac{\partial^k f}{\partial x_{i_1} \cdots \partial x_{i_k}}(p) = 0$ for all i_j with $k \leq m$.

Alternatively, one can think of $\mathrm{ord}_p(f)$ as the least degree of a term of f when expressed in coordinates centered at p. So, for example, if $p = (a_1, \dots, a_n)$, then let $X_i = x_i - a_i$, and substitute $x_i = X_i + a_i$ into f to get $g = f(X_1 + a_i, \dots, X_n + a_n)$. Then $\mathrm{ord}_p(f)$ is the least degree of a nonzero term of g, regarded as a polynomial in the X_i. This removes having to deal with partial derivatives, which can be problematic when working over arbitrary algebraically closed fields, i.e., when considering $f \in K[x_1, \dots, x_n]$.

To further algebraize the interpolation problem, we note that $\mathrm{ord}_p(f) \geq m$ if and only if $f \in I(p)^m$, where $I(p)$ is the ideal of all polynomials that vanish at p. Thus $\mathrm{ord}_{p_i}(f) \geq m_i$ for points $p_1, \dots, p_s$ and orders of vanishing m_i if and only if f is in the ideal $\cap I(p_i)^{m_i}$. It is convenient to use 0-cycle notation to specify the given data, so we write $Z = m_1 p_1 + \cdots + m_s p_s$, which we refer to as a fat point *scheme*, and we denote $\cap I(p_i)^{m_i}$ by $I(Z)$. (Readers familiar with schemes in the algebraic geometric sense can just regard Z as the subscheme of $\mathbb{A}^n_K$ defined by $I(Z) \subseteq K[x_1, \dots, x_n]$.)

Given $p_1 \dots, p_s \in K^n$ and nonnegative integers $m_1, \dots, m_s$, we have the ideal $I = I(m_1 p_1 + \cdots + m_s p_s) \subseteq K[x_1, \dots, x_n] = A$. Let $A_{\leq t}$ be the K-vector space span of all $f \in A$ with $\deg(f) \leq t$, and let $I_{\leq t} = I \cap A_{\leq t}$. Then we refer to the function $H_I^{\leq}(t) = \dim_K(I_{\leq t})$ as the Hilbert function of I. Also, given any ideal $0 \neq I \subseteq A$, define $\alpha(I)$ to be the minimum degree of nonzero elements of I.

(If $0 \neq J \subseteq R = K[x_0, \ldots, x_n]$ is a homogeneous ideal, then $\alpha(J)$ is in fact the degree of a nonzero homogeneous element of J of least degree.) We can now raise the following open problems:

Problem 1.1. Consider the following problems.

1. Find $\alpha(I)$.
2. Find the Hilbert function $H_I^{\leq}$ of I.
3. Find the maximum value of $\alpha(I(m_1q_1 + \cdots + m_sq_s))$ as the q_i range over all choices of s distinct points of $\mathbb{A}^n$. (The maximum occurs on a Zariski open subset of $(\mathbb{A}^n)^s$, so we can ask: What is the maximum value of $\alpha(I(m_1q_1+\cdots+m_sq_s))$ for general points $q_1, \ldots, q_s$?)

Example 1.2. All of the problems above are easy if $n = 1$, using the fact that $K[x_1]$ is a principal ideal domain. This is the case of Lagrange interpolation. In this case, a point p_i is just an element of K, so, for example, $I = I(m_1p_1 + \cdots + m_sp_s) = ((x_1-p_1)^{m_1} \cdots (x_1-p_s)^{m_s})$. The positions of the points p_i do not matter: we always have $\alpha(I) = m_1 + \cdots + m_s$ and $H_I^{\leq}(t) = \min(0, t + 1 - \sum_i m_i)$.

We end this introduction with an advisory to the reader. It is common to refer to the data $m_1p_1 + \cdots + m_sp_s$ as being points p_i with *multiplicities* m_i. This grows out of the universal terminology that a root of a polynomial in a single variable can be a multiple root; for example, $x = 1$ is a root of multiplicity 2 for $f(x) = x^2-2x+1$. This terminology is quite old (see [45,51], for example). More recently, commutative algebraists have used multiplicity to refer to what can also be called the degree of a fat point subscheme. In this sense, the multiplicity of mp for a point $p \in \mathbb{A}^n$ is $\binom{m+n-1}{n}$ (see [15, p. 66], for example). Regardless of priority, the term multiplicity has a multiplicity of well-established usage, so one should check what usage any given author is employing.

2 Affine Space and Projective Space

Let K be an algebraically closed field. For $n \geq 0$, let $\mathbb{A}^n$ denote K^n, and let $A = K[\mathbb{A}^n]$ denote $K[X_1, \ldots, X_n]$. We refer to $\mathbb{A}^n$ as *affine n-space*. For any subset $S \subseteq \mathbb{A}^n$, let $I(S) \subseteq A$ denote the ideal of all polynomials that vanish on S. (For those familiar with Spec, the *affine scheme* associated with S is $\mathrm{Spec}(A/I(S))$. Note that any ideal $I \subseteq A$ defines an affine subscheme of $\mathrm{Spec}(A)$, and ideals I and J define the same affine subscheme if and only if $I = J$.)

For $n \geq 0$, let $\mathbb{P}^n$ denote equivalence classes of nonzero $(n+1)$-tuples, where $(a_0, \ldots, a_n)$ and $(b_0, \ldots, b_n)$ are equivalent if there is a nonzero $t \in K$ such that $(a_0, \ldots, a_n) = t(b_0, \ldots, b_n)$. Let $R = K[\mathbb{P}^n]$ denote $K[x_0, \ldots, x_n]$. We refer to $\mathbb{P}^n$ as *projective n-space*. For any subset $S \subseteq \mathbb{P}^n$, we obtain an associated homogeneous ideal (i.e., an ideal generated by homogeneous polynomials, also called forms) $I(S) \subseteq R$, the ideal generated by all homogeneous polynomials that vanish on S, where we regard R as being a graded ring with each variable having degree 1 and constants having degree 0. For those familiar with Proj, the *projective scheme* associated to S is $\mathrm{Proj}(R/I(S))$. If $M = (x_0, \ldots, x_n)$, any homogeneous ideal $I \subseteq M \subset R$ defines a subscheme $\mathrm{Proj}(R/I) \subseteq \mathrm{Proj}(R) = \mathbb{P}^n$, and homogeneous ideals $I \subseteq M$ and $J \subseteq M$ define the same subscheme if and only if $I_t = J_t$ for $t \gg 0$ (or equivalently, if and only if $I \cap M^t = J \cap M^t$ for $t \gg 0$), where I_t and J_t are the homogeneous components of the ideals of degree t. (Thus I_t is the vector space span of the elements of I of degree t. This applies in particular to R, so R_t is the K-vector space span of the homogeneous polynomials in R of degree t, and we have $I_t = R_t \cap I$.) Given a homogeneous ideal I, among all homogeneous ideals J such that $I_t = J_t$ for $t \gg 0$ there is a largest such ideal contained in M which contains all of the others, called the *saturation* of I, denoted $\mathrm{sat}(I)$. Thus given homogeneous ideals $I \subseteq M$ and $J \subseteq M$, we have $\mathrm{Proj}(R/I) = \mathrm{Proj}(R/J)$ if and only if $\mathrm{sat}(I) = \mathrm{sat}(J)$. We say an ideal is *saturated* if it is equal to its saturation. Thus geometrically we are most interested in homogeneous ideals which are saturated, since projective schemes are in bijective correspondence with the saturated homogeneous ideals. (Indeed, readers uncomfortable with Proj can get by just thinking about saturated homogeneous ideals.)

We can regard $\mathbb{A}^n \subset \mathbb{P}^n$ via the inclusion $(a_1, \ldots, a_n) \mapsto (1, a_1, \ldots, a_n)$. We have an isomorphism of function fields

$$K(X_1, \ldots, X_n) = K(\mathbb{A}^n) \cong K(\mathbb{P}^n) = K(x_1/x_0, \ldots, x_n/x_0)$$

defined by $X_i \mapsto \frac{x_i}{x_0}$.

Remark 2.1. Some authors use $\mathbb{A}^n$ to denote $\mathrm{Spec}(K[x_1, \ldots, x_n])$. Since we are assuming K is algebraically closed, our usage is (by the Nullstellensatz) equivalent to taking $\mathbb{A}^n$ to be the set of closed points (i.e., of points corresponding to maximal ideals) of $\mathrm{Spec}(K[x_1, \ldots, x_n])$. Likewise, some authors use $\mathbb{P}^n$ to denote $\mathrm{Proj}(K[x_0, \ldots, x_n])$. In our definition, $\mathbb{P}^n$ denotes the set of closed points of $\mathrm{Proj}(K[x_0, \ldots, x_n])$.

As discussed in the previous section, we will denote the span of all polynomials of degree at most t by $A_{\leq t}$. Given an ideal $I \subseteq A$, let $I_{\leq t}$ denote $A_{\leq t} \cap I$, so $I_{\leq t}$ is the subspace of I spanned by all $f \in I$ of degree at most t. Given an ideal $I \subseteq A$, the *Hilbert function* of I is the function $H_I^{\leq}$ where $H_I^{\leq}(t) = \dim_K(I_{\leq t})$; i.e., $H_I^{\leq}(t)$ is the K-vector space dimension of the vector space spanned by all $f \in I$ with $\deg(f) \leq t$. The *Hilbert function* of A/I (or of the scheme $\mathrm{Spec}(A/I)$) is $H_{A/I}^{\leq}(t) = \dim_K(A_{\leq t}/I_{\leq t}) = \binom{n+t}{n} - H_I^{\leq}(t)$. Given a homogeneous ideal $I \subseteq R$, the *Hilbert function* H_I of I is the function $H_I(t) = \dim_K(I_t)$; i.e., $H_I(t)$ is the K-vector space dimension of the vector space spanned by all homogeneous $f \in I$ with $\deg(f) = t$. The *Hilbert function* of R/I (or of the scheme $\mathrm{Proj}(R/I)$) is $H_{R/I}(t) = \dim_K(R_t/I_t) = \binom{t+n}{n} - H_I(t)$.

It is known that $H_I^{\leq}$ and $H_{A/I}^{\leq}$ become polynomials for $t \gg 0$ (see Exercise 3.8 for an example). This polynomial is called the *Hilbert polynomial* of I or A/I, respectively. (We will see in the next section that the Hilbert polynomial for the ideal I of the fat point subscheme $m_1p_1 + \cdots + m_rp_r$ is $\binom{t+n}{n} - \sum_i \binom{m_i+n-1}{n}$. Similarly, $\sum_i \binom{m_i+n-1}{n}$ is the Hilbert polynomial for A/I.) Likewise, if $I \subseteq R$ is a homogeneous ideal, H_I and $H_{R/I}$ become polynomials for $t \gg 0$, called the *Hilbert polynomial* of I or R/I as the case may be. Note that $H_I^{\leq}(t) = H_A^{\leq}(t) - H_{A/I}^{\leq}(t) = \binom{t+n}{n} - H_{A/I}^{\leq}(t)$ for all $t \geq 0$. Using Exercise 2.1 we also see that $H_I(t) = H_R(t) - H_{R/I}(t) = \binom{t+n}{n} - H_{R/I}(t)$ for all $t \geq 0$.

It is a significant and often difficult problem to determine the least value i such that the Hilbert polynomial and Hilbert function become equal for all $t \geq i$. (For an ideal of fat points, this value is sometimes called the *regularity index* of I, and $i + 1$ in the case of an ideal of fat points is known as the *Castelnuovo–Mumford regularity* $\operatorname{reg}(I)$ of I.)

Exercises

Exercise 2.1. Show that there is a bijection between the set $\mathcal{M}_{\leq t}(A)$ of monomials of degree at most t in $A = K[x_1, \ldots, x_n]$ and the set $\mathcal{M}_t(R)$ of monomials of degree exactly t in $R = K[x_0, \ldots, x_n]$ for every $t \geq 0$. (This shows that $H_A^{\leq}(t) = H_R(t)$ for all $t \geq 0$.)

Exercise 2.2. If $0 \neq I \subseteq A$ is an ideal, show that $\alpha(I^m) \leq m\alpha(I)$, but if $0 \neq J \subseteq R$ is homogeneous, then $\alpha(J^m) = m\alpha(J)$. (See Exercise 3.6 for an example where equality in $\alpha(I^m) \leq m\alpha(I)$ fails.)

Exercise 2.3. Let $I \subseteq M \subset R$ be a homogeneous ideal. Let P be the ideal generated by all homogeneous $f \in R$ such that $fM^i \subseteq I$ for some $i > 0$. Show that $I \subseteq P$, that P contains every homogeneous ideal $J \subseteq M$ such that $I_t = J_t$ for $t \gg 0$, and that $I_t = P_t$ for $t \gg 0$. Conclude that P is the saturation of I and that $P = \operatorname{sat}(P)$. (In terms of colon ideals, $\operatorname{sat}(I) = \cup_{i \geq 1} I : M^i$.)

3 Fat Points in Affine Space

A *fat point* subscheme of affine n-space is the scheme corresponding to an ideal of the form $I = \cap_{i=1}^r I(p_i)^{m_i} \subset A$ for a finite set of points $p_1, \ldots, p_r \in \mathbb{A}^n$ and positive integers m_i. We denote $\operatorname{Spec}(A/I)$ in this case by $m_1p_1 + \cdots + m_rp_r$, and we denote the ideal $\cap_{i=1}^r I(p_i)^{m_i}$ by $I(m_1p_1 + \cdots + m_rp_r)$.

Given distinct points $p_1, \ldots, p_r \in \mathbb{A}^n$, let $I = \cap_{i=1}^r I(p_i)$; following Waldschmidt [52] we define a constant we denote by $\gamma(I)$ as the following limit

$$\gamma(I) = \lim_{m \to \infty} \frac{\alpha(\cap_{i=1}^r (I(p_i)^m))}{m}.$$

By Exercise 3.1, $\cap_{i=1}^r(I(p_i)^m) = I^m$, so

$$\gamma(I) = \lim_{m\to\infty} \frac{\alpha(I^m)}{m},$$

but for a unified treatment, whether the points p_i are in affine space or projective space, it is better to take

$$\gamma(I) = \lim_{m\to\infty} \frac{\alpha(\cap_{i=1}^r(I(p_i)^m))}{m}$$

as the definition of $\gamma(I)$.

We say the points $p_1, \ldots, p_r \in \mathbb{A}^n$ are *generic* points if the coordinates of the points are algebraically independent over the prime field Π_K of K. (This is possible only if the transcendence degree of K over Π_K is at least rn.) The following problem is open for $n > 1$ and $r \gg 0$.

Problem 3.1. Let I be the ideal of r generic points of $\mathbb{A}^n$. Determine $\gamma(I)$.

There is a conjectural solution to the problem above, when $r \gg 0$, due to Nagata [44] for $n = 2$ and Iarrobino [40] for $n > 2$:

Conjecture 3.2 (Nagata/Iarrobino Conjecture). Let I be the ideal of $r \gg 0$ generic points of $\mathbb{A}^n$. Then $\gamma(I) = \sqrt[n]{r}$ for $r \gg 0$.

Remark 3.3. The value of $\gamma(I)$ is known for r generic points of $\mathbb{A}^2$ for $1 \le r \le 9$ (see, for example, [9, Appendix 1] and [45, Theorem 7]) or when r is a square [44]. In particular, $\gamma(I) = 1$ if $r = 1, 2$, while $\gamma(I) = 3/2$ if $r = 3$, $\gamma(I) = 2$ if $r = 4, 5$, $\gamma(I) = 12/5$ if $r = 6$, $\gamma(I) = 21/8$ if $r = 7$, $\gamma(I) = 48/17$ if $r = 8$, and $\gamma(I) = \sqrt{r}$ if $r \ge 9$ is a square. Moreover, when $n > 2$ and $\sqrt[n]{r}$ is an integer, then again $\gamma(I) = \sqrt[n]{r}$ (see [17, Theorem 6]).

We will for now just verify that the values given in Remark 3.3 are upper bounds. By Exercise 3.3, the Hilbert polynomial of the ideal of a fat point subscheme $m_1p_1 + \cdots + m_rp_r \subset \mathbb{A}^n$ is $\binom{t+n}{n} - \sum_i \binom{m_i+n-1}{n}$, and so $\sum_i \binom{m_i+n-1}{n}$ is the *Hilbert polynomial* for A/I or equivalently for the scheme $m_1p_1 + \cdots + m_rp_r$.

Proposition 3.4. *Consider the ideal I of r distinct points of $\mathbb{A}^n$. Then $\gamma(I) \le \sqrt[n]{r}$. Moreover, when $n = 2$, we have: $\gamma(I) = 1$ if $r = 1, 2$; $\gamma(I) \le 3/2$ if $r = 3$; $\gamma(I) \le 2$ if $r = 4, 5$; $\gamma(I) \le 12/5$ if $r = 6$; $\gamma(I) \le 21/8$ if $r = 7$; and $\gamma(I) \le 48/17$ if $r = 8$.*

Proof. For $\gamma(I) \le \sqrt[n]{r}$, see Exercise 3.9. Now let $n = 2$. Say $r = 1$. Then by Exercise 3.5, $H^{\le}_{I^m}(t) = 0$ for $t < m$ (so $\alpha(I^m) \ge m$) and clearly I^m has elements of degree m (so $\alpha(I^m) \le m$), hence $\alpha(I^m) = m$. Thus $\gamma(I) = 1$ by definition.

Now let $r = 2$; let p_1 and p_2 be the $r = 2$ points. Then $I^m \subseteq I(p_1)^m$, so $\alpha(I(p_1)^m) \le \alpha(I^m)$, hence $1 = \gamma(I(p_1)) \le \gamma(I^m)$, but again I^m clearly has elements of degree m (take the mth power of the linear polynomial defining the line through p_1 and p_2), so $\alpha(I^m) \le m$, hence $\gamma(I) \le 1$ so we have $\gamma(I) = 1$.

Now let $r = 3$. If the points p_1, p_2, p_3 are collinear, then as for two points we have $\gamma(I) = 1$. Otherwise, consider the cubic polynomial $L_{12}L_{13}L_{23}$ defining the union of the three lines L_{ij} through pairs $\{p_i, p_j\}$ of the $r = 3$ points. But $L_{ij} \in I(p_i) \cap I(p_j)$ and $I(p_i) \cap I(p_j) = I(p_i)I(p_j)$ by Exercise 3.1, so $L_{12}L_{13}L_{23} \in (I(p_1)I(p_2))(I(p_1)I(p_3))(I(p_2)I(p_3)) = (I(p_1)I(p_2)I(p_3))^2$, which (again by Exercise 3.1) is I^2. Thus $L_{12}L_{13}L_{23}$ is in I^2 and has degree 3, so Exercise 3.2(c) shows that $\gamma(I) \leq \alpha(I^2)/2 \leq 3/2$.

For $r = 4$, it's easy to see that $\alpha(I) \leq 2$, so $\gamma(I) \leq \alpha(I)/1 \leq 2$.

For $r = 5$, $H_I^{\leq}(2) \geq \binom{2+2}{2} - 5\binom{1+2-1}{2} = 1$, so $\alpha(I) \leq 2$ and $\gamma(I) \leq \alpha(I)/1 \leq 2$.

For $r = 6$, through every subset of 5 of the 6 points there is (as we just saw) a conic, hence I^5 contains a nonzero polynomial of degree 12 (coming from the conics through the 6 subsets of 5 of the 6 points), so $\alpha(I^5) \leq 12$ and $\gamma(I) \leq \alpha(I^5)/5 \leq 12/5$.

For $r = 7$, there is a cubic which has a point of multiplicity at least 2 at any one of the points and multiplicity at least 1 at the other 6 points, since $H_I^{\leq}(3) \geq \binom{3+2}{2} - \binom{2+2-1}{2} - 6\binom{1+2-1}{2} = 1$. Multiplying together the seven cubics (one having a point of multiplicity at least 2 at the first point, the next having a point of multiplicity 2 at the second point, etc.) gives a polynomial of degree 21 having multiplicity at least 8 at each of the points, so $\gamma(I) \leq \alpha(I^8)/8 \leq 21/8$.

For $r = 8$, there is a sextic which has a point of multiplicity at least 3 at any one of the points and multiplicity at least 2 at the other 7 points, since $H_I^{\leq}(6) \geq \binom{6+2}{2} - \binom{3+2-1}{2} - 7\binom{2+2-1}{2} = 1$. Multiplying together the eight sextics gives a polynomial of degree 48 having multiplicity at least 17 at each of the points, so $\gamma(I) \leq \alpha(I^{17})/17 \leq 48/17$.

We will see in Sect. 7 and its exercises and Sect. 8 why equality holds above for $r < 9$ when $n = 2$ if the points are sufficiently general.

Exercises

Exercise 3.1. Let $p_1, \ldots, p_r$ be distinct points of $\mathbb{A}^n$. Show that $\cap_{i=1}^r I(p_i)^{m_i} = I(p_1)^{m_1} \cdots I(p_r)^{m_r}$.

Exercise 3.2. [Waldschmidt's Constant, [52, 53]] Let $p_1, \ldots, p_r$ be distinct points of $\mathbb{A}^n$ and let $I = \cap_{i=1}^r I(p_i)$. Let b and c be positive integers.

(a) Show that

$$\frac{\alpha(I^{bc})}{bc} \leq \frac{\alpha(I^b)}{b}.$$

(b) Show that

$$\lim_{m\to\infty} \frac{\alpha(I^{m!})}{m!}$$

exists.

(c) Show that

$$\lim_{m\to\infty}\frac{\alpha(I^m)}{m}$$

exists is equal to the limit given in (b) and satisfies

$$\lim_{m\to\infty}\frac{\alpha(I^m)}{m}\leq\frac{\alpha(I^t)}{t}$$

for all $t \geq 1$.

Exercise 3.3. Show that the K-vector space dimension of $A_{\leq t}$ is $\dim_K(A_{\leq t}) = \binom{t+n}{n}$.

Exercise 3.4. Show that there are $\binom{t+n}{n}$ monomials of degree t in $n+1$ variables.

Exercise 3.5. Let I be the ideal of the point $p = (a_1, \ldots, a_n) \in \mathbb{A}^n$. Show that $H^{\leq}_{I^m}(t) \geq \binom{t+n}{n} - \binom{m+n-1}{n}$, with equality for $t \geq m-1$.

Exercise 3.6. Let p_1, p_2, p_3 be distinct noncollinear points of $\mathbb{A}^2$. If $I = I(p_1) \cap I(p_2)$, show that $\alpha(I^m) = m\alpha(I)$. If $J = I(p_1) \cap I(p_2) \cap I(p_3)$ and $m > 1$, show that $\alpha(J^m) < m\alpha(J)$.

Exercise 3.7. Let $I \subseteq A$ be an ideal. Show that $H^{\leq}_{A/I}$ is nondecreasing.

The following exercise is a version of the Chinese Remainder Theorem.

Exercise 3.8. Let I be the ideal of $m_1 p_1 + \cdots + m_r p_r$ for r distinct points $p_i \in \mathbb{A}^n$. Show that $H^{\leq}_I(t) \geq \binom{t+n}{n} - \sum_i \binom{m_i+n-1}{n}$, with equality if $t \gg 0$.

Exercise 3.9. Let I be the ideal of r distinct points of $\mathbb{A}^n$. Show that $\gamma(I) \leq \sqrt[n]{r}$. If $1 \leq r \leq n$, show that $\gamma(I) = 1$.

Exercise 3.10. If $s \geq 9$ and $n = 2$, show that $\inf\{\frac{t}{m} : \binom{t+n}{n} - s\binom{m+n-1}{n} > 0; m, t \geq 1\} = \sqrt[n]{s}$. (The same fact is true for $n > 2$ with $s \gg 0$ replacing $s \geq 9$. This is part of the motivation for the Conjecture 3.2.)

Exercise 3.11. Let $p \in \mathbb{A}^n$ and let $m > 0$. Show that every element $\overline{f} \in A/(I(p))^m$ is the image of a polynomial $f \in A$ of degree at most $m-1$, and that $\overline{f}$ is a unit if and only if $f(p) \neq 0$.

Exercise 3.12. For any nonzero element $f \in K[\mathbb{A}^n]$, show there exists a point $p \in \mathbb{A}^n$ such that $f(p) \neq 0$.

Exercise 3.13. Let $n \geq 1$ and let $p_1, \ldots, p_r$ be distinct points of $\mathbb{A}^n$. Show that there is a linear form $f \in K[\mathbb{A}^n]$ such that $f(p_i) \neq f(p_j)$ whenever $p_i \neq p_j$.

Here is a more explicit version of Exercise 3.8, one solution of which applies Exercises 3.11–3.13.

Exercise 3.14. Let I be the ideal of $m_1p_1+\cdots+m_rp_r$ for r distinct points $p_i \in \mathbb{A}^n$. Show that $H_I^{\leq}(t) = \binom{t+n}{n} - \sum_i \binom{m_i+n-1}{n}$ if $t \geq m_1 + \cdots + m_r - 1$. If the points are collinear, show that $H_I^{\leq}(t) > \binom{t+n}{n} - \sum_i \binom{m_i+n-1}{n}$ if $t < m_1 + \cdots + m_r - 1$.

4 Fat Points in Projective Space

A *fat point* subscheme of projective n-space is the scheme corresponding to an ideal of the form $I = \cap_{i=1}^r I(p_i)^{m_i} \subset R$ for a finite set of distinct points $p_1, \ldots, p_r \in \mathbb{P}^n$ and positive integers m_i. We again denote the subscheme defined by I by $m_1p_1 + \cdots + m_rp_r$ (in this case the subscheme is $\mathrm{Proj}(R/I)$), and we denote the ideal $\cap_{i=1}^r I(p_i)^{m_i}$ by $I(m_1p_1 + \cdots + m_rp_r)$.

Remark 4.1. If $p_1, \ldots, p_r \subset \mathbb{A}^n \subset \mathbb{P}^n$, then there is no ambiguity in the notation $m_1p_1+\cdots+m_rp_r$, since there is a canonical isomorphism from $m_1p_1+\cdots+m_rp_r$ regarded as a subscheme of $\mathbb{A}^n$ and $m_1p_1+\cdots+m_rp_r$ regarded as a subscheme of $\mathbb{P}^n$. However, there is ambiguity in the notation $I(m_1p_1 + \cdots + m_rp_r)$, so we will sometimes use $I_A(m_1p_1 + \cdots + m_rp_r)$ to denote the ideal in A and $I_R(m_1p_1 + \cdots + m_rp_r)$ to denote the homogeneous ideal in R of $m_1p_1 + \cdots + m_rp_r$.

Remark 4.2. If $I_R = \cap_{i=1}^r I_R(p_i)$, it can sometimes happen that $I_R^m = \cap_{i=1}^r (I_R(p_i)^m)$, but $I_R(p_1)^{m_1} \cdots I_R(p_r)^{m_r} = \cap_{i=1}^r I_R(p_i)^{m_i}$ essentially never happens (see Exercise 4.1), and in general the most one can say about I_R^m is that $I_R^m \subseteq \cap_{i=1}^r (I_R(p_i)^m)$. Thus, we define the mth *symbolic* power $I_R^{(m)}$ of $I_R = \cap_{i=1}^r I_R(p_i)$ to be $I_R^{(m)} = \cap_{i=1}^r (I_R(p_i)^m)$. One can see the difference between the ideals $I_R(p_1)^{m_1} \cdots I_R(p_r)^{m_r}$ and $\cap_{i=1}^r I_R(p_i)^{m_i}$ and between I_R^m and $I_R^{(m)}$ by looking at primary decompositions. The intersection $\cap_{i=1}^r (I_R(p_i)^m)$ is the primary decomposition of $I_R^{(m)}$, but I_R^m has a primary decomposition of the form $I_R^{(m)} \cap J$ where J is M-primary (possibly $J = M$, in which case we have $I_R^m = I_R^{(m)} \cap M = I_R^{(m)}$), M being the irrelevant ideal (the ideal generated by the coordinate variables in $K[\mathbb{P}^n]$). Similarly, the primary decomposition of $I_R(p_1)^{m_1} \cdots I_R(p_r)^{m_r}$ also has the form $I_R^{(m)} \cap J$ where J is M-primary. In any case, we see that $I_R^m \subseteq I_R^{(m)}$ for all $m \geq 1$. We also see that $(I_R^m)_t = (I_R^{(m)})_t$ for $t \gg 0$, since for large t, any M-primary ideal J contains M^t and thus has $J_t = M_t$.

By Exercise 4.7, we have $I^r \subseteq I^{(m)}$ if and only if $r \geq m$. However, it is a hard problem to determine for which m and r we have $I^{(m)} \subseteq I^r$. See, for example, [11, 14, 32, 39] and the references therein.

Problem 4.3. Let $p_1, \ldots, p_s \in \mathbb{P}^n$ be distinct points. Let $I = I_R(p_1 + \cdots + p_s)$. Is it true that $I^{(ns-n+1)} \subseteq I^s$ for all $s \geq 1$? In particular, is it true that $I^{(3)} \subseteq I^2$ always holds when $n = 2$?

Remark 4.4. Problem 4.3 was open when the course these notes are based on was given in 2012. The situation changed shortly thereafter. An example with

$I^{(3)} \not\subseteq I^2$ was posted to the arXiv early in 2013 [13]. This inspired another example [5, Remark 3.11]; see also [34] for further discussion. Thus the problem now seems to be to classify the configurations of points in the plane for which we have $I^{(3)} \not\subseteq I^2$. So far, they seem to be quite rare.

Let $\delta_t : R_t \to A_{\leq t}$ be the map defined for any $F \in R_t$ by $\delta_t(F) = F(1, X_1, \ldots, X_n)$ and let $\eta_t : A_{\leq t} \to R_t$ be the map $\eta_t(f) = x_0^t f(x_1/x_0, \ldots, x_n/x_0)$. Note that these are K-linear maps, each being the inverse of the other. In particular, $\dim(R_t) = \dim(A_{\leq t}) = \binom{t+n}{n}$.

If $p \in \mathbb{A}^n \subset \mathbb{P}^n$, so $p = (a_1, \ldots, a_n) \in \mathbb{A}^n$ and can be represented in projective coordinates by $p = (1, a_1, \ldots, a_n) \in \mathbb{P}^n$, let $I = (X_1 - a_1, \ldots, X_n - a_n)$ be the ideal of p in A and let $J = (x_1 - a_1x_0, \ldots, x_n - a_nx_0)$ be the ideal of $p \in \mathbb{P}^n$ in R. Then $\eta_t((I^m)_{\leq t}) \subseteq (J^m)_t$ and $\delta_t((J^m)_t) \subseteq (I^m)_{\leq t}$, so we have K-linear vector space isomorphisms $(I^m)_{\leq t} \to (J^m)_t$ given by η_t, hence $H^{\leq}_{I^m}(t) = H_{J^m}(t)$ and $H^{\leq}_{A/I^m}(t) = H_{R/J^m}(t)$ for all t. Similarly, if $p_1, \ldots, p_r \in \mathbb{A}^n \subset \mathbb{P}^n$, and if $I = I_A(m_1p_1 + \cdots + m_rp_r) \subset A$ and $J = I_R(m_1p_1 + \cdots + m_rp_r) \subset R$, then again we have K-linear isomorphisms $I_{\leq t} \to J_t$ given by η_t, hence

$$H^{\leq}_I(t) = H_J(t) \text{ and } H^{\leq}_{A/I}(t) = H_{R/J}(t) \tag{1}$$

for all t. Hence the Hilbert functions and Hilbert polynomials for $m_1p_1 + \cdots + m_rp_r$ are the same whether we regard them as affine or projective subschemes. In particular, if $p_1, \ldots, p_r \subset \mathbb{A}^n \subset \mathbb{P}^n$ and if $I_A = I_A(p_1 + \cdots + p_r)$ and $I_R = I_R(p_1 + \cdots + p_r)$, then $\alpha(I_R^{(m)}) = \alpha(I_A^m)$ for all $m \geq 1$ and $\gamma(I_A) = \gamma(I_R)$. By Exercise 3.8, we also have $H_{I_R}(t) \geq \binom{t+n}{n} - \sum_i \binom{m_i+n-1}{n}$ and hence clearly

$$H_{I_R}(t) \geq \max\left\{\binom{t+n}{n} - \sum_i \binom{m_i+n-1}{n}, 0\right\}.$$

This is an equality for $t \gg 0$. There is a conjecture, known as the SHGH Conjecture, that gives a conjectural value for $H_{I_R}(t)$ when $n = 2$ and the points p_i are generic. Here is a simple to state special case of the SHGH Conjecture, named for various people who published what turns out to be equivalent conjectures: Segre [48] in 1961, Harbourne [28] in 1986, Gimigliano [22] in 1987 (also see [23]) and Hirschowitz [38] in 1989.

Conjecture 4.5 (SHGH Conjecture (Special Case)). Given $r \geq 9$ generic points $p_i \in \mathbb{P}^2$ and any nonnegative integers m and t, let $I = I_R(m(p_1 + \cdots + p_r))$. Then

$$H_I(t) = \max\left\{\binom{t+2}{2} - r\binom{m+1}{2}, 0\right\}.$$

There has been a lot of work done on this conjecture (see, for example, [2,10,33], but there are many more papers than this). The SHGH Conjecture is, however, only a starting point: one might also want to know the graded Betti numbers for a minimal free resolution. There are conjectures and results here too, mostly for $\mathbb{P}^2$. See, for example, [30] for some conjectures, and [3,7,18,24,25,29,35,41] for various results.

Most questions about fat points can be studied either from the point of view of subschemes of affine space or of subschemes of projective space. It can be more convenient to work with homogeneous ideals, so we will focus on the latter point of view.

We now mention some bounds on $\gamma(I)$ for an ideal $I = I_R(p_1 + \cdots + p_r)$ of distinct points $p_i \in \mathbb{P}^n$. Waldschmidt and Skoda [49, 52, 53] showed that $\gamma(I) \geq \frac{\alpha(I^{(m)})}{m+n-1}$ holds over the complex numbers for all positive integers m, and in particular that $\gamma(I) \geq \frac{\alpha(I)}{n}$. The proof involved some hard complex analysis. Easier and more general proofs which hold for any field K in any characteristic can be given using recent results on containments of symbolic powers in ordinary powers of I: we know by [14, 39] that $I^{(nm)} \subseteq I^m$ holds for all $m \geq 1$. Thus $m\alpha(I) = \alpha(I^m) \leq \alpha(I^{(nm)})$, so dividing by mn and taking the limit as $m \to \infty$ gives

$$\frac{\alpha(I)}{n} \leq \gamma(I).$$

(See [47] for a different specifically characteristic $p > 0$ argument.)

Chudnovsky [9] showed $\frac{\alpha(I)+1}{2} \leq \gamma(I)$ in case $n = 2$ and conjectured $\frac{\alpha(I)+n-1}{n} \leq \gamma(I)$ in general; this conjecture is still open. By Exercise 4.6 we know

$$\frac{\alpha(I^{(m)})}{n+m-1} \leq \gamma(I).$$

Esnault and Viehweg [16] obtained $\frac{\alpha(I^{(m)})+1}{m+n-1} \leq \gamma(I)$ in characteristic 0. It seems reasonable to extend Chudnovsky's conjecture [32, Question 4.2.1]:

Conjecture 4.6. For an ideal $I = I_R(p_1 + \cdots + p_r)$ of distinct points $p_i \in \mathbb{P}^n$ and for all $m \geq 1$,

$$\frac{\alpha(I^{(m)}) + n - 1}{n + m - 1} \leq \gamma(I).$$

If this conjecture is correct, it is sharp, since there are configurations of points (so-called star configurations) for which equality holds (apply [4, Lemma 8.4.7] with $j = 1$).

Exercises

Exercise 4.1. Given $r > 1$ and distinct points $p_1, \ldots, p_r \in \mathbb{P}^n$ with $m_i > 0$ for all i, show that $I(p_1)^{m_1} \cdots I(p_r)^{m_r} \subsetneq \cap_{i=1}^r I(p_i)^{m_i}$.

Exercise 4.2. Let $p_1, \ldots, p_r \in \mathbb{P}^n$ be distinct points. Let $I = I_R = I(m_1p_1 + \cdots + m_rp_r) \subset R$. Show that multiplication by a linear form F that does not vanish at any of the points p_i induces injective vector space homomorphisms $R_t/I_t \to R_{t+1}/I_{t+1}$. Conclude that $H_{R/I}$ is a nondecreasing function of t.

Exercise 4.3. Let $p_1, \ldots, p_r \in \mathbb{P}^n$ be distinct points. Let $I = I_R = I(m_1p_1 + \cdots + m_rp_r) \subset R$. Show that $H_{R/I}(t)$ is strictly increasing until it becomes constant (i.e., if c is the least t such that $H_{R/I}(c) = H_{R/I}(c+1)$, show that $H_{R/I}(t)$ is a strictly increasing function for $0 \le t \le c$, and that $H_{R/I}(t) = H_{R/I}(c)$ for all $t \ge c$).

Exercise 4.4. Give an example of a monomial ideal $J \subset K[x, y]$ such that $H_{R/J}$ is eventually constant but is not nondecreasing.

Exercise 4.5. Show that Conjecture 4.5 implies the $n = 2$ case of Conjecture 3.2.

Exercise 4.6. If $I \subset R$ is the radical ideal of a finite set of points in $\mathbb{P}^n$, then $I^{((m-1+n)t)} \subseteq (I^{(m)})^t$ [14, 39]. Use this to show

$$\frac{\alpha(I^{(m)})}{n+m-1} \le \gamma(I).$$

Exercise 4.7. Let $r, m \ge 1$. If $I = I(p_1 + \cdots + p_s) \subset R$ is the radical ideal of a finite set of distinct points $p_i \in \mathbb{P}^n$, show $I^r \subseteq I^{(m)}$ if and only if $r \ge m$.

5 Examples: Bounds on the Hilbert Function of Fat Point Subschemes of $\mathbb{P}^2$

Let $p_1, \ldots, p_r \in \mathbb{P}^2$ be distinct points. Let $m_1, \ldots, m_r$ be positive integers. Let $L_0, \ldots, L_{s-1}$ be lines, repeats allowed, such that every point p_i is on at least m_i of the lines L_j. Let $Z_0 = Z = m_1p_1 + \cdots + m_rp_r$. Define Z_{j+1}, for $j = 0, \ldots, s-1$, recursively as follows. We set $m_{i0} = m_i$ for all i and $Z_j = m_{1j}p_1 + \cdots + m_{rj}p_r$. Then $Z_{j+1} = m_{1\,j+1}p_1 + \cdots + m_{r\,j+1}p_r$ where $m_{i\,j+1} = m_{ij}$ if $p_i \notin L_j$, $m_{i\,j+1} = 0$ if $m_{ij} = 0$, and $m_{i\,j+1} = m_{ij} - 1$ if $p_i \in L_j$ and $m_{ij} > 0$. We get a sequence of fat point subschemes $Z = Z_0 \supseteq Z_1 \supseteq \cdots \supseteq Z_s = \varnothing$. Geometrically, Z_{j+1} is the fat point subscheme residual to Z_j with respect to the line L_j. Algebraically, $I(Z_{j+1}) = I(Z_j) : (F_j)$, where F_j is the form defining the line L_j.

Define a reduction vector $\mathbf{d} = (d_0, \ldots, d_{s-1})$, where $d_j = \sum_{p_i \in L_j} m_{i\,j-1}$, so d_j is the sum of the multiplicities $m_{i\,j-1}$ for points $p_i \in L_j$. From the reduction vector we construct a new vector, diag($\mathbf{d}$). The entries of diag($\mathbf{d}$) are obtained as follows. Make an arrangement of dots in s rows, the first row at the bottom, the next row above it (aligned at the left), and so on, one row for each entry of $\mathbf{d}$, where the number of dots in each row is given by the corresponding entry of $\mathbf{d}$ and where the dots are placed at integer lattice points. The entries of diag($\mathbf{d}$) are obtained by counting the number of dots on each diagonal (of slope -1). Fig. 1 is Example 2.5.5 of [11], where $\mathbf{d} = (8, 6, 5, 2)$ and diag($\mathbf{d}$) $= (1, 2, 3, 4, 4, 3, 3, 1, 0, 0, \ldots)$.

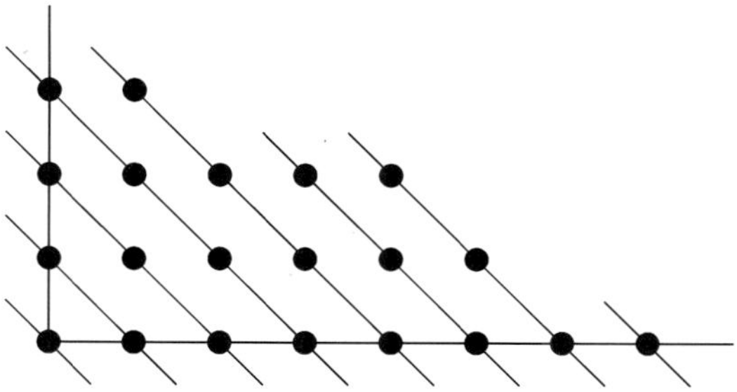

Fig. 1 Obtaining diag(**d**) from a reduction vector **d**

Theorem 5.1 ([11, Theorem 1.1]). *Let* **d** *be the reduction vector for a fat point scheme $Z \subset \mathbb{P}^2$ with respect to a given choice of lines L_i, and let v_{t+1} be the sum of the first $t+1$ entries of* diag(**d**)*. Then $H_{R/I(Z)}(t) \geq v_{t+1}$, and equality holds for all t if the entries of* **d** *are strictly decreasing.*

For example, choose distinct lines L_0, L_1, L_2, and L_3. Now choose any eight points on L_0 (possibly including points of intersection of L_0 with the other lines), then any six additional points on L_1 (again possibly including points of intersection of L_1 with the other lines but avoiding points already chosen, so now we have 14 distinct points), five on L_2 (possibly including points of intersection of L_2 with the other lines but avoiding points already chosen, so now we have 19 distinct points), and two on L_3 (as before possibly including points of intersection of L_3 with the other lines but avoiding points already chosen, so we end up with 21 distinct points). Then Z_0 is the reduced scheme consisting of all 21 points; removing the first 8 gives Z_1, removing from Z_1 the next 6 gives Z_2, removing from Z_2 the next 5 gives Z_3, and removing the last 2 gives $Z_4 = \varnothing$. The corresponding reduction vector is $\mathbf{d} = (8, 6, 5, 2)$, and (regarding a function of the nonnegative integers as a sequence) $H_{R/I(Z)}$ is $(1, 3, 6, 10, 14, 17, 20, 21, 21, 21, \ldots)$.

It is sometimes convenient to give not $H_{R/I(Z)}$ itself, but its first difference $\Delta H_{R/I(Z)}$, defined as $\Delta H_{R/I(Z)}(0) = 1$ and $\Delta H_{R/I(Z)}(t) = H_{R/I(Z)}(t) - H_{R/I(Z)}(t-1)$ for $t > 0$. In the preceding example, $\Delta H_{R/I(Z)}$ is $(1, 2, 3, 4, 4, 3, 3, 1, 0, 0, \ldots)$. In particular, when the entries of **d** are strictly decreasing, then $\Delta H_{R/I(Z)} = \text{diag}(\mathbf{d})$.

Sketch of proof (Theorem 5.1). We content ourselves here with merely obtaining an upper bound on $H_{R/I}(t)$. The fact that this bound agrees with the statement given in the theorem involves some combinatorial analysis, for which we refer you to the original paper.

We pause for a notational comment. Given a line $L \subset \mathbb{P}^2$ and a point $p \in L \subset \mathbb{P}^2$, it can be ambiguous whether by $I(p)$ we mean the ideal of p in $K[L]$ or in $K[\mathbb{P}^2]$. Thus we use $I(p)$ for the ideal in $K[\mathbb{P}^2]$ and we use $I_L(p)$ to indicate the ideal of p in $K[L]$.

Let $Z = Z_0$ be the original fat point scheme and let $Z_1, Z_2, \ldots, Z_s = \varnothing$ be the successive residuals with respect to the lines $L_0, L_1, \ldots, L_{s-1}$. Let $I = I(Z) \subset K[\mathbb{P}^2]$ be the ideal defining Z. Let $\mathbf{d} = (d_0, \ldots, d_{s-1})$. Let F_i be a linear form defining L_i. Given any fat point subscheme $X = a_1 q_1 + \cdots + a_u q_u \subsetneq \mathbb{P}^2$, we have the ideal $I(X) \subset K[\mathbb{P}^2]$ as usual. Given a line $L \subset \mathbb{P}^2$ defined by a linear form F, the scheme theoretic intersection $X \cap L = \sum_{q_i \in L} a_i q_i$ is the fat point subscheme of $L \cong \mathbb{P}^1$ defined by the ideal $I_L(X \cap L) = \cap_{q_i \in L} I_L(q_i)^{a_i} \subset K[L] = K[\mathbb{P}^2]/(F) \cong K[\mathbb{P}^1]$, where for a point $q \in L \subset \mathbb{P}^2$, $I_L(q) \subset K[L]$ is the principal ideal defining q as a point of $L \cong \mathbb{P}^1$. Specifically, $I_L(q) = I(q)/(F) \subset K[L] = K[\mathbb{P}^2]/(F)$.

We have canonical inclusions $I(Z_{i+1}) \to I(Z_i)$ given by multiplying by F_i. The quotient $I(Z_i)/F_i I(Z_{i+1})$ is an ideal of $K[L_i]$ whose saturation is $I_L(Z_i \cap L_i)$. Thus we have an inclusion $I(Z_i)/F_i I(Z_{i+1}) \subseteq I_{L_i}(Z_i \cap L_i)$ which need not be an equality. Thus for all t we have $I(Z_i)_t/F_i(I(Z_{i+1}))_{t-1} = (I(Z_i)/F_i I(Z_{i+1}))_t \subseteq (I_{L_i}(Z_i \cap L_i))_t$, but for $t \gg 0$ this becomes

$$I(Z_i)_t/F_i(I(Z_{i+1}))_{t-1} = (I(Z_i)/F_i I(Z_{i+1}))_t = (I_{L_i}(Z_i \cap L_i))_t.$$

Thus for each i and t we have an exact sequence

$$0 \to (I(Z_{i+1}))_{t-1} \to (I(Z_i))_t \to (I_{L_i}(Z_i \cap L_i))_t.$$

By definition of the reduction vector, $Z_i \cap L_i$ has degree d_i. Since $I_{L_i}(Z_i \cap L_i)$ is a principal ideal, we have $\dim_K((I_{L_i}(Z_i \cap L_i))_t) = \binom{t-d_i+1}{1} = \max\{t - d_i + 1, 0\}$, since there are $t - d_i + 1$ monomials in two variables of degree $t - d_i$ whenever $t - d_i \geq 0$. Thus for each i we get an inequality: for $i = 0$ we have

$$\dim_K((I(Z_0))_t) \leq \dim_K((I(Z_1))_{t-1}) + \max\{t - d_0 + 1, 0\};$$

for $i = 1$ we have

$$\dim_K((I(Z_1))_{t-1}) \leq \dim_K((I(Z_2))_{t-2}) + \max\{t - 1 - d_1 + 1, 0\};$$

and continuing in this way we eventually obtain

$$\dim_K((I(Z_{s-1}))_{t-(s-1)}) \leq \dim_K((I(Z_s))_{t-s}) + \max\{t - (s-1) - d_{s-1} + 1, 0\}.$$

Note that $(I(Z_s))_{t-s} = M_{t-s}$, M being the irrelevant ideal (so generated by the variables), hence $\dim_K((I(Z_s))_{t-s}) = \binom{t-s+2}{2}$.

By back substitution, we get

$$\dim_K((I(Z_0))_t) \leq \binom{t-s+2}{2} + \sum_{0 \leq i \leq s-1} \max\{t - i - d_i + 1, 0\}.$$

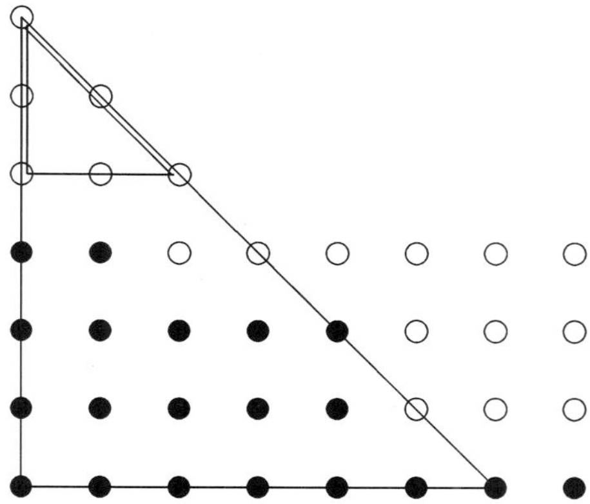

Fig. 2 Obtaining upper bounds on Hilbert functions

Thus

$$H_{R/I}(t) = \binom{t+2}{2} - \dim_K((I(Z_0))_t)$$
$$\geq \binom{t+2}{2} - \binom{t-s+2}{2} - \sum_{0 \leq i \leq s-1} \max\{t - i - d_i + 1, 0\}.$$

A combinatorial analysis shows this bound is what is claimed in the statement of the theorem. Basically, if you arrange the dots as specified by the reduction vector $\mathbf{d}$ (for Fig. 2, $\mathbf{d} = (8, 5, 5, 2)$), then $\binom{t+2}{2} - \binom{t-s+2}{2} - \sum_{0 \leq i \leq s-1} \max\{t - i - d_i + 1, 0\}$ will for each t count the number of black dots in an isosceles right triangle with legs of length t; in Fig. 2 this triangle is the big triangle, which has $t = 6$. The term $\binom{t+2}{2}$ counts the total number of dots in the big triangle, black and open (giving 28 in Fig. 2). To get the number of black dots, you must first subtract the open dots in the little triangle; there are $\binom{t-s+2}{2}$ of these (where, in Fig. 2, $t = 6$ and $s = 4$, giving 6 open dots). The remaining terms then subtract off the number of open dots in the big triangle where each term accounts for each horizontal line on which there is a black dot (these terms would be $\max\{t - 0 - d_0 + 1, 0\} = \max\{6 - 8 + 1, 0\} = 0$ for the bottom row, $\max\{t - 1 - d_1 + 1, 0\} = \max\{6 - 1 - 5 + 1, 0\} = 1$ for the next row up, $\max\{t - 2 - d_2 + 1, 0\} = \max\{6 - 2 - 5 + 1, 0\} = 0$ for the row above that, and $\max\{t - 3 - d_3 + 1, 0\} = \max\{6 - 3 - 2 + 1, 0\} = 2$ for the top row below the little triangle).

The fact that the bound is an equality when the entries of the reduction vector are decreasing involves showing that the third map in the sequence

$$0 \to (I(Z_{i+1}))_{t-1} \to (I(Z_i))_t \to (I_{L_i}(Z_i \cap L_i))_t \qquad (*)$$

is surjective for every i and t. This is done using the long exact sequence in cohomology, where the terms in $(*)$ become modules of global sections of ideal sheaves, and where the lack of surjectivity on the right is controlled by an h^1 term. Working back from the last sequence, one shows for each i and t that either the controlling h^1 term is 0 (and hence we have surjectivity for that i and t) or $(I_{L_i}(Z_i \cap L_i))_t = 0$, hence again we have surjectivity for the given i and t.

Exercises

Exercise 5.1. Let $r_1 > \cdots > r_s > 0$ be integers. Pick s distinct lines, and on line i pick any r_i points, such that none of the points chosen is a point of intersection of the ith line with another of the s lines. Let Z be the reduced scheme consisting of all of the chosen points. Show that $\Delta H_{R/I(Z)}$ is the sequence $(1, 2, \ldots, s, {}^{r_s-1}s, {}^{r_{s-1}-r_s-1}(s-1), {}^{r_{s-2}-r_{s-1}-1}(s-2), \ldots)$, where ${}^{i}j$ denotes a sequence consisting of i repetitions of j.

Exercise 5.2. Take any four distinct lines L_0, L_1, L_2, L_3, no three of which contain a point. There are six points, $p_1, \ldots, p_6$, where pairs of the lines intersect. Let $Z = 3p_1 + \cdots + 3p_6$. Determine the Hilbert function of $R/I(Z)$. (This generalizes to s lines, no 3 of which are coincident at a point; see [11].)

Exercise 5.3. Let $p_1, \ldots, p_r$ be distinct points of $\mathbb{P}^2$. Let $Z = m_1 p_1 + \cdots + m_r p_r$. Pick lines $L_0, \ldots, L_{r-1}$ such that L_{i-1} contains p_i but does not contain p_j for $j \neq i$. Let $\mathbf{d}$ be the reduction vector obtained by choosing m_1 copies of L_0, then m_2 copies of L_1, etc. Show that $\mathbf{d} = (m_1, m_1 - 1, m_1 - 2, \ldots, m_1 - (m_1 - 1), m_2, m_2 - 1, \ldots, m_2 - (m_2 - 1), \ldots, m_r, m_r - 1, \ldots, m_r - (m_r - 1))$; conclude that $H_{R/I(Z)}(t) = \sum_i \binom{m_i+1}{2}$ for all $t \geq m_1 + \cdots + m_r - 1$.

6 Hilbert Functions: Some Structural Results

By Exercises 4.2 and 4.3, we know the Hilbert function of a fat point subscheme is nondecreasing in a strong way (it is strictly increasing until it is constant). It is possible to characterize the functions that are Hilbert functions of fat point subschemes: the Hilbert function of every fat point subscheme of projective space is what is known as a *differentiable O-sequence* (defined below), and for every differentiable O-sequence f there is an n and a finite set of points $p_1, \ldots, p_r \in \mathbb{P}^n$ such that $f = H_{R/I}$ where $R = K[\mathbb{P}^n]$ and $I = I_R(p_1 + \cdots + p_r)$.

It is worth noting that this leads to a characterization of Hilbert functions of reduced zero-dimensional subschemes of projective space: a function f is $H_{R/I}$ for some homogeneous radical ideal I of a finite set of points of projective space if and only if f is a zero-dimensional differentiable O-sequence. It is also true that a function f is $H_{R/I}$ for the homogeneous ideal $I = I(Z)$ for some fat point subscheme Z of projective space if and only if f is a zero-dimensional differentiable O-sequence, but this is because reduced schemes of finite sets of points are special cases of fat point schemes. It is not known, for example, which zero-dimensional differentiable O-sequences occur as Hilbert functions $H_{R/I^{(2)}}$ for homogeneous radical ideals I defining finite sets of points in projective space. (A general reference for the material in this section is [6].)

Definition-Proposition 6.1 (See, For Example, [26]). Let h and d be positive integers. Then h can be expressed uniquely in the form

$$\binom{m_d}{d} + \binom{m_{d-1}}{d-1} + \cdots + \binom{m_j}{j}$$

where $m_d > m_{d-1} > \cdots > m_j \geq j \geq 1$. This expression for h is called the *d-binomial expansion of h*. Given the d-binomial expansion of h, we also define

$$h^{\langle d \rangle} = \binom{m_d + 1}{d + 1} + \binom{m_{d-1} + 1}{d} + \cdots + \binom{m_j + 1}{j + 1}.$$

Example 6.1. The 3-binomial expansion of 15 is

$$15 = \binom{5}{3} + \binom{3}{2} + \binom{2}{1} = 10 + 3 + 2.$$

It is convenient to relate this to Pascal's triangle. The binomial coefficients $\binom{m}{d}$ with d fixed lie on a diagonal of slope 1 say in Pascal's triangle. So to obtain the d-binomial expansion of h, one picks the largest $\binom{m_d}{d}$ on this line less than or equal to h. Then one makes up as much of the remainder $h - \binom{m}{d}$ as possible by choosing a coefficient $\binom{m_{d-1}}{d-1}$ on the next line up of slope 1, etc. To obtain $h^{\langle 3 \rangle}$, one just slides the choices made for h down and to the right. Thus

$$15^{\langle 3 \rangle} = \binom{6}{4} + \binom{4}{3} + \binom{3}{2} = 15 + 4 + 3 = 22.$$

Definition 6.2. A sequence of nonnegative integers $\{h_d\}_{d \geq 0}$ is called an *O-sequence* if

- $h_0 = 1$, and
- $h_{d+1} \leq h_d^{\langle d \rangle}$ for all $d \geq 1$, where $0^{\langle d \rangle} = 0$ for all i.

With these definitions we can state a well-known theorem of Macaulay (see [42] and [50] for full details):

Theorem 6.2 (Macaulay's Theorem). *The following are equivalent:*

(a) $\{h_d\}_{d \geq 0}$ is an O-sequence;
(b) $\{h_d\}_{d \geq 0}$ is the Hilbert function $H_{R/I}$ for some homogeneous ideal $I \subsetneq R$; and
(c) $\{h_d\}_{d \geq 0}$ is the Hilbert function $H_{R/J}$ for some monomial ideal $J \subsetneq R$.

Definition 6.3. Let $\mathscr{H} = \{h_d\}_{d\geq 0}$ be an O-sequence and $\Delta\mathscr{H} = \{e_d\}_{d\geq 0}$ be defined by $e_0 = h_0$ and $e_d = h_d - h_{d-1}$ for $d \geq 1$. We say that $\mathscr{H}$ is a *differentiable O-sequence* if $\Delta\mathscr{H}$ is also an O-sequence. We say $\mathscr{H}$ is *zero-dimensional* if $\Delta\mathscr{H}$ is 0 for all $t \gg 0$.

Proposition 6.3. *Let $p_1, \ldots, p_s \in \mathbb{P}^n$ be distinct points, let $m_1, \ldots, m_s$ be positive integers, and let $I = I(m_1 p_1 + \cdots + m_s p_s)$ be the ideal of the fat point subscheme $m_1 p_1 + \cdots + m_s p_s \subset \mathbb{P}^n$. Then the Hilbert function $H_{R/I}$ is a differentiable zero-dimensional O-sequence.*

Proof. By Macaulay's Theorem, $H_{R/I}$ is an O-sequence. By Exercise 4.3, $H_{R/I}$ is zero-dimensional. But if $x \in R$ is a linear form that does not vanish at any of the points, and if $J = I + (x)$, then

$$\frac{R}{J} \cong \frac{R/I}{J/I} = \frac{R/I}{((x)+I)/I} \cong \frac{R/I}{x(R/I)}$$

so we have $H_{R/J} = H_{\frac{R/I}{x(R/I)}}$ and since x maps to a unit in R/I, we obtain $H_{\frac{R/I}{x(R/I)}} = \Delta H_{R/I}$. But by Macaulay's Theorem again, $H_{R/J}$ is an O-sequence, hence $H_{R/I}$ is a differentiable O-sequence.

There is also a converse:

Theorem 6.4 ([21]). *Let $\mathscr{H} = \{h_d\}_{d\geq 0}$ be a differentiable zero-dimensional O-sequence with $h_1 \leq n+1$. Then there is a finite set of points in $\mathbb{P}^n$ and the ideal $I \subseteq R$ of those points is a radical ideal such that $\mathscr{H} = H_{R/I}$. In case $n = 2$, those points can be chosen as in Exercise 5.1 and hence $\Delta\mathscr{H} = \mathrm{diag}(\mathbf{d})$ for some decreasing sequence $\mathbf{d}$ of positive integers.*

We give some idea how one can prove this, involving monomial ideals and their liftings. The original proof, given in [21], is somewhat different.

Definition 6.4. Let $J \subseteq K[x_1, x_2]$ be a homogeneous ideal and let $\phi : K[x_0, x_1, x_2] \to K[x_1, x_2]$ be defined by $\phi(x_0) = 0$ and $\phi(x_i) = x_i$ for $i > 0$. We say that J *lifts to* $I \subseteq K[x_0, x_1, x_2]$ if

- I is a radical ideal in $K[x_0, x_1, x_2]$;
- x_0 is not a zero-divisor on $K[x_0, x_1, x_2]/I$; and
- $\phi(I) = J$.

If $\mathscr{H} = \{h_d\}_{d\geq 0}$ is a differentiable zero-dimensional O-sequence (with $n = 2$), let $\Delta\mathscr{H} = \{e_d\}_{d\geq 0}$ be defined by $e_0 = 1, e_d = h_d - h_{d-1}$ for $d \geq 1$. By Macaulay's Theorem, we know there exists an ideal $J \subseteq K[x_1, x_2]$ generated by some monomials $\{x_1^{m_{10}} x_2^{m_{20}}, \ldots, x_1^{m_{1r}} x_2^{m_{2r}}\}$ such that $H_{K[x_1,x_2]/J} = \Delta\mathscr{H}$. Since the O-sequence is zero-dimensional, we know that among the generators are pure powers of x_1 and x_2. In fact, Macaulay proved more than the statement we gave above of Macaulay's Theorem; he showed that J can be taken to be a lex ideal, which means that whenever $x_1^i x_2^j \in J$ with $i > 0$, then $x_1^{i-1} x_2^{j+1} \in J$. (Here we mean lex with respect to the monomial ordering with $x_2 > x_1$, which is nonstandard, but which is needed to be consistent with the exposition in [20].) Since in our case

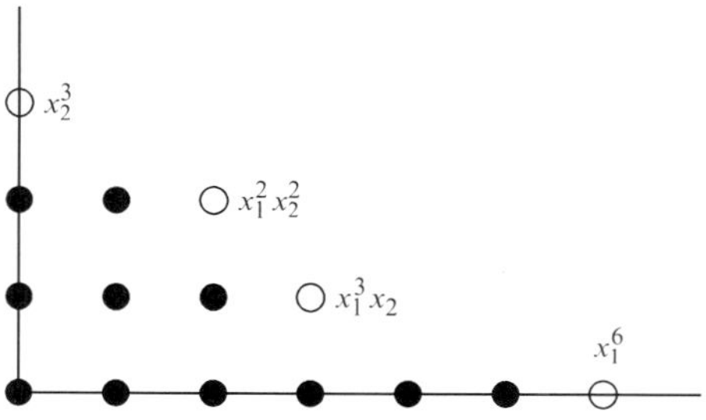

Fig. 3 A monomial ideal

J is not only lex but also contains pure powers of x_1 and x_2, we may assume that $m_{2i} = i$ and $m_{1i} > m_{1i+1}$ for all i, with $m_{1r} = 0$. Geramita–Gregory–Roberts [20] and Hartshorne [36] showed that J lifts to an ideal I which is the ideal of a finite set of points whose coordinates are given by the exponent vectors (m_{1i}, m_{2i}). To explain this in more detail we introduce some notation and bijections.

To an element $\alpha = (a_1, a_2) \in \mathbb{N}^2$ we associate the point $\overline{\alpha} = [1 : a_1 : a_2] \in \mathbb{P}^2$. Further, for each monomial $g = x^\alpha = x_1^{a_1}x_2^{a_2}$ we associate

$$\overline{g} = \prod_{j=1}^{2}\left(\prod_{i=0}^{a_j-1}(x_j - ix_0)\right).$$

Observe that $\overline{g}$ is homogeneous.

Now, since J is a monomial ideal, the set $\mathscr{M} \setminus N$, where $\mathscr{M}$ denotes the monomials in $K[x_1, x_2]$ (including 1) and N denotes the set of monomials in J, gives representatives for a K-basis of $K[x_1, x_2]/J$. Let $\overline{\mathscr{M}}$ denote the set of all points $\overline{\alpha} = \overline{(a_1, a_2)} \in \mathbb{P}^2$ such that $x_1^{a_1}x_2^{a_2} \in \mathscr{M}$. It can then be shown (see [20] for full details) that J lifts to $I = (\overline{g_i})$, where $\{g_i\}$ is the minimal generating set for J. The key step in the proof is to show that

$$I = \{f \in K[x_0, x_1, x_2] : f(\overline{\alpha}) = 0 \text{ for all } \overline{\alpha} \in \overline{\mathscr{M}}\}.$$

Note that I is the ideal of a finite set of points which can be chosen as in Exercise 5.1.

Example 6.5. Consider $\mathscr{H} = (1, 3, 6, 9, 10, 11, 11, 11, \ldots)$. This is a differentiable zero-dimensional O-sequence with $\Delta\mathscr{H} = (1, 2, 3, 3, 1, 1, 0, 0, \ldots)$. To find a finite set of points $\mathbb{X}$ where $H_{R/I(\mathbb{X})} = \mathscr{H}$ we consider the monomial ideal $J = (x_2^3, x_1^2x_2^2, x_1^3x_2, x_1^6)$. We can visualize the monomials in $\mathscr{M} \setminus N$ as the circles in the x_1x_2-plane in Fig. 3, where the monomial $x_1^{a_1}x_2^{a_2}$ is represented by the pair (a_1, a_2). The open circles represent the generators of J.

Let $\mathbb{X}$ be the set consisting of the points in $\mathbb{P}^2$ which are in $\overline{\mathscr{M}}$; these points are $[1 : 0 : 0]$, $[1 : 1 : 0]$, $[1 : 2 : 0]$, $[1 : 3 : 0]$, $[1 : 4 : 0]$, $[1 : 5 : 0]$, $[1 : 0 : 1]$, $[1 : 1 : 1]$, $[1 : 2 : 1]$, $[1 : 0 : 2]$, $[1 : 1 : 2]$. The ideal $I = I(\mathbb{X})$ is generated by:

$$\overline{x_2^3} = x_2(x_2 - x_0)(x_2 - 2x_0)$$
$$\overline{x_1^2x_2^2} = x_1(x_1 - x_0)x_2(x_2 - x_0)$$
$$\overline{x_1^3x_2} = x_1(x_1 - x_0)(x_1 - 2x_0)x_2$$
$$\overline{x_1^6} = x_1(x_1 - x_0)(x_1 - 2x_0)(x_1 - 3x_0)(x_1 - 4x_0)(x_1 - 5x_0).$$

We have that J lifts to I. Observe that $\mathbb{X}$ is a configuration of points contained in a union of three "horizontal" lines in $\mathbb{P}^2$, with six points on the bottom line, three on the middle line and two on the top line.

The method used in the above example will work in general. Given a differentiable zero-dimensional O-sequence $\mathscr{H}$ where $\Delta\mathscr{H} = (h_0, h_1, h_2, \ldots)$, then one applies the steps above using the ideal J found by setting the degree t monomials of $\mathscr{M} \setminus N$ to be the first h_t monomials in R using lexicographic ordering.

Example 6.6. Suppose $h = (1, 3, 5, 5, 5, \ldots)$. This is a differentiable zero-dimensional O-sequence. Using the methods of the previous section, one can check that it is the Hilbert function of five points in $\mathbb{P}^2$, two on one line, and three on another line, none where the lines meet.

Example 6.7. Suppose $h = (1, 3, 2, 0, 0 \ldots)$. This is a zero-dimensional O-sequence but it is not differentiable. It is the Hilbert function of R/I for $R = K[x, y, z]$ and $I = (x^2, xy, x^2, y^2) + (x, y, z)^3$.

Exercises

Exercise 6.1. Let $I = I(3p)$ for a point $p \in \mathbb{P}^2$. Find a set of points $p_1, \ldots, p_r \in \mathbb{P}^2$ such that $H_{R/I} = H_{R/J}$ where $J = I(p_1 + \cdots + p_r)$.

Exercise 6.2. Show that **d** in the statement of Theorem 6.4 is unique.

7 Bézout's Theorem in $\mathbb{P}^2$ and Applications

We start with some intuition as to what Bézout's Theorem is all about. One way to think about it is as a generalization of the Fundamental Theorem of Algebra (FTA). One can state FTA as follows:

Theorem 7.1 (FTA). *A nonconstant polynomial $f \in \mathbf{C}[x]$ of degree d has exactly d roots, counted with multiplicity, where $\mathbf{C}$ is the field of complex numbers.*

Replacing $\mathbf{C}$ by any algebraically closed field K, a simplified version of Bézout's Theorem says the following:

Theorem 7.2 (Baby Bézout). *Let $F \in K[\mathbb{P}^2]$ be a nonconstant form of degree d and let L be a linear form (and also, by abuse of notation, the line it defines). Then either the restriction of F to L has exactly d roots (counted with multiplicity), or L divides F.*

The full version of Bézout's Theorem (see below) says that forms $F, G \in K[\mathbb{P}^2]$ of degrees $d_1, d_2 > 0$ have exactly $d_1 d_2$ common zeros (counted with multiplicity) unless F and G have a common factor of positive degree. The rigorous statement requires dealing with how to count common zeros correctly. So let $0 \neq F \in K[\mathbb{P}^2] = K[x_0, x_1, x_2]$ be homogeneous. The multiplicity $\mathrm{mult}_p(F)$ of F at a point $p \in \mathbb{P}^2$ is the largest m such that $F \in I(p)^m$, where we regard $I(p)^0$ as being R. If projective coordinates are chosen so that $p = (1, 0, 0)$, then $\mathrm{mult}_p(F)$ is the degree of a term of least degree in $F(1, x_1, x_2)$. The homogeneous component h of $F(1, x_1, x_2)$ of least degree factors as a product of powers of homogeneous linear factors l_i; i.e., $h = l_1^{m_1} \cdots l_s^{m_s}$. The factors l_i are the *tangents* to F at p, and the exponent m_i is the multiplicity of l_i.

Suppose F and G are homogeneous polynomials which do not have a common factor vanishing at p. For each $m \geq 1$, the K-vector space dimension of the tth homogeneous component of $R/((F, G) + I(p)^m)$ is equal to some limiting value $\Lambda_m(F, G, p)$ for all $t \gg 0$. For all $m \gg 0$, $\Lambda_m(F, G, p)$ also attains a limiting value, $\Lambda(F, G, p)$. We define the *intersection multiplicity* $I_p(F, G)$ to be $\Lambda(F, G, p)$. Since F and G determine one-dimensional subschemes $C_F, C_G \subset \mathbb{P}^2$ which in turn determine F and G, we also will refer to $I_p(F, G)$ as $I_p(C_F, C_G)$.

Assume that F, G, and H are homogeneous polynomials which do not have a common factor vanishing at p. Then some facts about intersection multiplicities are (see [37] or [19]):

(a) $I_p(F, G) \geq \mathrm{mult}_p(F)\,\mathrm{mult}_p(G)$, where equality holds if and only if F and G have no tangent in common at p;
(b) $I_p(F, GH) = I_p(F, G) + I_p(F, H)$;
(c) intersection multiplicities are invariant under projective linear homogeneous changes of coordinates; and:

Theorem 7.3 (Bézout's Theorem). *If $F, G \in K[\mathbb{P}^2]$ are forms which have no common factor of positive degree, then*

$$(\deg(F))(\deg(G)) = \sum_{p \in \mathbb{P}^2} I_p(F, G).$$

Example 7.4. Let $Z = m_1 p_1 + \cdots + m_r p_r$, where $p_1, \ldots, p_r \in \mathbb{P}^2$ are distinct points and each m_i is a positive integer. Let $C \subset \mathbb{P}^2$ be an irreducible curve of degree d such that $\mathrm{mult}_{p_i}(C) = e_i$ for each i (i.e., $\mathrm{mult}_{p_i}(G) = e_i$ where G is the form defining C). Say $0 \neq F \in I(Z)_t$, so $\mathrm{mult}_{p_i}(F) \geq m_i$ for all i. If $\sum_i m_i e_i > td$, then $\sum_i I_{p_i}(F, G) \geq \sum_i \mathrm{mult}_{p_i}(F)\,\mathrm{mult}_{p_i}(G) \geq \sum_i m_i e_i > td$ so by Bézout's Theorem, G and F have a common factor, but G is irreducible, so $G|F$. Thus $H \in I((m_1 - e_1)p_1 + \cdots + (m_r - e_r)p_r)$, where $H = F/G$.

We can apply this to get bounds on $\alpha(I(Z))$. For example, let $L_1, L_2, L_3, L_4 \subset \mathbb{P}^2$ be lines no three of which meet at a point. We will regard L_i as denoting either the line itself or the linear homogeneous form that defines the line, depending on context. Let $p_{ij} = L_i \cap L_j$ for $i \neq j$, so $\{p_{ij}\}$ are the six points of pair-wise intersections of the lines. Let $Z = \sum_{ij} 3p_{ij}$. It is easy to check that $(L_1L_2L_3)^2L_4$ is in $I(p_{ij})^3$ for each of the six points. Thus $(L_1L_2L_3)^2L_4 \in I(Z)_7$ so $\alpha(I(Z)) \leq 7$. On the other hand, assume we have $0 \neq F \in I(Z)_6$. There are three points where both F and L_i vanish, with F having multiplicity at least 3 at each and L_i having multiplicity 1. Since $3 \cdot (3 \cdot 1) > \deg(F)\deg(L_i) = 6$, then $L_i|F$. This is true for all i, so $L_1L_2L_3L_4|F$. Let $H = F/(L_1L_2L_3L_4)$. Then $\deg(H) = 2$ and $\mathrm{mult}_{p_{ij}}(H) \geq 1$. Now $3 \cdot (1 \cdot 1) > \deg(H)\deg(L_i) = 2$, so again $L_1 \cdots L_4|H$, but this is impossible since $\deg(H) < \deg(L_1L_2L_3L_4)$. Thus H and therefore F must be 0, so $\alpha(I(Z)) > 6$ and hence $\alpha(I(Z)) = 7$. (Note that this is in agreement with the result of Exercise 5.2.)

Example 7.5. Let $I = I(p_1 + p_2 + p_3)$ for three noncollinear points of $\mathbb{P}^2$. We show that $\gamma(I) = 3/2$. Consider $I^{(m)} = I(m(p_1 + p_2 + p_3))$. Assume $m = 2s$ is even, and suppose $0 \neq F \in (I^{(m)})_{3s-1}$. Note that F vanishes to order at least m at each of two points for any line L_{ij} through two of the points $p_i, p_j, i \neq j$. Since $2m = 4s > 3s - 1$, this means by Bézout that the linear forms (also denoted L_{ij}) defining the lines are factors of F. Dividing F by $L_{12}L_{13}L_{23}$ we obtain a form G of degree $3(s-1) - 1$ in $I^{(m-2)}$. The same argument applies: $L_{12}L_{13}L_{23}$ must divide G. Eventually we obtain a form of degree 2 divisible by $L_{12}L_{13}L_{23}$, which is impossible. Thus $F = 0$, and $\alpha(I^{(m)}) > \frac{3m}{2} - 1$. Since $(L_{12}L_{13}L_{23})^s \in I^{(m)}$, we see that $\alpha(I^{(m)}) \leq \frac{3m}{2}$, thus $\alpha(I^{(m)}) = \frac{3m}{2}$, and hence $\gamma(I) = \lim_{m\to\infty} \alpha(I^{(m)})/m = 3/2$.

Exercises

Exercise 7.1. Show that $I_p(F, G) = 0$ if either F or G does not vanish at p.

Exercise 7.2. Let $p = (1, 0, 0)$, $F = x_1x_0 - x_2^2$, and $G = x_1x_0^2 - x_2^3$. Compute $I_p(F, G)$ and verify that $\sum_{p\in\mathbb{P}^2} I_p(F, G) = \deg(F)\deg(G)$ by explicit computation.

Exercise 7.3. Consider the $\binom{s}{2}$ points of pairwise intersection of s distinct lines in $\mathbb{P}^2$, no three of which meet at a point. Let I be the radical ideal of the points. Mimic Example 7.4 to show that $\alpha(I^{(m)}) = ms/2$ if m is even, and $\alpha(I^{(m)}) = (m+1)s/2 - 1$ if m is odd.

Exercise 7.4. Let $I = I(p_1 + p_2 + p_3 + p_4)$ for four points of $\mathbb{P}^2$, no three of which are collinear. Show that $\gamma(I) = 2$.

Exercise 7.5. Let $I = I(p_1 + p_2 + p_3 + p_4 + p_5)$ for five points of $\mathbb{P}^2$, no three of which are collinear. Show that $\gamma(I) = 2$.

Exercise 7.6. Show that there exist six points of $\mathbb{P}^2$ which do not all lie on any conic, and no three of which are collinear.

Exercise 7.7. Let $I = I(p_1 + \cdots + p_6)$ for six points of $\mathbb{P}^2$, no three of which are collinear and which do not all lie on a conic (such point sets exist by Exercise 7.6). Show that $\gamma(I) = 12/5$.

Exercise 7.8. Show that there exist seven points of $\mathbb{P}^2$, no three of which are collinear and no six of which lie on a conic.

Exercise 7.9. Let $I = I(p_1 + \cdots + p_7)$ for seven points of $\mathbb{P}^2$, no three of which are collinear and no six of which lie on a conic (such point sets exist by Exercise 7.8). Show that $\gamma(I) = 21/8$.

Exercise 7.10. Given nine distinct points $p_i \in \mathbb{P}^2$ on an irreducible cubic C such that $\text{mult}_{p_i}(C) = 1$ for all i, show that $\gamma(I) = 3$ for $I = I(p_1 + \cdots + p_9)$.

8 Divisors, Global Sections, the Divisor Class Group and Fat Points

For this section, our references are [12,27,31,37,45]. Given any finite set of distinct points $p_1, \ldots, p_r \in \mathbb{P}^2$, there is a projective algebraic surface X, a projective morphism $\pi : X \to \mathbb{P}^2$ (obtained by blowing up the points p_i) such that each $\pi^{-1}(p_i) = E_i$ is a smooth rational curve and such that π induces an isomorphism $X \setminus \cup_i E_i \to \mathbb{P}^2 \setminus \{p_1, \ldots, p_r\}$.

The divisor class group $\text{Cl}(X)$ (of divisors modulo linear equivalence, where a divisor is an element of the free abelian group on the irreducible curves on X) is the free group with basis $e_0, e_1, \ldots, e_r$, where $e_0 = [E_0]$ is the class of the pullback E_0 to X of a line $L \subset \mathbb{P}^2$, and $e_i = [E_i]$ for $i > 0$ is the class of the curve E_i. The group $\text{Cl}(X)$ comes with a bilinear form, called the intersection form, defined as $-e_0^2 = e_i^2 = -1$ for all $i > 0$, and $e_i \cdot e_j = 0$ for $i \neq j$. An important element, known as the canonical class, is $K_X = -3e_0 + e_1 + \cdots + e_r$. If C and D are divisors, we define $C \cdot D = [C] \cdot [D]$. If C and D are prime divisors meeting transversely, then $C \cdot D$ is just the number of points of intersection of C with D.

If D is a divisor on X, its class can be written as $[D] = de_0 - \sum_i m_i e_i$ for some integers d and m_i. Associated with D is an invertible sheaf $\mathcal{O}_X(D)$. The space of global sections of this sheaf is a finite dimensional K-vector space, denoted $\Gamma(\mathcal{O}_X(D))$ and also $H^0(X, \mathcal{O}_X(D))$. The dimension of this vector space is denoted $h^0(X, \mathcal{O}_X(D))$; if $[D] = [D']$, then $h^0(X, \mathcal{O}_X(D)) = h^0(X, \mathcal{O}_X(D'))$.

In case $D = dE_0 - \sum_i m_i E_i$ such that each $m_i \geq 0$, then there is a canonical identification of $H^0(X, \mathcal{O}_X(D))$ with $I(m_1 p_1 + \cdots + m_r p_r)_d$ [31, Proposition IV.1.1]. Thus techniques for computing $h^0(X, \mathcal{O}_X(D))$ can be applied to computing the Hilbert function of $m_1 p_1 + \cdots + m_r p_r$. One important tool is the theorem of Riemann–Roch for surfaces; see Exercise 8.2. Bézout's Theorem also has a natural interpretation in this context. If C and D are effective divisors such that $[C] = c_0e_0 - c_1e_1 - \cdots - c_re_r$ and $[D] = d_0e_0 - d_1e_1 - \cdots - d_re_r$, then $C \cdot D = c_0d_0 - c_1d_1 - \cdots - c_rd_r$; if this is negative, then C and D have a common component. In particular, if C is a prime divisor, then C itself is the common component, hence $D - C$ is effective.

Another important technique involves a group action on $\mathrm{Cl}(X)$ related to the Cremona group of birational transformations of the plane. Given $\pi : X \to \mathbb{P}^2$ as above, there can exist morphisms $\pi' : X \to \mathbb{P}^2$ obtained by blowing up other points (possibly infinitely near) $p'_1, \ldots, p'_r \in \mathbb{P}^2$. The composition $\pi'\pi^{-1}$, defined away from the points p_i, is a birational transformation of $\mathbb{P}^2$, hence an element of the Cremona group (named for Luigi Cremona, after whom there is named a street in Rome near the Colosseum). We thus have a second basis $e'_0, e'_1, \ldots, e'_r$ of $\mathrm{Cl}(X)$ corresponding to curves E'_i. In particular, we can write $dE_0 - \sum_i m_i E_i$ as $d'E'_0 - \sum_i m'_i E'_i$. The change of basis transformation from the basis e_i to the basis e'_i is always an element of a particular group, now known as the Weyl group, W_r (we give generators s_i for W_r below). For $r < 9$, W_r is finite, but it is infinite for all $r \geq 9$.

Example 8.1. Consider the quadratic Cremona transformation on $\mathbb{P}^2$, defined away from $x_0x_1x_2 = 0$ as $Q : (a, b, c) \mapsto (1/a, 1/b, 1/c)$. Alternatively, one can define it at all points of $\mathbb{P}^2$ except $(1, 0, 0)$, $(0, 1, 0)$ and $(0, 0, 1)$ as $(a, b, c) \mapsto (bc, ac, ab)$. It can also be obtained by as $\pi'\pi^{-1}$, where $\pi : X \to \mathbb{P}^2$ is the morphism given by blowing up the points $(1, 0, 0)$, $(0, 1, 0)$, and $(0, 0, 1)$ and $\pi' : X \to \mathbb{P}^2$ contracts the proper transforms of the lines through pairs of those points. More generally one can define the quadratic transform at any three noncollinear points, by blowing them up and blowing down the proper transforms of the lines through pairs of the three points. An important theorem (now fully proved, but originally announced by Noether with a proof felt to be incomplete) is that the Cremona group for $\mathbb{P}^2$ is generated by invertible linear transformations of the plane and quadratic transformations [1].

Let $n_0 = e_0 - e_1 - e_2 - e_3$ and let $n_i = e_1 - e_{i+1}$ for $i = 1, \ldots, r - 1$. For any $x \in \mathrm{Cl}(X)$ and any $0 \leq i < r$, let $s_i(x) = x + (x \cdot n_i)n_i$. Then W_r is defined to be the group generated by $s_i \in W_r$. When $i > 0$, the element s_i just transposes e_i and e_{i+1}, so $\{s_1, \ldots, s_{r-1}\}$ generates the group of permutations on the set $\{e_1, \ldots, e_r\}$. When the points p_1, p_2, p_3 are not collinear, the element s_0 corresponds to the quadratic transformation $Q : (a, b, c) \mapsto (\frac{1}{a}, \frac{1}{b}, \frac{1}{c})$. Note that $s_0(e_1) = e_0 - e_2 - e_3$, $s_0(e_2) = e_0 - e_1 - e_3$, and $s_0(e_3) = e_0 - e_1 - e_2$: blowing up p_1, p_2, and p_3, to get E_1, E_2, E_3 and blowing down the proper transforms of the line through p_2 and p_3, the line through p_1 and p_3 and the line through p_1 and p_2 is precisely Q. (Note also that $s_0(e_0) = 2e_0 - e_1 - e_2 - e_3$ and a line $a_0x_0 + a_1x_1 + a_2x_2 = 0$ pulls back under Q to $a_0/x_0 + a_1/x_1 + a_2/x_2 = 0$ which, by multiplying through by $x_0x_1x_2$ to clear the denominators is the same as $a_0x_1x_2 + a_1x_0x_2 + a_2x_0x_1 = 0$; i.e., on the surface X obtained by blowing up the coordinate vertices we have $e'_0 = 2e_0 - e_1 - e_2 - e_3$.)

When the points p_i are sufficiently general (such as being generic, meaning, say, that the projective coordinates a_{ij} for each point $p_i = (a_{i0}, a_{i1}, a_{i2})$ are all nonzero, and the ratios $\frac{a_{11}}{a_{10}}, \frac{a_{12}}{a_{10}}, \frac{a_{21}}{a_{20}}, \frac{a_{22}}{a_{20}}, \ldots, \frac{a_{r1}}{a_{r0}}, \frac{a_{r2}}{a_{r0}}$ are algebraically independent over the prime field of K) and given the surface $\pi : X \to \mathbb{P}^2$ obtained by blowing up the points p_i, the birational morphisms $X \to \mathbb{P}^2$ (up to projective equivalence) are in one-to-one correspondence with the elements of W_r. We denote by π_w the morphism corresponding to w. The identity element w corresponds to the basis $\{e_0, e_1, \ldots, e_r\}$ obtained by blowing up the points p_i, and this gives π since for $i > 0$, E_i is the unique effective divisor whose class is e_i. Contracting $E_r, E_{r-1}, \ldots, E_1$ in order gives π. Likewise, for any $w \in W_r$, the basis $e'_i = w(e_i)$ gives the sequence of curves E'_i which must be contracted to define π_w.

Given a divisor $F = dE_0 - \sum_i m_i E_i$, we denote by wF the divisor $d'E'_0 - \sum_i m'_i E'_i$ where $w(de_0 - \sum_i m_i e_i) = d'e'_0 - \sum_i m'_i e'_i$. Since w represents a change of basis, we have $H^0(X, \mathcal{O}_X(F)) = H^0(X, \mathcal{O}_X(wF))$ and thus $\dim I(\sum_i m_i p_i)_d = \dim I(\sum_i m'_i p'_i)_{d'}$. (The fact that $H^0(X, \mathcal{O}_X(F)) = H^0(X, \mathcal{O}_X(wF))$ also shows that $I(\sum_i m_i p_i)_d$ has an irreducible element if and only if $I(\sum_i m'_i p'_i)_{d'}$ does.) But if the points p_i are generic, so are the points p'_i (up to projective equivalence), so $\dim I(\sum_i m_i p_i)_d = \dim I(\sum_i m'_i p_i)_{d'}$. (There is an automorphism $\phi : K \to K$ such that the coordinates of the points p_i map to the coordinates of the points p'_i. This induces an invertible map $\Phi : I(\sum_i m'_i p_i)_{d'} \to I(\sum_i m'_i p'_i)_{d'}$ such that if $a_i \in K$ and $F_i \in I(\sum_i m'_i p_i)_{d'}$, then $\Phi(\sum_i a_i F_i) = \sum_i \phi(a_i)\Phi(F_i)$, from which it follows that $\dim I(\sum_i m'_i p_i)_{d'} = \dim I(\sum_i m'_i p_i)_{d'}$ and hence that $\dim I(\sum_i m_i p_i)_d = \dim I(\sum_i m'_i p_i)_{d'}$.)

Example 8.2. Let $p_1, \ldots, p_9$ be generic points of $\mathbb{P}^2$. We show that $I(p_1 + \cdots + p_5)_2$, $I(2p_1 + p_2 + \cdots + p_7)_3$, and $I(3p_1 + 2p_2 + \cdots + 2p_8)_6$ each are one-dimensional, with basis given by an irreducible form. In each case we have a homogeneous component of the form $I(\sum_i m_i p_i)_d$. It is enough to show that there is an element $w \in W_8$ such that $w[F] = e_0 - e_1 - e_2$, where $[F] = de_0 - \sum_i m_i e_i$. But $s_0(2e_0 - e_1 - \cdots - e_5) = e_0 - e_4 - e_5$ and we apply a permutation σ to obtain $\sigma(e_0 - e_4 - e_5) = e_0 - e_1 - e_2$. Thus $\dim I(p_1 + \cdots + p_5)_2 = \dim I(p_1 + p_2)_1$ and since $I(p_1 + p_2)_1$ clearly has an irreducible element so does $I(p_1 + \cdots + p_5)_2$. The other cases with $r < 9$ are similar. The case that $r = 9$ is also similar if we show that $I(p_1 + \cdots + p_9)_3$ has an irreducible element.

Exercises

Exercise 8.1. Let X be the blow up of $\mathbb{P}^2$ at r distinct points. Show that $w(x) \cdot w(y) = x \cdot y$ for all $x, y \in \mathrm{Cl}(X)$ and all $w \in W_r$, and show that $w(K_X) = K_X$ for all $w \in W_r$, where $K_X = -3e_0 + e_1 + \cdots + e_r$.

Exercise 8.2. Let X be the blow up of $\mathbb{P}^2$ at s distinct points $p_i \in \mathbb{P}^2$. Let $F = tE_0 - m_1E_1 - \cdots - m_sE_s$. The theorem of Riemann–Roch for surfaces says that

$$h^0(X, \mathcal{O}_X(F)) - h^1(X, \mathcal{O}_X(F)) + h^2(X, \mathcal{O}_X(F)) = \frac{F^2 - K_X \cdot F}{2} + 1.$$

Serre duality says $h^2(X, \mathcal{O}_X(F)) = h^0(X, \mathcal{O}_X(K_X - F))$, and hence $h^2(X, \mathcal{O}_X(F)) = 0$ if $t \geq 0$. Thus for $t \geq 0$ and $m_i \geq 0$ for all i, taking $I = I(m_1p_1 + \cdots + m_sp_s)$, we have $H_I(t) = h^0(X, \mathcal{O}_X(F)) = \frac{F^2-K_X \cdot F}{2} + 1 + h^1(X, \mathcal{O}_X(F))$. Show that

$$\frac{F^2 - K_X \cdot F}{2} + 1 = \binom{t+2}{2} - \sum_i \binom{m_i+1}{2}.$$

Conclude that $P_I(t) = \frac{F^2-K_X \cdot F}{2} + 1$ where P_I is the Hilbert polynomial for I, and that $h^1(X, \mathcal{O}_X(F)) = H_I(t) - P_I(t)$ is the difference between the Hilbert function and Hilbert polynomial for I.

Exercise 8.3. Let $p_1, \ldots, p_8$ be generic points of $\mathbb{P}^2$. Show that $\alpha(I(6p_1 + \cdots + 6p_8)) = 17$.

Exercise 8.4. Let $p_1, \ldots, p_r \in \mathbb{P}^2$ be generic points of $\mathbb{P}^2$. Let X be the surface obtained by blowing up the points. Let $w \in W_r$ and let $[C] = w(e_1)$. Show that C is a smooth rational curve with $C^2 = C \cdot K_X = 1$. Conclude that $((mC)^2 - K_X \cdot (mC))/2 + 1 \leq 0$ for all $m > 1$. Such a curve C is called an *exceptional curve*. (By [45, Theorem 2b], when $r \geq 3$, the set of classes of exceptional curves is precisely the orbit $W_r(e_1)$.)

Exercise 8.5. Let $p_1, \ldots, p_r \in \mathbb{P}^2$ be distinct points of $\mathbb{P}^2$. Let X be the surface obtained by blowing up the points. Let C be an exceptional curve on X, let D be an effective divisor, let $m = -C \cdot D > 0$, and let $F = D - mC$. If $m > 1$, show that $h^0(X, \mathcal{O}_X(D)) = h^0(X, \mathcal{O}_X(F))$ (hence C is a *fixed component* of $|D| = |F| + mC$ of multiplicity m, where $|D|$ is the linear system of all curves corresponding to elements of $H^0(X, \mathcal{O}_X(D))$), and that $(D^2 - K_X \cdot D)/2 < (F^2 - K_X \cdot F)/2$; conclude that $h^0(X, \mathcal{O}_X(D)) > (D^2 - K_X \cdot D)/2 + 1$.

9 The SHGH Conjecture

The SHGH Conjecture [22, 28, 38, 48] gives an explicit conjectural value for the Hilbert function of the ideal of a fat point subscheme of $\mathbb{P}^2$ supported at generic (or even just sufficiently general) points.

Consider I_4 where I is the ideal of the fat point subscheme $3p_1 + 3p_2 + p_3 + p_4 \subset \mathbb{P}^2$. Let $D = 4E_0 - 3E_1 - 3E_2 - E_3 - E_4$ and let $C = E_0 - E_1 - E_2$.

Note that $D \cdot C = -2$; let $F = D - 2C = 2E_0 - E_1 - \cdots - E_4$. We know $H_I(4) = h^0(X, \mathcal{O}_X(D)) \geq (D^2 - K_X \cdot D)/2 + 1 = \binom{4+2}{2} - 2\binom{3+1}{2} - 2\binom{1+1}{2} = 1$. But by Exercise 8.5 we also have

$$H_I(4) = h^0(X, \mathcal{O}_X(F)) \geq (F^2 - K_X \cdot F)/2 + 1 = 2.$$

The occurrence of C as a fixed component of $|D|$ of multiplicity more than 1 results in a strict inequality $h^0(X, \mathcal{O}_X(D)) > (D^2 - K_X \cdot D)/2 + 1$.

The SHGH Conjecture says that whenever we have a divisor $D = dE_0 - m_1E_1 - \cdots - m_rE_r$ with $d, m_1, \ldots, m_r \geq 0$, (assuming that the E_i were obtained by blowing up $r \geq 3$ generic points of $\mathbb{P}^2$) then either $h^0(X, \mathcal{O}_X(D)) = \max(0, (D^2 - K_X \cdot D)/2 + 1)$ or there is an exceptional curve C (i.e., an effective divisor whose class is an element of the W_r-orbit of E_1) such that $C \cdot D < -1$. If $h^0(X, \mathcal{O}_X(D)) > 0$, it is easy to find all such C and subtract them off, leaving one with F such that $h^0(X, \mathcal{O}_X(F)) = (F^2 - K_X \cdot F)/2 + 1$. (If $D \cdot C \geq 0$ for all C, one can show that $[D]$ can be reduced by W_r to a nonnegative linear combination of the classes e_0, $e_0 - e_1, 2e_0 - e_1 - e_2, 3e_0 - e_1 - e_2 - e_3, \cdots, 3e_0 - e_1 - \cdots - e_r$; see [27].)

The SHGH Conjecture is known to hold for $r \leq 9$.

Example 9.1. Consider the fat point subscheme $Z = 13p_1 + 13p_2 + 10p_3 + \cdots + 10p_7$ for generic points $p_i \in \mathbb{P}^2$. We determine the Hilbert function of $I = I(Z)$. First $H_I(28) = 0$. We have $H_I(28) = h^0(X, \mathcal{O}_X(D))$ for the divisor $D = 28E_0 - 13E_1 - 13E_2 - 10E_3 - \cdots - 10E_7$. But $[D]$ reduces via W_7 to $-2e_0 + 2e_4 + 2e_5 + 5e_6 + 5e_7$, so $h^0(X, \mathcal{O}_X(D)) = h^0(X, \mathcal{O}_X(D'))$, where $D' = -2E_0 + 2E_4 + 2E_5 + 5E_6 + 5E_7$. The occurrence of a negative coefficient for e_0 means $h^0(X, \mathcal{O}_X(D')) = 0$, hence $H_I(t) = 0$ for $t < 29$. Now consider $D = 29E_0 - 13E_1 - 13E_2 - 10E_3 - \cdots - 10E_7$. Then via the action of W_7 we obtain $D' = 4E_0 - E_1 - \cdots - E_5 + 2E_6 + 2E_7$. As in Exercise 8.5, we can subtract off $2E_6 + 2E_7$ to get $F = D - (2E_6 + 2E_7) = 4E_0 - E_1 - \cdots - E_5 = (E_0) + (3E_0 - E_1 - \cdots - E_5)$. Thus $F \cdot C \geq 0$ for all exceptional C, so by the SHGH Conjecture $H_I(29) = h^0(X, \mathcal{O}_X(D)) = h^0(X, \mathcal{O}_X(D')) = h^0(X, \mathcal{O}_X(F)) = (F^2 - K_X \cdot F)/2 + 1 = 10$. Finally consider $D = 30E_0 - 13E_1 - 13E_2 - 10E_3 - \cdots - 10E_7$. Here we get $F = D' = 12E_0 - 4(E_1 + \cdots + E_5) - E_6 - E_7 = 3(3E_0 - E_1 - \cdots - E_5) + (3E_0 - E_1 - \cdots - E_7)$. Thus $D' \cdot C \geq 0$ for all exceptional C, so we get $H_I(30) = h^0(X, \mathcal{O}_X(D)) = h^0(X, \mathcal{O}_X(D')) = h^0(X, \mathcal{O}_X(F)) = (F^2 - K_X \cdot F)/2 + 1 = 39$. For $t \geq 30$ and $D = tE_0 - 13E_1 - 13E_2 - 10E_3 - \cdots - 10E_7$, we have $D = (t - 30)E_0 + (30E_0 - 13E_1 - 13E_2 - 10E_3 - \cdots - 10E_7)$ so $D \cdot C = (t - 30)E_0 \cdot C + C \cdot (30E_0 - 13E_1 - 13E_2 - 10E_3 - \cdots - 10E_7) \geq 0$ for all exceptional C, so $H_I(t) = h^0(X, \mathcal{O}_X(D)) = \max(0, (D^2 - K_X \cdot D)/2 + 1)$, but $(D^2 - K_X \cdot D)/2 + 1$ was positive for $t = 30$ and adding a nonnegative multiple of E_0 only makes it bigger so we have $H_I(t) = h^0(X, \mathcal{O}_X(D)) = (D^2 - K_X \cdot D)/2 + 1 = \binom{t+2}{2} - 2\binom{13+1}{2} - 5\binom{10+1}{2}$.

We close by relating the statement of the SHGH Conjecture given above to the special case stated in Conjecture 4.5. Consider $F = tE_0 - m(E_1 + \cdots + E_r)$, where $p_1, \ldots, p_r \in \mathbb{P}^2$ are $r \geq 9$ generic points of $\mathbb{P}^2$, X is the surface obtained by blowing up the points and E_i is the exceptional curve obtained by blowing up p_i. Let I be the radical ideal of the points. Then $H_{I^{(m)}}(t) = h^0(X, \mathcal{O}_X(F))$. For simplicity, we consider only the cases $t \geq 3m \geq 0$. Then $F = -mK_X + (t-3m)E_0$ with $t - 3m \geq 0$. But for any exceptional curve E we have $[E] = w([E_1])$ for some $w \in W_r$, so $-K_X \cdot E = -K_X \cdot E_1 = 1$ by Exercise 8.1. Since E is a curve on X, its image in $\mathbb{P}^2$ has nonnegative degree, so $E_0 \cdot E \geq 0$. Thus $F \cdot E \geq m \geq 0$. The SHGH Conjecture therefore asserts $H_{I^{(m)}}(t) = h^0(X, \mathcal{O}_X(F)) = \max\left(0, (D^2 - K_X \cdot D)/2 + 1\right) = \max\left(0, \binom{t+2}{2} - r\binom{m+1}{2}\right)$, as conjectured in Conjecture 4.5.

Exercises

Exercise 9.1. Find the Hilbert function of the ideal I of $Z = 12p_1 + 10p_2 + \cdots + 10p_8 \subset \mathbb{P}^2$, assuming the points are generic.

10 Solutions

2 *Affine Space and Projective Space*

Solution 2.1. Define a map * : $\mathcal{M}_{\leq t}(A) \to \mathcal{M}_t(R)$ by $x_1^{m_1} \cdots x_n^{m_n} \mapsto (x_1^{m_1} \cdots x_n^{m_n})^* = x_0^{m_0} x_1^{m_1} \cdots x_n^{m_n}$ where $m_0 = t - (m_1 + \cdots + m_n)$, and define a map $_*$: $\mathcal{M}_t(R) \to \mathcal{M}_{\leq t}(A)$ by evaluating x_0 at 1; i.e., by $x_0^{m_0} x_1^{m_1} \cdots x_n^{m_n} \mapsto (x_0^{m_0} \cdots x_n^{m_n})_* = x_1^{m_1} \cdots x_n^{m_n}$. If f is a monomial in $\mathcal{M}_{\leq t}(A)$, clearly $(f^*)_* = f$, while if $F \in \mathcal{M}_t(R)$, then just as clearly $(F_*)^* = F$. Thus * and $_*$ are inverse to each other and hence are bijections.

Solution 2.2. Pick $f \in I$ of degree $\alpha(I)$. Then $f^m \in I^m$, so $\alpha(I^m) \leq \deg(f^m) = m\alpha(I)$. Since J is homogeneous, J has a set of homogeneous generators $g_1, \ldots, g_r$, hence J^m is generated by products of m of the generators g_i (repeats allowed), the minimum degree of which is $m\alpha(J)$. But for any homogeneous elements $b_1, \ldots, b_t$ in R, where we assume (by reindexing if need be) that $\deg(b_1) \leq \deg(b_2) \leq \cdots \leq \deg(b_t)$, the ideal $(b_1, \ldots, b_t)$ is contained in M^s for $s = \deg(b_1)$, where M is the ideal generated by the variables. Since M^s is the span of the monomials of degree at least s, there are no elements in M^s (and hence none in $(b_1, \ldots, b_t)$) of degree less than s. Applied to J^m, we see that J^m has an element of degree $m\alpha(J)$ and no nonzero elements of degree less than that, hence $\alpha(J^m) = m\alpha(J)$.

Solution 2.3. Since $fM \subseteq I$ for all $f \in I$, we have $f \in P$ for all homogeneous $f \in I$. Thus $I \subseteq P$. If $J \subseteq M$ is any homogeneous ideal such that $J_t = I_t$ for all $t \gg 0$, then for any homogeneous $g \in J$ and for i large enough we have $gM^i \in J_t = I_t$, hence $g \in P$, so $J \subseteq P$. Thus P contains every nontrivial homogeneous ideal whose homogeneous components eventually coincide with those of I. Since P is finitely generated, there is an s large enough such that $fM^s \subset I$ for every generator f in a given finite set of homogeneous generators for P. Thus $PM^s \subseteq I$ for $s \gg 0$. But for degrees $t \geq \omega$, where ω is the maximum degree in a minimal set of generators of P, we have $P_tM_1 = P_{t+1}$, hence $(PM^i)_t = P_t$ for all $t \geq \omega + i$. Thus $P_t = (PM^i)_t \subseteq I_t \subset P_t$ for $t \gg 0$. Hence P is the largest ideal among all homogeneous ideals J such that $J_t = I_t$ for $t \gg 0$; i.e., $\text{sat}(I) = P$. Of course, by maximality of the saturation we always have $P \subseteq \text{sat}(P)$, but $(\text{sat}(P))_t = P_t = I_t$ for $t \gg 0$, hence $\text{sat}(P) \subseteq P$, so $P = \text{sat}(P)$.

3 Fat Points in Affine Space

Solution 3.1. Clearly $I(p_1)^{m_1} \cdots I(p_r)^{m_r} \subseteq \cap_{i=1}^r I(p_i)^{m_i}$. For the reverse inclusion, note that not every polynomial which vanishes at p_1 vanishes at p_2, so we can pick a polynomial f such that $f(p_1) = 0$ but $f(p_2) \neq 0$. Normalizing allows us to assume $f(p_2) = 1$. Let $g = 1 - f$. Then $f \in I(p_1)$, $g \in I(p_2)$ and $f + g = 1$. Writing $(f + g)^{m_1+m_2}$ as a linear combination of terms of the form $\binom{i+j}{j} f^i g^j$ with $i + j = m_1 + m_2$, each term is either in $I(p_1)^{m_1}$ or in $I(p_2)^{m_2}$. Thus we can write $1 = F + G$ where F is the sum of the terms in $I(p_1)^{m_1}$ and G is the sum of the terms in $I(p_2)^{m_2}$. Therefore every element $h \in I(p_1)^{m_1} \cap I(p_2)^{m_2}$ can be written $h = hF + hG \in I(p_1)^{m_1}I(p_2)^{m_2}$; i.e., $I(p_1)^{m_1} \cap I(p_2)^{m_2} = I(p_1)^{m_1}I(p_2)^{m_2}$. Similarly, $I(p_1)^{m_1} \cap I(p_2)^{m_2} \cap I(p_3)^{m_3} = I(p_1)^{m_1}I(p_2)^{m_2} \cap I(p_3)^{m_3} = I(p_1)^{m_1}I(p_2)^{m_2}I(p_3)^{m_3}$. Continuing in this way, we eventually have $I(p_1)^{m_1} \cap \cdots \cap I(p_r)^{m_r} = I(p_1)^{m_1} \cdots I(p_r)^{m_r}$.

Solution 3.2. (a) Since $I^{bc} = (I^b)^c$, we have $\alpha(I^{bc}) = \alpha((I^b)^c) \leq c\alpha(I^b)$ by Exercise 2.2. Now the result follows by dividing by bc.

(b) By (a), $\frac{\alpha(I^{m!})}{m!}$ is decreasing as m increases but is always positive, so it has a limit L.

(c) For any $\varepsilon > 0$, we will show for $t \gg 0$ that $L \leq \alpha(I^t)/t \leq L + \varepsilon$. For $m \gg 0$ we may assume that $L \leq \alpha(I^{m!})/m! \leq L + \varepsilon/2$. For $t \geq m!$ we can write $t = s \cdot m! + d$ for some $0 \leq d < m!$. Then $I^{(s+1)m!} \subseteq I^t$, so $\alpha(I^t) \leq \alpha(I^{(s+1)m!}) \leq (s+1)\alpha(I^{m!})$, so

$$L \leq \frac{\alpha(I^{t!})}{t!} \leq \frac{\alpha(I^t)}{t} \leq \frac{(s+1)\alpha(I^{m!})}{s \cdot m! + d}$$

$$= \frac{s\alpha(I^{m!})}{s \cdot m! + d} + \frac{\alpha(I^{m!})}{s \cdot m! + d} \leq \frac{\alpha(I^{m!})}{m!} + \frac{\alpha(I^{m!})}{s \cdot m!} \leq L + \frac{\varepsilon}{2} + \frac{\alpha(I^{m!})}{s \cdot m!},$$

but for $s \gg 0$ (i.e., for $t \gg m!$), we have $\alpha(I^{m!})/(s \cdot m!) \leq \varepsilon/2$. The fact that $\lim_{m\to\infty} \frac{\alpha(I^m)}{m} \leq \frac{\alpha(I^t)}{t}$ for all $t \geq 1$ follows from (a) and (b).

Solution 3.3. The vector space $A_{\leq t}$ has basis consisting of monomials μ of degree at most t in the n variables $X_1, \dots, X_n$. By introducing an extra variable X_0, we can create a bijection between the monomials of degree t in $X_0, \dots, X_n$ and the monomials μ of degree at most t in $X_1, \dots, X_n$ (given by multiplying each such μ by X_0^i where $i = t - \deg(\mu)$). Now see Exercise 3.4.

Solution 3.4. We must count the number of arrangements of n ones and t zeros, since such arrangements are in bijection with the monomials in $n+1$ variables of degree t (for example, 001011 is the monomial $x_0^2 x_1$, since there are 2 zeros before the first 1, giving x_0^2, 1 zero immediately before the second 1, giving x_1^1, and no zeros immediately before the third 1 or the fourth 1, giving x_2^0 and x_3^0, and so altogether $x_0^2 x_1^1 x_2^0 x_3^0$). But the number of arrangements of n ones and t zeros is $\binom{t+n}{n}$.

Solution 3.5. Let $q = (0, \dots, 0) \in \mathbb{A}^n$. There is an automorphism $\psi : \mathbb{A}^n \to \mathbb{A}^n$ taking p to q, given by translation $(b_1, \dots, b_n) \mapsto (b_1 - a_1, \dots, b_n - a_n)$. The corresponding automorphism on rings is $\psi^* : K[X_1, \dots, X_n] \to K[X_1, \dots, X_n]$ where $X_i \mapsto X_i + a_i$. Note that $\psi^*(I(q)^m) = I(p)^m$ and that ψ^* induces vector space bijections $A_{\leq t} \to A_{\leq t}$ and $(I(q)^m)_{\leq t} \to (I(p)^m)_{\leq t}$. Thus it is enough to consider the case that $a_i = 0$ for all i. In this case $I = (X_1, \dots, X_n)$ is a monomial ideal, and hence homogeneous. Thus $\alpha(I^m) = m\alpha(I) = m$. Therefore, $t < m$ implies $H^{\leq}_{I^m}(t) = 0$. If $t < m$, let $t + i = m$ for some $i > 0$. Then $\binom{t+n}{n} \leq \binom{m-1+n}{n}$ (look at Pascal's triangle) so $\binom{t+n}{n} - \binom{m+n-1}{n} \leq 0$, hence $H^{\leq}_{I^m}(t) \geq \binom{t+n}{n} - \binom{m+n-1}{n}$ with equality for $t = m - 1$. For $t \geq m$, $(I^m)_{\leq t}$ is spanned by the monomials of degree m through degree t. By introducing a variable X_0, we can regard these as being monomials of degree exactly t in $K[X_0, \dots, X_n]$ such that X_0 has exponent at most $t - m$: given any monomial μ in $X_1, \dots, X_n$ of degree $m \leq i \leq t$, $X_0^{t-i}\mu$ is a monomial in $X_0, \dots, X_n$ of degree t such that X_0 has exponent at most $t - m$. By Exercise 3.4, there are $\binom{t+n}{n}$ monomials in $X_0, \dots, X_n$ of degree t. The monomials in $X_0, \dots, X_n$ of degree t but for which X_0 has exponent more than $t - m$ are in bijective correspondence with the monomials in $X_0, \dots, X_n$ of degree $m-1$ (just multiply by X_0^{t-m+1}). There are thus $\binom{t+n}{n} - \binom{m-1+n}{n}$ monomials of degree t in $X_0, \dots, X_n$ for which X_0 has exponent at most $t - m$, hence $H^{\leq}_{I^m}(t) = \binom{t+n}{n} - \binom{m+n-1}{n}$.

Solution 3.6. There is a linear polynomial f defining the line through p_1 and p_2. Thus $f^m \in I^m$ so $\alpha(I^m) \leq m = m\alpha(I)$ (see Exercise 2.2). By Exercise 3.5, $H^{\leq}_{I(p_1)^m}(m-1) = 0$ and $H^{\leq}_{I(p_1)^m}(m) > 0$, so $\alpha(I(p_1)^m) = m$. But $I \subset I(p_1)$ so $I^m \subset I(p_1)^m$ hence $m = \alpha(I(p_1)^m) \leq \alpha(I^m)$ so $\alpha(I^m) = m = m\alpha(I)$. Now consider the second statement. Since p_1, p_2, p_3 are noncollinear, no linear polynomial can vanish at all three points. Thus $\alpha(J) \geq 2$. Let f_1 be the linear polynomial defining the line through p_2 and p_3, f_2 the linear polynomial defining the line through p_1 and p_3, and f_3 the linear polynomial defining the line through p_1 and p_2. If $m = 2s$, then $(f_1 f_2 f_3)^s$ has degree $3s = 3m/2$ but is in $I(p_1)^m \cap I(p_2)^m \cap I(p_3)^m = J^m$ so $\alpha(J^m) \leq 3m/2 < 2m \leq m\alpha(J)$. If $m = 2s + 1$, then $(f_1 f_2 f_3)^s f_1 f_2 \in J^m$, hence $\alpha(J^m) \leq 3s + 2 < 4s + 2 = 2m \leq m\alpha(J)$.

Solution 3.7. We have a vector space inclusion $\phi : A_{\leq t} \to A_{\leq t+1}$. Compose with the quotient $A_{\leq t+1}/I_{\leq t+1}$; the kernel is $I_{\leq t}$, hence ϕ induces an injective map $A_{\leq t}/I_{\leq t} \to A_{\leq t+1}/I_{\leq t+1}$.

Solution 3.8. The polynomials in $I(p_i)^{m_i}$ of degree at most t form a linear subspace of $A_{\leq t}$ defined by $\binom{m_i+n-1}{n}$ homogeneous linear equations. Thus $(I(m_1p_1 + \cdots + m_rp_r))_{\leq t}$ is a linear subspace defined by $\sum_i \binom{m_i+n-1}{n}$ homogeneous linear equations. Therefore $H_I^{\leq}(t) \geq \binom{t+n}{n} - \sum_i \binom{m_i+n-1}{n}$, with the inequality (as opposed to equality) arising since the equations need not be independent.

For the rest, note that by the Chinese Remainder Theorem we have an isomorphism $A/I \to \bigoplus_i A/I(p_i)^{m_i}$ in which $f + I \mapsto (f + I(p_1)^{m_1}, \ldots, f + I(p_r)^{m_r})$. But $A/I(p_i)^{m_i}$ is finite dimensional for each i (of dimension $\binom{m_i+n-1}{n}$ in fact), so for some d we have a surjection $A_{\leq d} \to \oplus_i A/I(p_i)^{m_i}$ and hence a surjection $A_{\leq t} \to \oplus_i A/I(p_i)^{m_i}$ for all $t \geq d$. Thus $A_{\leq t}/I_{\leq t} \cong \bigoplus_i A/I(p_i)^{m_i}$ for all $t \geq d$, so $H_{A/I}^{\leq}(t) = \dim_K(A_{\leq t}/I_{\leq t}) = \sum_i \binom{m_i+n-1}{n}$ and $H_I^{\leq}(t) = \binom{t+n}{n} - \sum_i \binom{m_i+n-1}{n}$.

Solution 3.9. By Exercise 3.8, $\alpha(I^m) \leq t$ if $\binom{t+n}{n} - r\binom{m+n-1}{n} > 0$. If we regard $\binom{t+n}{n}$ as being $(t+n)(t+n-1)\cdots(t+1)/n!$ and $\binom{m+n-1}{n}$ as being $(m+n-1)\cdots(m+1)m/n!$, then substitute $t = \lambda m$ (so $\lambda = t/m$); $\binom{t+n}{n} - r\binom{m+n-1}{n}$ becomes a polynomial in m of degree n with leading coefficient $\lambda^n - r$. Thus, for any integers t and m such that $t > m\sqrt[n]{r}$, $\binom{t'+n}{n} - r\binom{m'+n-1}{n}$ will be positive for $t' = ti$ and $m' = mi$ for $i \gg 0$. I.e., $\gamma(I) \leq \alpha(I^m)/m \leq t/m$, but we can choose integers t and m such that t/m is arbitrarily close to but bigger than $\sqrt[n]{r}$, hence $\gamma(I) \leq \sqrt[n]{r}$.

If $1 \leq r \leq n$, the points lie on a hyperplane, so $\gamma(I) \leq \alpha(I)/1 = 1$. But $\gamma(J) = 1$ if J is the ideal of any one of the points, so (as we saw for $r = n = 2$ in the proof of Proposition 3.4) $1 = \gamma(J) \leq \gamma(I)$ so $\gamma(I) = 1$.

Solution 3.10. Let $a = \inf\{\frac{t}{m} : \binom{t+n}{n} - s\binom{m+n-1}{n} > 0; m, t \geq 1\}$. We can rewrite $\binom{t+n}{n} - s\binom{m+n-1}{n} > 0$ as $\frac{(t+1)(t+2)}{2} - s\frac{m(m+1)}{2} > 0$, which is equivalent to $t^2 + 3t - s(m^2 + m) \geq 0$. This in turn becomes $m^2(l^2 - s) + m(3l - s) \geq 0$ if we substitute $t = lm$. If $l = t/m < \sqrt{s}$ and $s \geq 9$, then $l^2 - s < 0$ and $3l - s < 0$, hence $m^2(l^2 - s) + m(3l - s) < 0$, so $t^2 + 3t - s(m^2 + m) < 0$ and therefore $\binom{t+n}{n} - s\binom{m+n-1}{n} \leq 0$. It follows that $a \geq \sqrt{s}$. But if $l = t/m > \sqrt{s}$, then the leading coefficient $l^2 - s$ of $m^2(l^2 - s) + m(3l - s)$ is positive, hence for $i \gg 0$, $(im)^2(l^2 - s) + im(3l - s) > 0$. Therefore $\binom{it+n}{n} - s\binom{im+n-1}{n} > 0$, so $a \leq (ti)/(mi) = t/m = l$ for all rationals $l > \sqrt{s}$, hence $a \leq \sqrt{s}$ so $a = \sqrt{s}$.

Solution 3.11. Say $p = (a_1, \ldots, a_n)$. Then $A = K[X_1, \ldots, X_n] = K[Y_1, \ldots, Y_n]$, where $Y_i = X_i - a_i$, and $I(p) = (X_1 - a_1, \ldots, X_n - a_n) = (Y_1, \ldots, Y_n)$. Given any element $f \in A$, it has the same degree whether expressed in terms of the variables Y_i or in terms of the X_i, but $A/(I(p))^m = A/(Y_1, \ldots, Y_n)^m$, so every element of $A/(I(p))^m$ is the image of an element of degree at most $m - 1$. Moreover, if $f(p) = 0$, then $f \in I(p)$, so $\overline{f}$ is nilpotent (since $\overline{f}^m = 0$) hence not a unit.

And if $f(p) \neq 0$, let $g = (f - f(p))/f(p)$. Then $\overline{g}^m = 0$, so $(1+\overline{g})(1+(-\overline{g})+(-\overline{g})^2+\cdots+(-\overline{g})^{m-1}) = 1$. Thus $\overline{f} = f(p)(1+\overline{g})$ is a unit since $f(p)$ and $1+\overline{g}$ are units.

Solution 3.12. If $f(p) = 0$ for all $p \in \mathbb{A}^n$, then $f \in \sqrt{(0)}$ by the Nullstellensatz, hence $f = 0$, contrary to assumption.

Solution 3.13. If $n = 1$, this is clear, so assume $n > 1$. For each i and j, consider the vector v_{ij} from p_i to p_j. Then it suffices to find f such that $f(v_{ij}) \neq 0$ for all $i \neq j$; i.e., given finitely many points $[v_{ij}] \in \mathbb{P}^{n-1}$, we must find a linear form $f \in K[\mathbb{P}^{n-1}]$ such that $f(v_{ij}) \neq 0$ for all $i \neq j$. I.e., regarding linear forms as points in the dual space $(\mathbb{P}^{n-1})^*$ and points v_{ij} as hyperplanes in $(\mathbb{P}^{n-1})^*$, we must find a point in $(\mathbb{P}^{n-1})^*$ not on any of a finite set of hyperplanes. But we can think of a point of $(\mathbb{P}^{n-1})^*$ as giving a point of $\mathbb{A}^n$ (unique up to multiplication by nonzero scalars) and vice versa, and we can think of hyperplanes in $(\mathbb{P}^{n-1})^*$ as giving codimension 1 linear subspaces of $\mathbb{A}^n$ and vice versa, so the result follows from Exercise 3.12.

Solution 3.14. Let $s = t - (m_1 + \cdots + m_r - 1)$ and let g be a degree 1 polynomial that does not vanish at any of the points p_i (start with any g with $\deg(g) = 1$ and replace g by $g - c$, where $c \in K \setminus \{g(p_1), \ldots, g(p_r)\}$). By Exercise 3.13 we can pick a linear form f such that $f(p_i) \neq f(p_j)$ whenever $p_i \neq p_j$. Define $f_i = g^s \Pi_{j \neq i}(f - f(p_j))^{m_j}$. Note that $\deg(f_i) = m_1 + \cdots + m_r - m_i + s = t - (m_i - 1)$, and that $f_i \in I(p_j)^{m_j}$ for all $j \neq i$, but by Exercise 3.11 f_i maps to a unit $\overline{f_i}$ in $A/I(p_i)^{m_i}$ under the quotient homomorphism $\phi_i : A \to A/I(p_i)^{m_i}$.

Given any element $(\overline{a_1}, \ldots, \overline{a_r}) \in \bigoplus_j A/I(p_j)^{m_j}$, we can by Exercise 3.11 pick elements $b_j \in A_{\leq(m_j-1)}$ such that $\phi_j(b_j) = \overline{f_j}^{-1}\overline{a_j}$, for $j = 1, \ldots, r$. Consider the homomorphism $\phi : A \to \bigoplus_j A/I(p_j)^{m_j}$ defined by $\phi(h) = (\phi_1(h), \ldots, \phi_r(h))$. Taking $h = \sum_j f_j b_j$, we see $\phi(h) = (\overline{a_1}, \ldots, \overline{a_r})$, and since $\deg(h) \leq \max_j\{\deg(f_j b_j)\} = \max_j\{t - (m_j - 1) + m_j - 1\} = t$, we see that $\phi(A_{\leq t}) = \bigoplus_j A/I(p_j)^{m_j}$. This gives the result, since $\bigoplus_j A/I(p_j)^{m_j} = \phi(A_{\leq t}) \cong A_{\leq t}/I_{\leq t}$, hence $H_I^{\leq}(t) = \dim(I_{\leq t}) = \dim(A_{\leq t}) - \dim(\bigoplus_j A/I(p_j)^{m_j}) = \binom{t+n}{n} - \sum_j \binom{m_j+n-1}{n}$.

Note that $H_I^{\leq}(t) = \binom{t+n}{n} - \sum_i \binom{m_i+n-1}{n}$ is equivalent to $\phi|_{A_{\leq t}}$ being surjective. Thus to show that $H_I^{\leq}(t) > \binom{t+n}{n} - \sum_i \binom{m_i+n-1}{n}$ it is enough to show $\phi|_{A_{\leq t}}$ is not surjective. Suppose the points are collinear. Let L be the line containing the points. Then we have a commutative diagram

$$\begin{array}{ccc} A & \to & \bigoplus_j A/I(p_j)^{m_j} \\ \downarrow & & \downarrow \\ \overline{A} = K[L] = K[X] & \to & \bigoplus_j \overline{A}/\overline{I}(p_j)^{m_j} \end{array}$$

where $\overline{A} = K[L] = K[X]$ is the coordinate ring of the line L, hence a polynomial ring in a single variable X, and $\overline{I}(p_j)$ is the ideal in $\overline{A}$ of the point p_j. The upper horizontal arrow is ϕ, the lower one is the corresponding homomorphism $\overline{\phi}$ for dimension 1. The vertical arrows are the usual quotients, and are therefore surjective. Thus to show $\phi|_{A_{\leq t}}$ is not surjective for $t < m_1 + \cdots + m_r - 1$, it is enough to show $(K[X])_{\leq t} \to \bigoplus_j \overline{A}/\overline{I}(p_j)^{m_j}$ is not surjective; i.e., it is enough to consider the case $n = 1$. But then $H_I^{\leq}(t) \geq 0$, while $\binom{t+1}{1} - \sum_i \binom{m_i}{1} = t + 1 - \sum_i m_i < 0$ if $t < m_1 + \cdots + m_r - 1$.

4 Fat Points in Projective Space

Solution 4.1. First, $I(p_1)^{m_1} \cdots I(p_r)^{m_r} \subseteq \cap_{i=1}^r I(p_i)^{m_i}$ and $\alpha(I(p_1)^{m_1} \cdots I(p_r)^{m_r}) = m_1 + \cdots + m_r$. Now, by pairing the points up p_1 with p_2, p_3 with p_4, etc. (there will be a point left over if r is odd), we can pick a linear form that vanishes on p_1 and p_2, and a linear form that vanishes on p_3 and p_4, etc. (if r is odd, just pick any line through the leftover point). Raising the first to the power $\max(m_1, m_2)$, the second to the power $\max(m_3, m_4)$, etc., and then multiplying the results together we obtain a form of degree $\max(m_1, m_2) + \max(m_3, m_4) + \cdots$ in $\cap_{i=1}^r I(p_i)^{m_i}$. But $\max(m_1, m_2) + \max(m_3, m_4) + \cdots < m_1 + \cdots + m_r$, hence $I(p_1)^{m_1} \cdots I(p_r)^{m_r} \neq \cap_{i=1}^r I(p_i)^{m_i}$.

Solution 4.2. Choose a linear form F that does not vanish at any of the points p_i. (This is always possible if the field K is large enough, but might not be possible if K is finite.) Let $G \in R_t$. By a linear change of coordinates, we may assume $F = x_0$. Recall the map δ_t defined right after Remark 4.4. If $FG \in I_{t+1}$, then $\delta_t(G) = \delta_{t+1}(x_0 G) = \delta_{t+1}(FG) \in (I_A)_{\leq t+1}$, but $\deg(\delta_t(G)) \leq t$, so $\delta_t(G) \in (I_A)_{\leq t}$, hence $G = \eta_t(\delta_t(G)) \in (I_R)_t = I_t$. Thus multiplication by F gives an injection $(R/I)_t \to (R/I)_{t+1}$, hence $H_{R/I}(t) \leq H_{R/I}(t+1)$ for all $t \geq 0$. (Alternatively, one could also approach this via a primary decomposition $I = \cap_i Q_i$. The primes corresponding to the primary components of the primary decomposition of I are just the ideals $I(p_i)$ of the points; i.e., $\sqrt{Q_i} = I(p_i)$. By hypothesis, $F \notin I(p_i)$ for the points p_i, hence for each i we have $F^j \notin Q_i$ for all $j \geq 1$. But $FG \in I$ implies $FG \in Q_i$ for all i; since no power of F is in Q_i we must have $G \in Q_i$ for all i, hence $G \in I$. Thus multiplication by F gives an injection $R/I \to R/I$, and since F is homogeneous of degree 1, this means multiplication by F gives an injection $(R/I)_t \to (R/I)_{t+1}$ for each $t \geq 0$.)

Solution 4.3. By Exercise 4.2 or Exercise 3.7 we know that $H_{R/I}$ is nondecreasing. By Eq. (1) and Exercise 3.14, $H_{R/I}(t) = \sum_i \binom{m_i+n-1}{n}$ for $t \gg 0$. Thus it is enough to show $H_{R/I}(s) = H_{R/I}(s+1)$ implies $H_{R/I}(s+1) = H_{R/I}(s+2)$ (and hence by induction $H_{R/I}(t)$ is constant, and in fact equal to $\sum_i \binom{m_i+n-1}{n}$, for all $s \geq c$).

Choose linearly independent linear forms $F_0, \ldots, F_n$ such that none of the F_j vanish at any of the p_i. By Exercise 4.2, multiplication by any F_j gives injective vector space homomorphisms $\lambda_{j,t} : R_t/I_t \to R_{t+1}/I_{t+1}$ for all $t \geq 0$. If $H_{R/I}(s) = H_{R/I}(s+1)$, then $\lambda_{j,s}$ is an isomorphism for all j. Thus, since multiplication is commutative, for all i and j we have $\lambda_{i,s+1}(R_{s+1}/I_{s+1}) = \lambda_{i,s+1}\lambda_{j,s}(R_s/I_s) = \lambda_{j,s+1}\lambda_{i,s}(R_s/I_s) = \lambda_{j,s+1}(R_{s+1}/I_{s+1})$. But $F_0, \dots, F_n$ generate R; in particular, $F_0 R_t + \cdots + F_n R_t = R_{t+1}$ for all $t \geq 0$, so $\sum_i F_i(R_t/I_t) = R_{t+1}/I_{t+1}$, hence $R_{s+2}/I_{s+2} = \sum_j \lambda_{j,s+1}(R_{s+1}/I_{s+1}) = \sum_j \lambda_{i,s+1}(R_{s+1}/I_{s+1}) = \lambda_{i,s+1}(R_{s+1}/I_{s+1})$, so $H_{R/I}(s+2) = \dim \lambda_{i,s+1}(R_{s+1}/I_{s+1}) = \dim R_{s+1}/I_{s+1} = H_{R/I}(s+1)$. (Alternatively, let F be a linear form not vanishing at any of the points. By Exercise 4.2, F induces an injection $(R/I)_t \to (R/I)_{t+1}$. So we have an exact sequence

$$0 \to (R/I)_t \overset{\times F}{\to} (R/I)_{t+1} \to (R/(I,F))_{t+1} \to 0.$$

The module on the right is a standard graded algebra, so it cannot be zero in one degree and nonzero in the next.)

Solution 4.4. Let $J = (x^2y, xy^3)$. Then $H_{R/J} = (1, 2, 3, 3, 2, 2, \ldots)$.

Solution 4.5. Let $I = I_R(p_1 + \cdots + p_r) \subset K[\mathbb{P}^2]$ be the ideal of $r \geq 9$ generic points $p_i \in \mathbb{P}^2$. By Exercise 3.9, we have $\gamma(I) \leq \sqrt{r}$. By Conjecture 4.5, it is enough now to show for $r \geq 9$ and $m > 0$ that $t < m\sqrt{r}$ implies $\binom{t+2}{2} - r\binom{m+1}{2} < 1$, since then $\alpha(I^{(m)})/m > t/m$ for all $t/m < \sqrt{r}$ and hence $\gamma(I) = \lim_{m\to\infty} \alpha(I^{(m)})/m \geq \sqrt{r}$. But $\binom{t+2}{2} - r\binom{m+1}{2} = (t^2 + 3t + 2 - r(m^2 + m))/2$, and $(t^2 + 3t + 2 - r(m^2 + m))/2 < 0$ for $t = 0$, so (since $(t^2 + 3t + 2 - r(m^2 + m))/2$ is strictly increasing as a function of t for $t \geq 0$) it suffices now to show $(t^2 + 3t + 2 - r(m^2 + m))/2 \leq 1$ for $t = m\sqrt{r}$. But $(t^2+3t+2-r(m^2+m))/2 = 1-m(r-3\sqrt{r})/2$ for $t = m\sqrt{r}$, and $r-3\sqrt{r} \geq 0$ for $r \geq 9$, so the result follows.

Solution 4.6. Since $I^{((m-1+n)t)} \subseteq (I^{(m)})^t$ we have

$$t\alpha(I^{(m)}) = \alpha((I^{(m)})^t) \leq \alpha(I^{(t(n+m-1))}),$$

so dividing by $t(n+m-1)$ and taking the limit as $t \to \infty$ gives

$$\frac{\alpha(I^{(m)})}{n+m-1} \leq \gamma(I).$$

Solution 4.7. Say $r \geq m$; then $I(p_i)^r \subseteq I(p_i)^m$, hence

$$I^r \subseteq I^{(r)} = \cap_i I(p_i)^r \subseteq \cap_i I(p_i)^m = I^{(m)}.$$

Conversely, assume $I^r \subseteq I^{(m)}$. By Eq. (1), for all t we have $H_{I_R^{(m)}}(t) = H^{\leq}_{I_A^m}(t)$ and $H_{I_R^{(r)}}(t) = H^{\leq}_{I_A^r}(t)$. By Remark 4.2 we have $H_{I_R^r}(t) = H_{I_R^{(r)}}(t)$ for $t \gg 0$, and thus $H_{I_R^r}(t) = H^{\leq}_{I_A^r}(t)$. By Exercise 3.8 for $t \gg 0$ we have $H^{\leq}_{I_A^r}(t) = \binom{t+n}{n} - s\binom{r+n-1}{n}$

and $H^{\leq}_{I^m_A}(t) = \binom{t+n}{n} - s\binom{m+n-1}{n}$. But $I^r \subseteq I^{(m)}$ implies $(I^r)_t \subseteq (I^{(m)})_t$ and thus, for $t \gg 0$ we have

$$\binom{t+n}{n} - s\binom{m+n-1}{n} = H^{\leq}_{I^m_A}(t) = H_{I^{(m)}_R}(t)$$

$$\geq H_{I^r_R}(t) = H^{\leq}_{I^r_A}(t) = \binom{t+n}{n} - s\binom{r+n-1}{n},$$

hence $s\binom{m+n-1}{n} \leq s\binom{r+n-1}{n}$ for $t \gg 0$. But as is easy to see by looking at Pascal's triangle, $\binom{j+n-1}{n}$ is an increasing function of j, so we conclude $m \leq r$.

5 *Examples: Bounds on the Hilbert Function of Fat Point Subschemes of* $\mathbb{P}^2$

Solution 5.1. Note that $\Delta H_{R/I(Z)} = \text{diag}(\mathbf{d})$ where $\mathbf{d} = (r_1, \ldots, r_s)$; the given answer is just diag($\mathbf{d}$) for this $\mathbf{d}$. It is tedious to write this out; examine some dot diagrams.

Solution 5.2. Consider the reduction vector $\mathbf{d} = (9, 8, 7, 6, 3, 2, 1)$ obtained from the sequence of lines $L_0, L_1, L_2, L_3, L_0, L_1, L_2$. Then

$$\Delta H_{R/I(Z)} = \text{diag}(\mathbf{d}) = (1, 2, 3, 4, 5, 6, 7, 4, 4, 0, 0, \ldots).$$

Solution 5.3. We obtain $\mathbf{d}$ by construction. Every dot in the dot diagram which we use to compute diag($\mathbf{d}$) is on a diagonal line with x-intercept at most $m_1 + \cdots + m_r - 1$, hence for $t \geq m_1 + \cdots + m_r - 1$, $H_{R/I(Z)}(t) \geq N$, where N is the total number of dots, but the number of dots is $N = \sum_i \binom{m_i+1}{2}$, and we know $\sum_i \binom{m_i+1}{2} \geq H_{R/I(Z)}(t)$ for all $t \geq 0$. Thus $\sum_i \binom{m_i+1}{2} \geq H_{R/I(Z)}(t)$ for all $t \geq 0$ and $H_{R/I(Z)}(t) \geq \sum_i \binom{m_i+1}{2}$ for $t \geq m_1 + \cdots + m_r - 1$, so $H_{R/I(Z)}(t) = \sum_i \binom{m_i+1}{2}$ for all $t \geq m_1 + \cdots + m_r - 1$.

6 *Hilbert Functions: Some Structural Results*

Solution 6.1. One solution is to use Theorem 5.1 to find a reduction vector $\mathbf{d}$. Pick a line L through p and let $L_1 = L_2 = L_3 = L$. The corresponding reduction vector is $\mathbf{d} = (3, 2, 1)$. By the theorem, $\Delta H_{R/I} = (1, 2, 3, 0, 0, \ldots)$. But if we pick three distinct lines L'_1, L'_2, L'_3 and on L'_1 we pick three points, on L'_2 we pick two points and on L'_3 we pick one point, where we avoid ever picking a point where two lines cross, and if we let Z be the union of these six points, then by the theorem $\Delta H_{R/I(Z)} = \text{diag}(\mathbf{d})$. Hence, I and $I(Z)$ have the same Hilbert function. Alternatively, it is not hard to work out the Hilbert function of a power of the ideal of a single point. Doing so gives $H_{R/I} = (1, 3, 6, 6, 6, \ldots)$, so another solution is to work backwards to find $\mathbf{d}$, given the fact asserted by Theorem 6.4 that $\Delta H_{R/I} = \text{diag}(\mathbf{d})$. Thus $\Delta H_{R/I} = (1, 2, 3, 0, 0, \ldots)$ so $\mathbf{d} = (3, 2, 1)$. Now proceed as in the first solution to obtain Z with Hilbert function $H_{R/I}$.

Solution 6.2. It is enough to show that $\mathbf{d} \neq \mathbf{d}'$ implies that $\text{diag}(\mathbf{d}) \neq \text{diag}(\mathbf{d}')$. Say $\mathbf{d} = (d_1, \ldots, d_r)$ and $\mathbf{d}' = (d_1', \ldots, d_s')$. By assumption, $\mathbf{d}$ and $\mathbf{d}'$ are decreasing. We will prove the contrapositive, so assume $\text{diag}(\mathbf{d}) = \text{diag}(\mathbf{d}')$. If $d_1 < d_1'$, then $d_i < d_1 < d_1'$, for all $1 < i \leq s$, hence the entries of $\text{diag}(\mathbf{d})$ for degrees t with $d_1 \leq t < d_1'$ will be 0 but nonzero for $\text{diag}(\mathbf{d}')$ (where we note the degree 0 entry is the first entry, the degree 1 entry is the second entry, etc.). Thus $d_1 \leq d_1'$ and by symmetry we have $d_1 = d_1'$. Therefore after deleting the first entries of $\mathbf{d}$ and $\mathbf{d}'$ we get $\text{diag}((d_2, \ldots, d_r)) = \text{diag}((d_2', \ldots, d_s'))$ and we repeat the argument. Eventually we obtain $d_i = d_i'$ for all i, and so $r = s$, and thus $\mathbf{d} = \mathbf{d}'$.

7 *Bézout's Theorem in* $\mathbb{P}^2$ *and Applications*

Solution 7.1. Say $F(p) \neq 0$. Then $F \notin I(p)^m$ for $m \geq 1$, so

$$(x_0, x_1, x_2) \subseteq \sqrt{(F, G) + I(p)^m}$$

by the Nullstellensatz, hence for t large enough $(x_0, x_1, x_2)_t = ((F, G) + I(p)^m)_t$ so $\dim R_t/((F, G) + I(p)^m)_t = 0$.

Solution 7.2. Since x_0 does not vanish at p, we have $I_p(x_0 F, G) = I_p(F, G) + I_p(x_0, G) = I_p(F, G)$. But $(x_0 F, G) = (x_0 F - G, G)$, so $I_p(x_0 F - G, G) = I_p(x_0 F, G)$. But $x_0 F - G = x_2^2(x_2 - x_0)$ so $I_p(x_0 F - G, G) = I_p(x_2^2(x_2 - x_0), G) = I_p(x_2^2, G) + I_p(x_2 - x_0, G) = 2I_p(x_2, G) + 0 = 2 \cdot 1$.

To compute $\sum_{p \in \mathbb{P}^2} I_p(F, G)$, it's enough to consider only those points $p \in \mathbb{P}^2$ where both F and G vanish; i.e., $\sum_{p \in \mathbb{P}^2} I_p(F, G) = I_{(1,0,0)}(F, G) + I_{(0,1,0)}(F, G) + I_{(1,1,1)}(F, G)$. We just found $I_{(1,0,0)}(F, G) = 2$. Similarly, we find $I_{(0,1,0)}(F, G) = 3$. At $p = (1, 1, 1)$, the tangent to F at p is $x_1 - 2x_2$ and the tangent to G at p is $x_1 - 3x_2$. These are different, so $I_{(1,1,1)}(F, G) = \text{mult}_p(F)\,\text{mult}_p(G) = 1$, hence $\sum_{p \in \mathbb{P}^2} I_p(F, G) = 6 = \deg(F)\deg(G)$.

Solution 7.3. See [11, Example 4.2.3] or [4, Lemma 8.4.7].

Solution 7.4. Consider $I^{(m)} = I(m(p_1 + p_2 + p_3 + p_4))$. Suppose $0 \neq F \in (I^{(m)})_{2m-1}$. Note that F vanishes to order at least m at each of two points on any line L_{ij} through two of the points p_i, p_j. Since $2m > 2m - 1$, this means by Bézout that the linear forms (also denoted L_{ij}) defining the lines are factors of F.

Dividing F by $B = L_{12}L_{13}L_{14}L_{23}L_{24}L_{34}$ we obtain a form G of degree $2m-7$ in $(I^{(m-3)})_{2m-7}$. The same argument applies: B must divide G. Eventually we obtain a form of degree less than 6 divisible by B, which is impossible. Thus $F = 0$, and $\alpha(I^{(m)}) > 2m-1$. Since $(L_{12}L_{34})^m \in I^{(m)}$, we see that $\alpha(I^{(m)}) \leq 2m$, thus $\alpha(I^{(m)}) = 2m$.

Solution 7.5. Since $H_I(2) \geq \binom{2+2}{2} - 5 = 1$, there is a nonzero form $F \in I_2$, hence $\gamma(I) \leq \alpha(I)/1 = 2$. If F were reducible, it would be a product of two linear forms, and hence three of the points would be collinear. Thus F is irreducible. Now let $0 \neq G \in I^{(m)}_{2m-1}$. By Bézout, F and G have a common factor, but F is irreducible, so $F|G$; say $FB = G$, hence $B \in I^{(m-1)}_{2(m-1)-1}$. Again we see that $F|B$, etc. Eventually we find that F divides a form of degree less than 2, which is impossible. Thus $I^{(m)}_{2m-1} = 0$, so $\alpha(I^{(m)}) \geq 2m$, so $\gamma(I) \geq 2m/m = 2$.

Solution 7.6. Pick five points $p_1, \ldots, p_5$ on an irreducible conic C, defined by an irreducible form F. Note that no three of these five points are collinear (else the line through the three is a component of the conic, which can't happen since the conic is irreducible). Pick any point p_6 not on C and not on any line through any two of the other points. If there were a nonzero form G of degree 2 such that G vanished at all six points, then $F|G$ by Bézout, hence G is a constant times F, so F would also have to vanish at p_6.

Solution 7.7. By Proposition 3.4 we know $\gamma(I) \leq 12/5$. As in the solution to Exercise 7.5, there is an irreducible form of degree 2 which vanishes at any five of the six points, and by hypothesis each such form does not vanish at the sixth point. Let F_i be the degree 2 form that vanishes at all of the points but p_i. Thus $F = F_1 \cdots F_6 \in (I^{(5)})_{12}$. Say $0 \neq G \in (I^{(5m)})_{12m-1}$. Then Bézout implies that each F_i divides G, hence $F|G$, so $B = G/F \in (I^{(5(m-1))})_{12(m-1)-1}$. The argument can be repeated, and eventually we find that F divides a form of degree less than the degree of F. Hence $(I^{(5m)})_{12m-1} = 0$, so $\alpha(I^{(5m)}) \geq 12m$, so $\gamma(I) \geq 12/5$.

Solution 7.8. Pick six points $p_1, \ldots, p_6$ which do not all lie on any conic, and no three of which are collinear. Any conic through any five of the points is irreducible, otherwise there is a line through 3 or more of the points. There is also at most one conic through any given five of the points, by Bézout's Theorem. (Alternatively, if there are two conics through the same five points, some linear combination of the forms defining the conics would give a form vanishing at all six points.) Now pick any seventh point p_7 not on any conic through five of the points $p_1, \ldots, p_6$ and not on any line through any two of the points $p_1, \ldots, p_6$. Clearly, no three of the points $p_1, \ldots, p_7$ can be collinear (since no three of $p_1, \ldots, p_6$ are and since p_7 is not on any of the lines through two of the points $p_1, \ldots, p_6$), and by the same argument no six of the points $p_1, \ldots, p_7$ can be contained in any conic.

Solution 7.9. By Proposition 3.4 we know $\gamma(I) \leq 21/8$. Since $\binom{3+2}{2} - \binom{2+1}{2} - 6\binom{1+1}{2} > 0$, there is for each i a form F_i of degree 3 that vanishes at each point p_j but has multiplicity at least 2 at p_i. If F_i were reducible, it would either consist of a line and a conic with p_i at a point where the two meet, and then the remaining six points would have to be put on the line or the conic, so either the line would have 3 or the conic would have 6, contrary to hypothesis, or F_i would consist of three lines, with one point where two of the lines meet, and the other six points placed elsewhere on the three lines, but then one of the lines would have to contain at least three of the points. Since F_i is irreducible, it must have multiplicity exactly 2 at p_i and 1 at the other points, otherwise by Bézout the line through p_i and any p_j, $j \neq i$, would be a component.

Note $F = F_1 \cdots F_7 \in (I^{(8)})_{21}$. As usual, if there is a G with $0 \neq G \in (I^{(8m)})_{21m-1}$ we get a contradiction by repeated applications of Bézout. Thus $\alpha(I^{(8m)}) \geq 21m$, hence $\gamma(I) \geq 21/8$.

Solution 7.10. Clearly, $\alpha(I^{(m)}) \leq 3m$, since $F^m \in (I^{(m)})_{3m}$, where F is the cubic form defining C. If $0 \neq G \in (I^{(m)})_{3m-1}$, then $F|G$ by Bézout, and we get $B \in (I^{(m-1)})_{3(m-1)-1}$. Repeating this argument we eventually get a form of degree less than that of F which F divides. Hence $(I^{(m)})_{3m-1} = 0$, so $\alpha(I^{(m)}) \geq 3m$, so $\gamma(I) = 3$.

8 Divisors, Global Sections, the Divisor Class Group, and Fat Points

Solution 8.1. It is enough to check this for $w = s_i$ for all i, for the generators s_i of W_r given above. But $s_i(x) \cdot s_i(y) = (x + (x \cdot n_i)n_i) \cdot (y + (y \cdot n_i)n_i) = x \cdot y + 2(y \cdot n_i)(x \cdot n_i) + (x \cdot n_i)(y \cdot n_i)(n_i \cdot n_i) = x \cdot y$, and $n_i \cdot K_X = 0$ for all i, so $s_i(K_X) = K_X + (K_X \cdot n_i)n_i = K_X$.

Solution 8.2. We have

$$\frac{F^2 - K_X \cdot F}{2} + 1 = (t^2 - \sum_i m_i^2 + 3t - \sum_i m_i)/2 + 1 = (t^2 + 3t + 2)/2 - \sum_i (m_i^2 + m_i)/2$$

which is just $\binom{t+2}{2} - \sum_i \binom{m_i+1}{2}$. The rest is clear.

Solution 8.3. The class $16e_0 - 6e_1 - \cdots - 6e_8$ reduces by W_8 to $2e_0 - 6e_1 - 2e_2$, but a conic can't vanish to order more than 2 at a point. Thus $\dim I(6p_1 + \cdots + 6p_8)_{16} = \dim I(3p_1 + p_2)_1 = 0$, hence $\alpha(I(6p_1 + \cdots + 6p_8)) \geq 17$. However, $H_{I^{(6)}}(17) \geq \binom{19}{2} - 8\binom{7}{2} = 3 > 0$, thus $\alpha(I(6p_1 + \cdots + 6p_8)) \leq 17$, hence $\alpha(I(6p_1 + \cdots + 6p_8)) = 17$. Alternatively, $17e_0 - 6e_1 - \cdots - 6e_8$ reduces by W_8 to e_0, hence $\dim I(6p_1 + \cdots + 6p_8)_{17} = \dim I(0)_1 = \dim R_1 = 3$, and we achieve the same conclusion.

Solution 8.4. We have $C^2 = e_1^2 = -1 = e_1 \cdot K_X = C \cdot K_X$. Since E_1 is a smooth rational curve, so is C. Moreover, $((mC)^2 - K_X \cdot (mC))/2 + 1 = (-m^2 + m + 2)/2 \leq 0$ for all $m \geq 2$.

Solution 8.5. It follows from Bézout's Theorem that $h^0(X, \mathcal{O}_X(D)) = h^0(X, \mathcal{O}_X(F))$. But $C \cdot F = 0$, so $D^2 = (F + mC)^2 = F^2 - m^2$ and $-K_X \cdot D = -K_X \cdot (F + mC) = -K_X \cdot F + m$, so $(D^2 - K_X \cdot D)/2 = (F^2 - K_X \cdot F)/2 - (m^2 - m)/2 < (F^2 - K_X \cdot F)/2$. Thus $h^0(X, \mathcal{O}_X(D)) = h^0(X, \mathcal{O}_X(F)) \geq (F^2 - K_X \cdot F)/2 + 1 > (D^2 - K_X \cdot D)/2 + 1$.

9 *The SHGH Conjecture*

Solution 9.1. For $t < 29$, the class of $D = tE_0 - 12E_1 - 12E_2 - 10E_3 - \cdots - 10E_8$ reduces via W_8 to a divisor class where e_0 has a negative coefficient, so $H_I(t) = 0$ for $t \leq 28$. For $t = 29$ we get $D' = E_0 - 2E_8$, so $H_I(29) = h^0(X, \mathcal{O}_X(E_0)) = 3$. For $t > 29$, $D \cdot C \geq 0$ for all exceptional C, so $H_I(t) = \max(0, (D^2 - K_X \cdot D)/2 + 1)$ and $(D^2 - K_X \cdot D)/2 + 1$ turns out to be positive so we have $H_I(t) = (D^2 - K_X \cdot D)/2 + 1 = \binom{t+2}{2} - 2\binom{12+1}{2} - 6\binom{10+1}{2}$ for all $t \geq 30$.

Acknowledgments We thank the University of Regina for its thoughtful hospitality, the (other) organizers for their work to make the Workshop a reality, the funders for their financial support, the participants for their enthusiastic engagement at the Workshop, and T. Szemberg for his feedback on these notes. We also are very grateful for the referee's detailed comments and careful reading of these notes.

References

1. Alexander, J.W.: On the factorization of Cremona plane transformations. Trans. Am. Math. Soc. **17**, 295–300 (1916). http://www.ams.org/journals/tran/1916-017-03/S0002-9947-1916-1501043-X/S0002-9947-1916-1501043-X.pdf
2. Alexander, J., Hirschowitz, A.: An asymptotic vanishing theorem for generic unions of multiple points. Invent. Math. **140**, 303–325 (2000)
3. Ballico, E., Idà, M.: On the minimal free resolution for fat point schemes of multiplicity at most 3 in P2 Journal of Pure and Applied Algebra **212**(7), 1756–1769 (2008)
4. Bauer, T., Di Rocco, S., Harbourne, B., Kapustka, M., Knutsen, A., Syzdek, W., Szemberg, T.: A primer on Seshadri constants. Contemp. Math. **496**, 33–70 (2009). In: Bates, D.J., Besana, G.-M., Di Rocco, S., Wampler, C.W. (eds.) Interactions of Classical and Numerical Algebraic Geometry. Proceedings of a conference in honor of A. J. Sommese, held at Notre Dame, 22–24 May 2008, 362 pp. (arXiv:0810.0728)
5. Bocci, C., Cooper, S., Harbourne, B.: Containment results for ideals of various configurations of points in PN, Journal of Pure and Applied Algebra 218 (2014) 65–75
6. Bruns, W., Herzog, J.: Cohen–Macaulay Rings. Cambridge Studies in Advanced Mathematics, vol. 39. Cambridge University Press, Cambridge (1993)
7. Catalisano, M.V.: Fat points on a conic. Commun. Algebra **19**(8), 2153–2168 (1991)
8. Catalisano, M.V., Geramita, A.V., Gimigliano, A.: Ranks of tensors, secant varieties of Segre varieties and fat points. Linear Algebra Appl. **355**, 263–285 (2002)
9. Chudnovsky, G.V.: Singular points on complex hypersurfaces and multidimensional Schwarz Lemma. In: Bertin M.-J. (ed.) Séminaire de Théorie des Nombres, Paris 1979–80, Séminaire Delange-Pisot-Poitou. Progress in Mathematics, vol. 12. Birkhäuser, Boston (1981)

10. Ciliberto, C., Miranda, R.: Degenerations of planar linear systems. J. Reine Angew. Math. **501**, 191–220 (1998)
11. Cooper, S., Harbourne, B., Teitler, Z.: Combinatorial bounds on Hilbert functions of fat points in projective space. J. Pure Appl. Algebra **215**, 2165–2179 (2011). arXiv:0912.1915
12. Du Val, P.: On the Kantor group of a set of points in a plane. Proc. Lond. Math. Soc. **42**, 18–51 (1936)
13. Dumnicki, M., Szemberg, T., Tutaj-Gasińska, H.: A counter-example to a question by Huneke and Harbourne. J. Algebra **393**, 24–29 (2013). arXiv:1301.7440
14. Ein, L., Lazarsfeld, R., Smith, K.E.: Uniform behavior of symbolic powers of ideals. Invent. Math. **144**, 241–252 (2001). arXiv:math/0005098
15. Eisenbud, D., Harris, J.: The Geometry of Schemes, 294 pp. Springer, Berlin (2000)
16. Esnault, H., Viehweg, E.: Sur une minoration du degré d'hypersurfaces s'annulant en certains points. Math. Ann. **263**(1), 75–86 (1983)
17. Evain, L.: On the postulation of s^d fat points in $\mathbb{P}^d$. J. Algebra **285**, 516–530 (2005)
18. Fitchett, S., Harbourne, B., Holay, S.: Resolutions of fat point ideals involving eight general points of $\mathbb{P}^2$. J. Algebra **244**, 684–705 (2001)
19. Fulton, W.: Algebraic Curves: An Introduction to Algebraic Geometry, 121 pp. (2008). http://www.math.lsa.umich.edu/~wfulton/CurveBook.pdf
20. Geramita, A.V., Gregory, D., Roberts, L.G.: Monomial ideals and points in projective space. J. Pure Appl. Algebra **40**, 33–62 (1986)
21. Geramita, A.V., Maroscia, P., Roberts, L.G.: The Hilbert function of a reduced K-algebra. J. Lond. Math. Soc. **28**, 443–452 (1983)
22. Gimigliano, A.: On linear systems of plane curves. Thesis, Queen's University, Kingston (1987)
23. Gimigliano, A.: Our thin knowledge of fat points. In: Queen's Papers in Pure and Applied Mathematics, no. 83. The Curves Seminar at Queen's, vol. VI (1989)
24. Gimigliano, A., Idà, M.: The ideal resolution for generic 3-fat points in $\mathbb{P}^2$. J. Pure Appl. Algebra **187**(1–3), 99–128 (2004)
25. Gimigliano, A., Harbourne, B., Idà, M.: Betti numbers for fat point ideals in the plane: a geometric approach. Trans. Am. Math. Soc. **361**, 1103–1127 (2009)
26. Greene, C., Kleitman, D.J.: Proof Techniques in the Theory of Finite Sets. Studies in Mathematics, vol. 17, pp. 22–79. Mathematical Association of America, Washington (1978)
27. Harbourne, B.: Blowings-up of $\mathbb{P}^2$ and their blowings-down. Duke Math. J. **52**, 129–148 (1985)
28. Harbourne, B.: The geometry of rational surfaces and Hilbert functions of points in the plane. In: Proceedings of the 1984 Vancouver Conference in Algebraic Geometry, CMS Conference Proceedings, vol. 6, pp. 95–111. American Mathematical Society, Providence (1986)
29. Harbourne, B.: Free resolutions of fat point ideals on $\mathbb{P}^2$. J. Pure Appl. Algebra **125**, 213–234 (1998)
30. Harbourne, B.: The ideal generation problem for fat points. J. Pure Appl. Algebra **145**(2), 165–182 (2000)
31. Harbourne, B.: Global aspects of the geometry of surfaces. Ann. Univ. Paed. Cracov. Stud. Math. **9**, 5–41 (2010). arXiv:0907.4151
32. Harbourne, B., Huneke, C.: Are symbolic powers highly evolved? J. Ramanujan Math. Soc. **28**, 3 311–330 (2013) (arXiv:1103.5809)
33. Harbourne, B., Roé, J.: Linear systems with multiple base points in $\mathbb{P}^2$. Adv. Geom. **4**, 41–59 (2004)
34. Harbourne, B., Seceleanu, A.: Containment counterexamples for ideals of various configurations of points in $\mathbf{P}^N$, preprint (2013). arXiv:1306.3668
35. Harbourne, B., Holay, S., Fitchett, S.: Resolutions of ideals of quasiuniform fat point subschemes of $\mathbb{P}^2$. Trans. Am. Math. Soc. **355**(2), 593–608 (2003)
36. Hartshorne, R.: Connectedness of the Hilbert scheme. Math. Inst. Hautes Etud. Sci. **29**, 261–304 (1966)
37. Hartshorne, R.: Algebraic Geometry. Graduate Texts in Mathematics, vol. 52, 496 pp. Springer, New York (1977)

38. Hirschowitz, A.: Une conjecture pour la cohomologie des diviseurs sur les surfaces rationelles génériques. J. Reine Angew. Math. **397**, 208–213 (1989)
39. Hochster, M., Huneke, C.: Comparison of symbolic and ordinary powers of ideals. Invent. Math. **147**(2), 349–369 (2002). arXiv:math/0211174
40. Iarrobino, A.: Inverse system of a symbolic power III: thin algebras and fat points. Compos. Math. **108**, 319–356 (1997)
41. Idà, M.: The minimal free resolution for the first infinitesimal neighborhoods of n general points in the plane. J. Algebra **216**, 741–753 (1999)
42. Macaulay, F.S.: Some properties of enumeration in the theory of modular systems. Proc. Lond. Math. Soc. **26**, 531–555 (1927)
43. Migliore, J., Nagel, U.: Glicci ideals, preprint (2012). arXiv:1208.6517
44. Nagata, M.: On the 14-th problem of Hilbert. Am. J. Math. **81**, 766–772 (1959)
45. Nagata, M.: On rational surfaces II. Mem. Col. Sci. Univ. Kyoto Ser. A Math. **33**(2), 271–293 (1960)
46. Schenck, H.: Linear systems on a special rational surface. Math. Res. Lett. **11**, 697–713 (2004)
47. Schwede, K.: A canonical linear system associated to adjoint divisors in characteristic $p > 0$, 14 pp., preprint (2011). arXiv:1107.3833v2
48. Segre, B.: Alcune questioni su insiemi finiti di punti in geometria algebrica. In: Atti del convegno internazionale di geometria Algebrica di Torino, pp. 15–33 (1961)
49. Skoda, H.: Estimations L^2 pour l'opérateur $\hat{\partial}$ et applications arithmétiques. In: Séminaire P.Lelong (Analyse), 1975/76. Lecture Notes in Mathematics, vol. 578, pp. 314–323. Springer, Berlin (1977)
50. Stanley, R.P.: Hilbert functions of graded algebras. Adv. Math. **28**, 57–82 (1978)
51. Sylvester, J.J.: Observations on a new theory of multiplicity. Philos. Mag. **3**(20), 460–467 (1852)
52. Waldschmidt, M.: Propriétés arithmétiques de fonctions de plusieurs variables II. In: Séminaire P. Lelong (Analyse), 1975–76. Lecture Notes Mathematics, vol. 578, pp. 108–135. Springer, Berlin (1977)
53. Waldschmidt, M.: Nombres transcendants et groupes algébriques. Astérisque 69/70, Socéte Mathématiqué de France (1979)

Part II
Research/Survey Papers

This part of the volume consists of papers (part survey, part new research, and accessible to researchers and graduate students) focusing on interactions between Algebra, Combinatorics, and Geometry.

A Good Leaf Order on Simplicial Trees

Sara Faridi

Dedicated to Tony Geramita for his many contributions to Mathematics

Abstract Using the existence of a good leaf in every simplicial tree, we order the facets of a simplicial tree in order to find combinatorial information about the Betti numbers of its facet ideal. Applications include an Eliahou–Kervaire splitting of the ideal, as well as a refinement of a recursive formula of Hà and Van Tuyl for computing the graded Betti numbers of simplicial trees.

1 Introduction

Given a monomial ideal I in a polynomial ring $R = k[x_1, \dots, x_n]$ over a field k, a *minimal free resolution* of I is an exact sequence of free R-modules

$$0 \to \bigoplus_d R(-d)^{\beta_{p,d}} \to \cdots \to \bigoplus_d R(-d)^{\beta_{0,d}} \to I \to 0$$

of R/I in which $R(-d)$ denotes the graded free module obtained by shifting the degrees of elements in R by d. The numbers $\beta_{i,d}$, which we shall refer to as the i-th $\mathbb{N}$-*graded Betti numbers* of degree d of R/I, are independent of the choice of graded minimal finite free resolution.

Questions about Betti numbers—including when they vanish and when they do not, what bounds they have, how they relate to the base field k, and what are the

S. Faridi (✉)
Department of Mathematics and Statistics, Dalhousie University, 6316 Coburg Rd, PO BOX 15000, Halifax, NS, Canada B3H 4R2
e-mail: faridi@dal.ca

S.M. Cooper and S. Sather-Wagstaff (eds.), *Connections Between Algebra, Combinatorics, and Geometry*, Springer Proceedings in Mathematics & Statistics 76, DOI 10.1007/978-1-4939-0626-0_4,

most effective ways to compute them—are of particular interest in combinatorial commutative algebra. Via a method called polarization [6], it turns out that it is enough to consider such questions for square-free monomial ideals [7]; i.e. a monomial ideal in which the generators are square-free monomials.

To a square-free monomial ideal I one can associate a unique simplicial complex called its facet complex. Conversely, every simplicial complex has a unique monomial ideal assigned to it called its facet ideal [3]. Simplicial trees [3] and related structures were developed as a class of simplicial complexes that generalize graph-trees, so that their facet ideals have similar properties to those of edge ideals of graphs discovered in a series of works by Villarreal and his coauthors [11].

This paper offers an order on the monomials generating the facet ideal of a simplicial tree which uses the existence of a "good leaf" in every simplicial tree [10]. This order in itself is combinatorially interesting and useful, but it turns out that it also produces a "splitting" [2] of the facet ideal of a tree which gives bounds on the Betti numbers of the ideal.

Our good leaf order also makes it possible to refine a recursive formula of Hà and Van Tuyl [9] for computing Betti numbers of facet ideals of simplicial trees, and to apply it to classes of trees with strict good leaf orders. The idea here is that a good leaf order will split an ideal to some extent, and within each one of these split pieces, one can apply Hà and Van Tuyl's formula quite efficiently if the order is strict.

2 Simplicial Complexes, Trees, and Forests

Definition 2.1 (Simplicial Complexes). A *simplicial complex* Δ over a set of vertices $V(\Delta) = \{v_1, \dots, v_n\}$ is a collection of subsets of $V(\Delta)$, with the property that $\{v_i\} \in \Delta$ for all i, and if $F \in \Delta$, then all subsets of F are also in Δ. An element of Δ is called a *face* of Δ, and the *dimension* of a face F of Δ is defined as $|F| - 1$, where $|F|$ is the number of vertices of F. The faces of dimensions 0 and 1 are called *vertices* and *edges*, respectively, and dim $\emptyset = -1$. The maximal faces of Δ under inclusion are called *facets* of Δ. The dimension of the simplicial complex Δ is the maximal dimension of its facets. A *subcollection* of Δ is a simplicial complex whose facets are also facets of Δ.

A simplicial complex Δ is *connected* if for every pair of facets F, G of Δ, there exists a sequence of facets $F_1, \dots, F_r$ of Δ such that $F_1 = F$, $F_r = G$, and $F_s \cap F_{s+1} \neq \emptyset$ for $1 \leqslant s < r$.

We use the notation $\langle F_1, \dots, F_q \rangle$ to denote the simplicial complex with facets $F_1, \dots, F_q$, and we call it the simplicial complex *generated by* $F_1, \dots, F_q$. By *removing the facet* F_i from Δ we mean the simplicial complex $\Delta \setminus \langle F_i \rangle$ which is generated by $\{F_1, \dots, F_q\} \setminus \{F_i\}$.

Definition 2.2 (Leaf, Joint, Simplicial Trees, and Forests [3]). A facet F of a simplicial complex Δ is called a *leaf* if either F is the only facet of Δ, or F intersects $\Delta \setminus \langle F \rangle$ in a face of $\Delta \setminus \langle F \rangle$. If F is a leaf and Δ has more than one

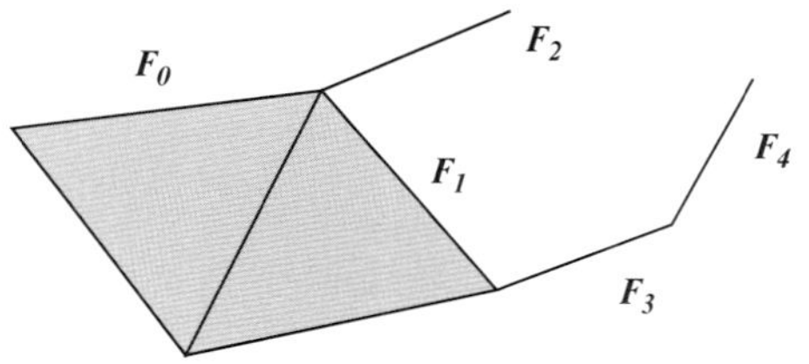

Fig. 1 Good leaves

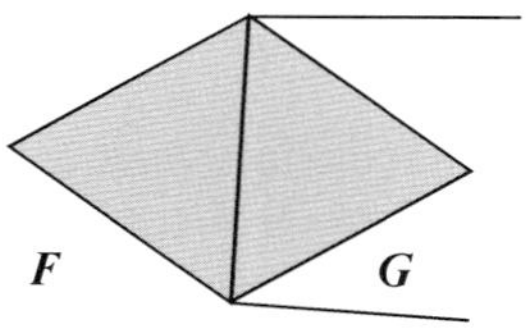

Fig. 2 A leaf that is not a good leaf

facet, then for some facet $G \in \Delta \setminus \langle F \rangle$ we have $F \cap H \subseteq G$ for all $H \in \Delta \setminus \langle F \rangle$. Such a facet G is called a *joint* of F.

A simplicial complex Δ is a *simplicial forest* if every nonempty subcollection of Δ has a leaf. A connected simplicial forest is called a *simplicial tree*.

It follows easily from the definition that a leaf must always contain at least one *free vertex*, that is a vertex that belongs to no other facet of Δ.

Example 2.3. The facets F_0, F_2 and F_4 are all leaves of the simplicial tree in Fig. 1. The first two have F_1 as a joint and F_4 has F_3 as a joint.

Definition 2.4 (Good Leaf [1, 12]). A facet F of a simplicial complex Δ is called a *good leaf* of Δ if F is a leaf of every subcollection of Δ which contains F.

All leaves of the simplicial tree in Fig. 1 are good leaves. Figure 2 contains an example of a leaf F in a simplicial tree which is not a good leaf: if we remove the facet G then F is no longer a leaf.

Good leaves were studied in [12] and then independently in [1] (where they were called "reducible leaves"). In both sources the existence of such a leaf in every tree was conjectured but not proved; the proof came later, using incidence matrices.

Theorem 2.5 ([10]). *Every simplicial tree contains a good leaf.*

Definition 2.6 (Facet Ideal, Facet Complex [3]). Let Δ be a simplicial complex with vertex set $\{x_1, \ldots, x_n\}$, and let $R = k[x_1, \ldots, x_n]$ be a polynomial ring over a field k with variables corresponding to the vertices of Δ. The *facet ideal* of Δ, denoted by $\mathscr{F}(\Delta)$, is an ideal of R whose generators are monomials, each of which is the products of the variables labeling the vertices of a facet of Δ. Given a square-free monomial ideal I in R, the *facet complex of* I is the simplicial complex whose facets are the set of variables appearing in each monomial generator of I.

Example 2.7. If $I = (xy, yzu, xz)$ is a monomial ideal in $R = k[x, y, z, u]$, its facet complex is the simplicial complex $\Delta = \langle\{x, y\}, \{y, z, u\}, \{x, z\}\rangle$. Similarly I is the facet ideal of Δ.

It is clear from the definition and example that every square-free monomial ideal has a unique facet complex, and every simplicial complex has a unique facet ideal. Because of this one-to-one correspondence we often abuse notation and use facets and monomials interchangeably. For example we say $F \cup G = \mathrm{lcm}(F, G)$ to imply the union of two facets F and G or the least common multiple of two monomials [corresponding to the facets] F and G.

Trees behave well under localization:

Lemma 2.8 (Localization of a Tree is a Forest [3]). *Let Δ be a simplicial tree with vertices $x_1, \ldots, x_n$, and let I be the facet ideal of Δ in the polynomial ring $R = k[x_1, \ldots, x_n]$ where k is a field. Then for any prime ideal p of R, I_p is the facet ideal of a simplicial forest.*

For a simplicial complex Δ with a facet F, we use the notation $\Delta_{\overline{F}}$ for facet complex of the localization $\mathscr{F}(\Delta)$ at the ideal generated by the complement of the facet F.

3 Good Leaf Orders

From its definition it is immediate that a good leaf F_0 of a tree Δ induces an order $F_0, F_1, \ldots, F_q$ on the facets of Δ so that

$$F_0 \cap F_1 \supseteq F_0 \cap F_2 \supseteq \cdots \supseteq F_0 \cap F_q.$$

Our goal in this section is to demonstrate that this order can be refined so that Δ is built leaf by leaf starting from the good leaf F_0. In other words, the order can be written so that for $i \leqslant q$, F_i is a leaf of $\Delta_i = \langle F_0, \ldots, F_i\rangle$. Such an order on the facets of Δ will be called a *good leaf order* on Δ.

Example 3.1. Let Δ be the simplicial tree in Fig. 1. Then F_0 is a good leaf and the labeling of facets $F_0, \ldots, F_4$ is a good leaf order on Δ, since $F_0 \cap F_1 \supseteq \cdots \supseteq F_0 \cap F_4$. Note that even though $F_0 \cap F_1 \supseteq F_0 \cap F_2 \supseteq F_0 \cap F_4 \supseteq F_0 \cap F_3$, this latter order $F_0, F_1, F_2, F_4.F_3$ is not a good leaf order since F_3 is not a leaf of Δ.

We show that every simplicial tree (forest) has a good leaf order.

Lemma 3.2. *Suppose $\Delta = \langle F, G, H\rangle$ is a simplicial tree with $F \cap G \not\subseteq H$ and $F \cap H \not\subseteq G$. Then G and H are the leaves of the tree Δ and F is the common joint so that $G \cap H \subseteq F$.*

Proof. If F is a leaf, then either $F \cap G \subseteq H$ or $F \cap H \subseteq G$. Either case is a contradiction, so the two leaves of the tree have to be G and H. If H is a joint of the leaf G then $F \cap G \subseteq H$ which is again a contradiction, so F is the joint of G. Similarly, F is the joint of H, and we have $G \cap H \subseteq F$. □

Proposition 3.3 (First Step to Build Good Leaf Order). *Let Δ be a simplicial tree with a good leaf F_0 and good leaf order*

$$F_0 \cap F_1 \supseteq F_0 \cap F_2 \supseteq \cdots \supseteq F_0 \cap F_q.$$

Let $1 \leqslant a \leqslant q$ and $0 \leqslant b < a$ and

$$F_0 \cap F_{a-b-1} \supsetneq F_0 \cap F_{a-b} = \cdots = F_0 \cap F_a.$$

Then one of $F_{a-b}, \dots, F_a$ is a leaf of $\langle F_0, \dots, F_a \rangle$.

Proof. Let $\Gamma = \langle F_0, \dots, F_a \rangle$. The subcollection $\Omega = \langle F_0, \dots, F_{a-b-1} \rangle$ of Γ is connected as all facets have nonempty intersection with F_0. If Γ is disconnected, Ω will be contained in one of the connected components of Γ, and there will be another connected component Σ whose facets are from $F_{a-b}, \dots, F_a$. Since Σ is a subcollection of a tree, it must have a leaf, and that leaf will be a leaf of Γ as well. So one of $F_{a-b}, \dots, F_a$ will be a leaf of Γ.

We now assume that Γ is connected and proceed by induction on a to prove our claim. If $a = 1$, then clearly F_1 is a leaf of the tree $\Gamma = \langle F_0, F_1 \rangle$. If $a = 2$ then since $F_2 \cap F_0 \subset F_1$, the facet F_2 must be a leaf with joint F_1.

Now suppose that $a > 2$ and the statement is true up to the $(a-1)$-st step. If $a - b = 1$, then

$$F_0 \cap F_1 = F_0 \cap F_2 = \cdots = F_0 \cap F_a.$$

By [4] Lemma 4.1 we know that Γ must have two leaves, and so one of the facets $F_1, \dots, F_a$ is a leaf.

We assume that $a - b \geqslant 2$ and neither one of $F_a, \dots, F_{a-b}$ is a leaf of Γ. There are two possible cases.

1. The case $b = 0$. Then $F_0 \cap F_{a-1} \supsetneq F_0 \cap F_a$. If $\Gamma' = \langle F_0, \dots, F_{a-2}, F_a \rangle$ then are two scenarios.

   a. If Γ' is disconnected, then the facet F_a alone is a connected component of Γ' (since all other facets intersect F_0) and therefore F_a is a leaf of Γ' and $F_a \cap F_i = \emptyset$ for $i = 0, \dots, a-2$. Since Γ is connected, $F_{a-1} \cap F_a \neq \emptyset$, and therefore F_a is a leaf of Γ with joint F_{a-1}.

   b. If Γ' is connected, we apply the induction hypothesis to the tree Γ' with good leaf F_0. In the ordering of the facets of Γ', F_a can only be at the right end of the sequence (since $F_0 \cap F_{a-2} \supsetneq F_0 \cap F_a$). So F_a is a leaf of Γ' and hence there is a joint $F_j \in \{F_0, \dots, F_{a-2}\}$ such that $F_a \cap F_k \subseteq F_j$ for all $F_k \in \{F_0, \dots, F_{a-2}\}$.

      If F_a is not a leaf of Γ then $F_a \cap F_{a-1} \not\subseteq F_j$. It also follows that $F_a \cap F_j \not\subseteq F_{a-1}$, as otherwise F_{a-1} would be a joint of F_a. Therefore, we can now apply Lemma 3.2 to the tree $\langle F_j, F_{a-1}, F_a \rangle$ to conclude that $F_j \cap F_{a-1} \subseteq F_a$. It follows that

$$F_0 \cap F_j \cap F_{a-1} \subseteq F_0 \cap F_a \Longrightarrow F_0 \cap F_{a-1} \subseteq F_0 \cap F_a \subsetneq F_{a-1} \cap F_0$$

which is a contradiction. So F_a has to be a leaf of Γ and we are done.

2. The case $b > 0$. We keep the good leaf F_0 and generate complexes $\Gamma_i = \Gamma \setminus \langle F_i \rangle$ for $i \in \{1, \dots, a\}$. By induction hypothesis each Γ_i has a leaf F_{u_i} where $u_i \in \{a-b, \dots, \hat{i}, \dots, a\}$. Since there are a total of $b+1$ facets that can be leaves of the Γ_i, and there are $a > b+1$ of the complexes Γ_i (recall that we are assuming $a - b \geqslant 2$), we must have $u_i = u_j = u$ for some distinct $i, j \in \{1, \dots, a\}$. Suppose F_{v_i} and F_{v_j} are the joints of F_u in Γ_i and Γ_j, respectively. So we have

$$F_u \cap F_h \subseteq F_{v_i} \text{ if } h \neq i$$
$$F_u \cap F_h \subseteq F_{v_j} \text{ if } h \neq j. \tag{1}$$

These two embeddings imply that

$$F_u \cap F_j \subseteq F_{v_i} \cap F_u \subseteq F_{v_j} \text{ if } v_i \neq j$$
$$F_u \cap F_i \subseteq F_{v_j} \cap F_u \subseteq F_{v_i} \text{ if } v_j \neq i. \tag{2}$$

Suppose $v_i \neq j$. Then from (1) and (2) we can see that F_u is a leaf of Γ with joint F_{v_j}. Similarly F_u is a leaf of Γ if $v_j \neq i$. So F_u is a leaf of Γ unless $v_i = j$ and $v_j = i$ are the only possible joints for F_u in Γ_i and Γ_j, respectively. In this case (1) turns into

$$F_u \cap F_h \subseteq F_j \text{ if } h \neq i$$
$$F_u \cap F_h \subseteq F_i \text{ if } h \neq j. \tag{3}$$

Now consider $\Gamma_u = \Delta \setminus \langle F_u \rangle$, which by induction hypothesis must have a leaf F_v with $v \in \{a-b, \dots, a\} \setminus \{u\}$ and a joint F_t. Since $F_i, F_j \in \Gamma_u$, we must have

$$F_i \cap F_v \subseteq F_t \text{ if } v \neq i$$
$$F_j \cap F_v \subseteq F_t \text{ if } v \neq j. \tag{4}$$

Once again, we consider two cases.

a. If v can be selected outside $\{i, j\}$, we combine (4) with (3) to get

$$F_u \cap F_v \subseteq F_j \cap F_v \subseteq F_t$$

meaning that F_v is a leaf of Γ.

b. If v must be in $\{i, j\}$, then the only leaves of Γ_u are F_i and F_j. As $F_0 \in \Gamma_u$ is a good leaf of Δ, one of i and j must be 0, say $j = 0$. But now we have

$$F_u \cap F_j = F_u \cap F_0 \subseteq F_i$$

which together with (3) implies that F_u is a leaf of Γ with joint F_i. □

Our main theorem is now just a direct consequence of Proposition 3.3, with a bit more added to it.

Theorem 3.4 (Main Theorem: Good Leaf Orders). *Let Δ be a simplicial tree with a good leaf F_0. Then there is an order $F_0, F_1, \ldots, F_q$ on the facets of Δ such that*

1. *$F_0 \cap F_1 \supseteq F_0 \cap F_2 \supseteq \cdots \supseteq F_0 \cap F_q$, and*
2. *The facet F_i is a leaf of $\Delta_i = \langle F_0, \ldots, F_i \rangle$ for $0 \leqslant i \leqslant q$.*
3. *The facet F_{i-1} is a either a leaf of Δ_i with the same joint as it has in Δ_{i-1}, or it is the unique joint of F_i in Δ_i, for $1 \leqslant i \leqslant q$.*
4. *$\Delta_i = \langle F_0, \ldots, F_i \rangle$ is connected for $0 \leqslant i \leqslant q$.*

Proof. The good leaf F_0 induces an order on the facets of Δ that satisfies the first property. We need to refine this order to achieve the second property. Let $i \in \{1, \ldots, q\}$. Starting from the beginning, here is how we proceed. For $i \in \{1, \ldots, q\}$ let c_i be the largest nonnegative integer such that $F_i \cap F_0 = F_{i+c_i} \cap F_0$ where $i + c_i \leqslant q$.

Set $i = 1$.

Step 1 If $c_i = 0$ then set $i := i + 1$ and go back to Step 1.

Step 2 If $c_i > 0$ then we reorder $F_i, \ldots, F_{i+c_i}$ as follows. By Proposition 3.3 there is a leaf $F_{\ell_{c_i}} \in \{F_i, \ldots, F_{i+c_i}\}$ of $\Gamma = \langle F_0, \ldots, F_{i+c_i} \rangle$. Applying the same proposition again there is a leaf $F_{\ell_{c_i - 1}} \in \{F_i, \ldots, F_{i+c_i}\} \setminus \{F_{\ell_{c_i}}\}$ of $\Gamma \setminus \langle F_{\ell_{c_i}} \rangle$. We continue this way $c_i + 1$ times and in the end we have a sequence

$$F_{\ell_0}, F_{\ell_1}, \ldots, F_{\ell_{c_i}}$$

which is a reordering of the facets $F_i, \ldots, F_{i+c_i}$ that satisfies both properties (1) and (2) in the statement of the theorem. We relabel $F_i, \ldots, F_{i+c_i}$ with this new order and set $i := i + c_i + 1$.

Step 3 If $i > q$ we stop and otherwise we go back to Step 1.

At the end of this algorithm, the facets of Δ have the desired order.

To prove the third part of the theorem, note that as F_{i-1} is a leaf in Δ_{i-1}, it has a set of free vertices in Δ_{i-1} which we call A. There are two scenarios.

- If $F_i \cap A \neq \emptyset$, then F_{i-1} has to be the unique joint of F_i in Δ_i, as no other facet of Δ_i would contain any element of A.
- If $F_i \cap A = \emptyset$ then $F_i \cap F_{i-1} \subseteq \Delta_{i-2} \cap F_{i-1} \subseteq F_\alpha$, where F_α is the joint of F_{i-1} in Δ_{i-1}. Therefore, F_{i-1} is a leaf of Δ_i.

Finally to see that Δ_i is connected for every i, we consider two situations.

1. $F_i \cap F_0 \neq \emptyset$. In this case Δ_i is connected as all facets of Δ_i intersect F_0.
2. $F_i \cap F_0 = \emptyset$. If $i = q$ then $\Delta_i = \Delta$ which is connected. Now we assume that i is the smallest index with $F_i \cap F_0 = \emptyset$, and $c_i > 0$, and we consider how $\Delta_i, \ldots, \Delta_q = \Delta$ are built in Step 2. We start from Δ, and pick a leaf for Δ from

$F_i, \ldots, F_q$. We call this facet F_q and we know already that $\Delta_q = \Delta$ must be connected. To pick Δ_{q-1} we remove the leaf F_q from Δ, and so Δ_{q-1} has to be connected. To build Δ_{q-2} we again remove a leaf from Δ_{q-1}, which forces Δ_{q-2} to be connected, and so on until we reach Δ_i, which by the same reasoning has to be connected. □

4 The Effect of Good Leaf Orders on Resolutions

Recall that for a monomial ideal I, the notation $\mathcal{G}(I)$ denotes the unique minimal monomial generating set for I.

Definition 4.1 (Splitting [2]). A monomial ideal I is called *splittable* if one can write $I = J + K$ for two nonzero monomial ideals J and K, such that

1. $\mathcal{G}(I)$ is the disjoint union of $\mathcal{G}(J)$ and $\mathcal{G}(K)$;
2. There is a *splitting function* $\mathcal{G}(J \cap K) \to \mathcal{G}(J) \times \mathcal{G}(K)$ taking each $w \in \mathcal{G}(J \cap K)$ to $(\phi(w), \psi(w))$ satisfying
   a. For each $w \in \mathcal{G}(J \cap K)$, $w = \mathrm{lcm}(\phi(w), \psi(w))$
   b. For each $S \subseteq \mathcal{G}(J \cap K)$, $\mathrm{lcm}(\phi(S))$ and $\mathrm{lcm}(\psi(S))$ strictly divide $\mathrm{lcm}(S)$.

If a monomial ideal is splittable, then its Betti numbers can be broken down into those of sub-ideals.

Theorem 4.2 ([2, 5]). *If I is a monomial ideal with a splitting $I = J + K$, then for all $i, j \geqslant 0$*

$$\beta_{i,j}(I) = \beta_{i,j}(J) + \beta_{i,j}(K) + \beta_{i-1,j}(J \cap K).$$

Our next observation is that a good leaf order on a simplicial tree provides a basic splitting of its facet ideal.

Theorem 4.3 (Splitting Using a Good Leaf Order). *If I is the facet ideal of a simplicial tree Δ with a good leaf F_0 and good leaf order*

$$F_0 \cap F_1 \supseteq F_0 \cap F_2 \supseteq \ldots \supseteq F_0 \cap F_t \supsetneq F_0 \cap F_{t+1} = \ldots = F_0 \cap F_q = \emptyset$$

and $J = (F_0, \ldots, F_t)$ and $K = (F_{t+1}, \ldots, F_q)$, then $I = J + K$ is a splitting of I.

Proof. It is clear that $I = J + K$. We number the vertices of $F_0, \ldots, F_t$ in some order as $x_1, \ldots, x_m$. We will build ϕ and ψ as in Definition 4.1. Suppose $L \in \mathcal{G}(J \cap K)$. Then there are facets F_i and F_j such that $i \leqslant t < j$ such that $L = \mathrm{lcm}(F_i, F_j)$. Of all choices of such F_i we pick one minimal with respect to lex order and call it G_L, and there is only one choice for F_j (since each F_j adds one or more new vertices to the sequence $F_0, \ldots, F_{j-1}$); call this facet H_L. So we have $L = \mathrm{lcm}(G_L, H_L)$. Let $\phi(L) = G_L$ and $\psi(L) = H_L$ so that we have a map

$$\begin{array}{ll}\mathscr{G}(J \cap K) & \to \mathscr{G}(J) \times \mathscr{G}(K) \\ L & \to (\phi(L), \psi(L)) = (G_L, H_L)\end{array}$$

We only need to show that Condition (b) in Definition 4.1 holds. Suppose $S = \{L_1, \dots, L_r\} \subseteq \mathscr{G}(J \cap K)$. Suppose, as before, for each i we can write $L_i = \mathrm{lcm}(G_{L_i}, H_{L_i}) = G_{L_i} \cup H_{L_i}$ where $G_{L_i} \in \mathscr{G}(K)$ and $H_{L_i} \in \mathscr{G}(K)$. We need to show

1. $G_{L_1} \cup \dots \cup G_{L_r} \subsetneq L_1 \cup \dots \cup L_r$.
   This is clear since each of the L_i contains vertices that are in $\mathscr{G}(K)$ but not in $\mathscr{G}(J)$.
2. $H_{L_1} \cup \dots \cup H_{L_r} \subsetneq L_1 \cup \dots \cup L_r$
   Each of the L_i has a nonempty intersection with F_0, but $H_{L_i} \cap F_0 = \emptyset$, which makes the inclusion above strict.

So we have shown that we have a splitting which completes the proof. □

As a result, we can use good leaf orders to bound invariants related to resolutions of trees. Recall that the *regularity* of an ideal I, denoted by $\mathrm{reg}(I)$, is the maximum value of $j - i$ where $\beta_{i,j}(I) \neq 0$. The *projective dimension* of I, denoted by $\mathrm{projdim}(I)$, is the maximum value of i where $\beta_{i,j}(I) \neq 0$ for some j. The projective dimension and regularity measure the "length" and the "width" of a minimal free resolution, as can be seen in the Betti diagram of the ideal; see Example 4.5 below. For a simplicial complex Γ we often use the notation $\beta_{i,j}(\Gamma)$, $\mathrm{reg}(\Gamma)$ and $\mathrm{projdim}(\Gamma)$ to indicate the Betti numbers, regularity and projective dimension of $\mathscr{F}(\Gamma)$.

The following statement is a direct application of Theorems 4.2 and 4.3.

Corollary 4.4 (Bounds on Betti Numbers of Trees). *Suppose Δ is a simplicial tree that can be partitioned into subcollections $\Delta_0, \dots, \Delta_s$, each of which is a tree, and such that for each $i = 0, \dots, s$, setting $a_0 = 0$ we have:*

1. *$\Delta_i = \langle F_{a_i}, F_{a_i+1}, \dots, F_{a_{i+1}-1} \rangle$ with good leaf F_{a_i}.*
2. *$F_{a_i+1} \cap F_{a_i} \supseteq \dots \supseteq F_{a_{i+1}-1} \cap F_{a_i} \neq \emptyset$ is a good leaf order on Δ_i;*
3. *$F_{a_i} \cap F_j = \emptyset$ for $j \geqslant a_{i+1}$.*

Then

$$\beta_{i,j}(\Delta) \geqslant \beta_{i,j}(\Delta_0) + \dots + \beta_{i,j}(\Delta_s).$$

In particular

$$\mathit{projdim}(\Delta) \geqslant \max\{\mathit{projdim}(\Delta_0), \dots, \mathit{projdim}(\Delta_s)\}$$

and

$$\mathit{reg}(\Delta) \geqslant \max\{\mathit{reg}(\Delta_0), \dots, \mathit{reg}(\Delta_s)\}.$$

We demonstrate the effect via the example of Fig. 1 which we will label below.

Example 4.5. For the ideal $I = (xyz, yzv, yu, vw, wt)$ the facet complex Δ is

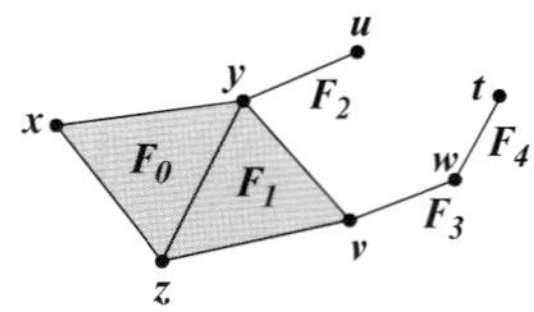

Following the statement of the corollary, we can find a splitting for I by partitioning the facets of Δ into two trees with the written good leaf orders

$$\langle F_0, F_1, F_2\rangle \qquad \text{and} \qquad \langle F_3, F_4\rangle$$

which correspond, respectively, to the two ideals

$$J = (xyz, yzv, yu) \qquad \text{and} \qquad K = (vw, wt).$$

We copy the Betti diagrams of I, J, and K (in that order) using Macaulay2 [8].

I	0	1	2	3
1	3	1	.	.
2	2	6	3	.
3	.	1	2	1

J	0	1	2
1	1	.	.
2	2	3	1

K	0	1
1	2	1

The bounds presented in Corollary 4.4 are now evident from the Betti diagrams:

$$\text{reg}(I) = 3 \geqslant \max\{\text{reg}(J), \text{reg}(K)\} = \max\{2, 1\} = 2$$

and

$$\text{projdim}(I) = 3 \geqslant \max\{\text{projdim}(J), \text{projdim}(K)\} = \max\{2, 1\} = 2$$

There can be different good leaf orders on a simplicial tree. It would be interesting to know which one gives a "better" splitting and better bounds for the resolution invariants.

4.1 *Recursive Calculations of Betti Numbers*

In [9] Hà and Van Tuyl used Eliahou–Kervaire splittings to reduce the computation of the Betti numbers of a given simplicial forest to that of smaller ones. Our goal here is to show that their formula can be refined in certain cases and be used to compute the Betti numbers of a given simplicial tree in terms of intersections of the faces. The method used is essentially a repeated application of a splitting formula due to Hà and Van Tuyl [9] to a good leaf order on a given tree, along with an argument that, at every stage, we know what the next splitting to consider should be.

Definition 4.6 ([9] Definition 5.1). Let F be a facet of a simplicial complex Δ. The *connected component of* F *in* Δ, denoted $\mathrm{conn}_\Delta(F)$, is defined to be the connected component of Δ containing F. If $\mathrm{conn}_\Delta(F) = \langle G_1, \dots, G_p \rangle$, then we define the *reduced connected component of* F *in* Δ, denoted by $\overline{\mathrm{conn}}_\Delta(F)$, to be the simplicial complex whose facets are a subset of $\{G_1 \setminus F, \dots, G_p \setminus F\}$, chosen so that if there exist G_i and G_j such that $\emptyset \neq G_i \setminus F \subseteq G_j \setminus F$, then we shall disregard the bigger facet $G_j \setminus F$ in $\overline{\mathrm{conn}}_\Delta(F)$.

Note that in the Definition 4.6, $\overline{\mathrm{conn}}_\Delta(F)$ is the localization of $\mathrm{conn}_\Delta(F)$ at the ideal generated by the complement of the facet F. Therefore if Δ is a tree then $\overline{\mathrm{conn}}_\Delta(F)$ is always a forest ([3]). Hà and Van Tuyl ([9] Lemma 5.7) prove this directly in their paper.

A facet F of Δ is called a *splitting facet of* Δ if $\mathscr{F}(\Delta) = (F) + \mathscr{F}(\Delta \setminus \langle F \rangle)$ is a splitting of $\mathscr{F}(\Delta)$ (here we are thinking of F as a monomial).

Theorem 4.7 ([9] Theorem 5.5). *If* F *is a splitting facet of a simplicial complex* Δ*, then for all* $i \geqslant 1$ *and* $j \geqslant 0$ *we have* $\beta_{i,j}(\mathscr{F}(\Delta))$ *is equal to*

$$\beta_{i,j}(\mathscr{F}(\Delta \setminus \langle F \rangle)) + \sum_{l_1=0}^{i} \sum_{l_2=0}^{j-|F|} \beta_{l_1-1,l_2}(\mathscr{F}(\overline{conn}_\Delta(F)) \beta_{i-l_1-1,j-|F|-l_2}(\mathscr{F}(\Delta \setminus conn_\Delta(F)). \quad (5)$$

So now the question is what is a good choice for a splitting facet. In their paper ([9] Theorem 5.6) Hà and Van Tuyl show that a leaf of a simplicial complex is a splitting facet. Their proof in fact only requires the facet to have a free vertex.

Proposition 4.8. *Let* Δ *be a simplicial complex. If* F *is a facet of* Δ *with a free vertex, then* F *is a splitting facet of* Δ.

Proof. The proof is identical to the proof of Theorem 5.6 in [9]. □

We use the convention that for any ideal I

$$\beta_{-1,j}(I) = \begin{cases} 1 & j = 0 \\ 0 & \text{otherwise.} \end{cases} \quad (6)$$

Suppose we have a simplicial tree Δ with good leaf order described as in Theorem 3.4. We apply (5) to $\Delta = \langle F_0, \dots, F_q \rangle$ peeling off leaves in the following order: $F_q, F_{q-1}, \dots, F_0$.

Suppose we are in step u, peeling off the leaf F_u from the tree $\Delta_u = \langle F_0, \dots, F_u \rangle$. Then $\mathrm{conn}_{\Delta_u}(F_u) = \Delta_u$ and so $\mathscr{F}(\Delta_u \setminus \mathrm{conn}_{\Delta_u}(F_u)) = 0$ and therefore

$$\beta_{a,b}(\mathscr{F}(\Delta_u \setminus \mathrm{conn}_{\Delta_u}(F_u)) = \begin{cases} 1 & a = -1,\ b = 0 \\ 0 & \text{otherwise.} \end{cases}$$

Applying this to (5), we solve $i - l_1 - 1 = -1$ and $j - |F_u| - l_2 = 0$ to find $l_1 = i$ and $l_2 = j - |F_u|$.

Moreover, we have $\overline{\mathrm{conn}}_{\Delta_u}(F_u) = (\Delta_{u-1})_{\overline{F_u}}$, that is Δ_{u-1} localized at the ideal generated by the complement of the facet F_u using notation as in Lemma 2.8. So (5) turns into

$$\begin{aligned}
\beta_{i,j}(\mathscr{F}(\Delta)) &= \beta_{i,j}(\mathscr{F}(\Delta_{q-1})) + \beta_{i-1,j-|F_q|}(\mathscr{F}((\Delta_{q-1})_{\overline{F_q}})) \\
&= \beta_{i,j}(\mathscr{F}(\Delta_{q-2})) + \beta_{i-1,j-|F_{q-1}|}(\mathscr{F}((\Delta_{q-2})_{\overline{F_{q-1}}})) + \beta_{i-1,j-|F_q|}(\mathscr{F}((\Delta_{q-1})_{\overline{F_q}})) \\
&\ \vdots \\
&= \beta_{i,j}(\mathscr{F}(\langle F_0\rangle)) + \sum_{u=1}^{q} \beta_{i-1,j-|F_u|}(\mathscr{F}((\Delta_{u-1})_{\overline{F_u}}))
\end{aligned}$$

Note that we did not use the fact that F_0 is a good leaf here, just that each F_u is a leaf if Δ_u. We have therefore justified the following statement.

Proposition 4.9. *Let Δ be a simplicial tree with a good leaf order $F_0, F_1, \ldots, F_q$ such that each F_u is a leaf of $\Delta_u = \langle F_0, \ldots, F_u\rangle$ for $u \leqslant q$. Then for all $i \geqslant 1$ and $j \geqslant 0$*

$$\beta_{i,j}(\mathscr{F}(\Delta)) = \beta_{i,j}(\mathscr{F}(\langle F_0\rangle)) + \sum_{u=1}^{q} \beta_{i-1,j-|F_u|}(\mathscr{F}((\Delta_{u-1})_{\overline{F_u}})). \tag{7}$$

By introducing an appropriate "δ" function we can say

$$\beta_{i,j}(\mathscr{F}(\langle F_0\rangle)) = \delta_{(i,j),(0,|F_0|)} = \begin{cases} 1 & i = 0,\ j = |F_0| \\ 0 & \text{otherwise.} \end{cases} \tag{8}$$

So now we focus on the structure of $(\Delta_{u-1})_{\overline{F_u}}$. The main point that we would like to make is that $(\Delta_{u-1})_{\overline{F_u}}$ behaves well, in other words, it satisfies the same kind of inclusion sequence enforced in Theorem 3.4, and the same "leaf-peeling" property. Note that though F_0 need not even survive the localization, its role is that of a virtual glue that forces facets to always stick together and have an appropriate order.

Proposition 4.10. *Let Δ be a simplicial tree with a good leaf F_0 and good leaf order*

$$F_0 \cap F_1 \supsetneq F_0 \cap F_2 \supsetneq \cdots \supsetneq F_0 \cap F_q.$$

Suppose $u \in \{1, \ldots, q\}$ *and* $\Delta_u = \langle F_1, \ldots, F_u \rangle$, *and suppose the facets of* $(\Delta_{u-1})_{\overline{F_u}}$ *are* $F_{a_1} \setminus F_u, \ldots, F_{a_s} \setminus F_u$ *with* $0 \leqslant a_1 < \ldots < a_s \leqslant u-1$. *Then*

1. $a_s = u - 1$,
2. $(F_0 \cap F_{a_1}) \setminus F_u \supsetneq \ldots \supsetneq (F_0 \cap F_{a_s}) \setminus F_u$,
3. $(\Delta_{u-1})_{\overline{F_u}}$ *is a simplicial tree,*
4. *If* F_v *is a joint of* F_u *in* Δ_u *then* $F_v \setminus F_u \in (\Delta_{u-1})_{\overline{F_u}}$,
5. $F_{u-1} \setminus F_u$ *has a free vertex in* $(\Delta_{u-1})_{\overline{F_u}}$.

Proof. To prove 1, suppose there is an $i < u-1$ such that $(F_i \setminus F_u) \subset (F_{u-1} \setminus F_u)$. By assumption there exists $y \in (F_0 \cap F_i) \setminus (F_0 \cap F_{u-1})$. As $(F_0 \cap F_{u-1}) \supset (F_0 \cap F_u)$, it follows that $y \in (F_i \setminus F_u)$ and $y \notin (F_{u-1} \setminus F_u)$, which contradicts the inclusion $(F_i \setminus F_u) \subset (F_{u-1} \setminus F_q)$.

The strict inclusions in 2 follow from the same observation, that for every i there is always an element in $F_0 \cap F_{a_i}$ which is not in $F_{a_{i+1}}$ or F_u.

Since $(\Delta_{u-1})_{\overline{F_u}}$ is a localization of the tree Δ_{u-1}, it is clear that it is a forest, and by 2, since $(F_0 \cap F_{a_s}) \setminus F_u \neq \emptyset$, it must be connected and therefore a simplicial tree. This settles 3.

For 4, suppose for some $j < u$ we have $F_j \setminus F_u \subseteq F_v \setminus F_u$. Then we will have

$$F_j = (F_j \cap F_u) \cup (F_j \setminus F_u) \subseteq F_v$$

which implies that $F_j = F_v$.

Finally to prove 5 we use induction on u. If $u = 1$ or 2, then $(\Delta_{u-1})_{\overline{F_u}}$ will have one or two facets, and in each case $F_{u-1} \setminus F_u$ clearly must have a free vertex. If $u = 3$, then F_2 is a leaf of Δ_2 with a joint F_i for some $i < 2$. If $F_i \setminus F_3 \in (\Delta_2)_{\overline{F_3}}$, then it acts as a joint of $F_2 \setminus F_3$ so $F_2 \setminus F_3$ is a leaf and must therefore have a free vertex. If $F_i \setminus F_3 \notin (\Delta_2)_{\overline{F_3}}$, then $(\Delta_2)_{\overline{F_3}}$ has at most two facets including $F_2 \setminus F_3$, each of which must have a free vertex. This settles the base cases for induction.

Now suppose $u \geqslant 4$ and $F_{u-1} \setminus F_u$ has no free vertex in $(\Delta_{u-1})_{\overline{F_u}}$.

By induction hypothesis, if we consider $\Gamma = \Delta \setminus \langle F_{u-2} \rangle$, then $F_{u-1} \setminus F_u$ will have a free vertex x in $(\Gamma_{u-1})_{\overline{F_u}}$. If x is not a free vertex in $(\Delta_{u-1})_{\overline{F_u}}$, then for some $j < u-1$ we have $x \in F_j \setminus F_u \in (\Delta_{u-1})_{\overline{F_u}}$ and $F_j \setminus F_u \notin (\Gamma_{u-1})_{\overline{F_u}}$. The only possible such index j is $j = u-2$. In other words, $x \in F_{u-1} \cap F_{u-2}$ and $x \notin F_i$ for any other $i \leqslant u$.

Similarly, if we remove F_{u-3} from Δ we will find a vertex $y \in F_{u-1} \cap F_{u-3}$ and $y \notin F_i$ for any other $i \leqslant u$.

By Lemma 3.2, we must then have $F_{u-3} \cap F_{u-2} \subseteq F_{u-1}$. Intersecting both sides with F_0 we obtain

$$F_{u-1} \cap F_0 \subseteq F_{u-2} \cap F_0 = F_{u-3} \cap F_{u-2} \cap F_0 \subseteq F_{u-1} \cap F_0$$

which means that $F_{u-1} \cap F_0 = F_{u-2} \cap F_0$; a contradiction. □

Proposition 4.10 now allows us to continue solving (7) by applying Theorem 4.7 once again, since we have a splitting facet for each $(\Delta_{u-1})_{\overline{F_u}}$. Consider the tree Δ

as described above with the good leaf order described in Theorem 3.4 and for some $u \in \{1,\ldots,q\}$, let $(\Delta_{u-1})_{\overline{F_u}} = \langle F_{a_1} \setminus F_u, \ldots, F_{a_s} \setminus F_u\rangle$ where $0 \leqslant a_1 < \ldots < a_s < u$. By Proposition 4.10 $(\Delta_{u-1})_{\overline{F_u}}$ is a simplicial tree with an order of the facets induced by the good leaf order of Δ, and with splitting facet $F_{a_s} \setminus F_u$.

We continue in the same spirit. Let $u_1 = u, u_2 = a_s$ and

$$\mathscr{C}_{u_1,u_2} = ((\Delta_{u_1-1})_{F_{u_1}})_{F_{u_2}\setminus F_{u_1}} = \langle F_{d_1} \setminus (F_{u_1} \cup F_{u_2}), \ldots, F_{d_w} \setminus (F_{u_1} \cup F_{u_2})\rangle$$

where $0 \leqslant d_1 < \ldots < d_w < u_2 < u_1$.

Similarly, we can build $\mathscr{C}_{u_1,\ldots,u_m}$ which is the localization of

$$C_{u_1,\ldots,u_{m-1}} = \langle F_{c_1} \setminus (F_{u_1} \cup \ldots \cup F_{u_{m-1}}), \ldots, F_{c_r} \setminus (F_{u_1} \cup \ldots \cup F_{u_{m-1}})\rangle \tag{9}$$

at the ideal generated by the complement of the facet $F_{u_m} \setminus (F_{u_1} \cup \ldots \cup F_{u_{m-1}})$ where $u_m = c_r$. So we have

$$\mathscr{C}_{u_1,\ldots,u_m} = \langle F_{b_1} \setminus (F_{u_1} \cup \ldots \cup F_{u_m}), \ldots, F_{b_t} \setminus (F_{u_1} \cup \ldots \cup F_{u_m})\rangle \tag{10}$$

where $b_1, \ldots, b_t \in \{c_1, \ldots, c_{r-1}\}$, and

$$0 \leqslant b_1 < b_2 < \ldots < b_t < c_r = u_m < u_{m-1} < \ldots < u_1.$$

Proposition 4.11. *Let Δ be a simplicial tree with a good leaf F_0 and good leaf order*

$$F_0 \cap F_1 \supsetneq F_0 \cap F_2 \supsetneq \cdots \supsetneq F_0 \cap F_q.$$

With notation as in (9) and (10) above, we have

1. $b_t = c_{r-1}$,
2. $(F_0 \cap F_{b_1}) \setminus (F_{u_1} \cup \ldots \cup F_{u_m}) \supsetneq \ldots \supsetneq (F_0 \cap F_{b_t}) \setminus (F_{u_1} \cup \ldots \cup F_{u_m})$,
3. *$\mathscr{C}_{u_1,\ldots,u_m}$ is a simplicial tree,*
4. *$F_{b_t} \setminus (F_{u_1} \cup \ldots \cup F_{u_m})$ has a free vertex in $\mathscr{C}_{u_1,\ldots,u_m}$ and is therefore a splitting facet of $\mathscr{C}_{u_1,\ldots,u_m}$.*

Proof. Let $A = F_{u_1} \cup \ldots \cup F_{u_m}$. To show 1, suppose there is an $i < r-1$ such that $F_{c_i} \setminus A \subset F_{c_{r-1}} \setminus A$. By the strict inclusions assumed there exists $y \in (F_0 \cap F_{c_i}) \setminus (F_0 \cap F_{c_{r-1}})$. As

$$F_0 \cap F_{c_{r-1}} \supsetneq F_0 \cap F_{u_m} \supsetneq \ldots \supsetneq F_0 \cap F_{u_1},$$

it follows that $y \in F_{c_i} \setminus A$ and $y \notin F_{c_{r-1}} \setminus A$, which is a contradiction.

For 2 it is easy to see that

$$(F_0 \cap F_{b_1}) \setminus A \supseteq \ldots \supseteq (F_0 \cap F_{b_t}) \setminus A.$$

To show that these inclusions are strict pick $1 \leqslant i < j < t$, we know that

$$F_0 \cap F_{b_i} \supsetneq F_0 \cap F_{b_j} \supsetneq F_0 \cap F_{u_m} \supsetneq F_0 \cap F_{u_{m-1}} \supsetneq \ldots \supsetneq F_0 \cap F_{u_1},$$

and therefore there exists $y \in (F_{b_i} \cap F_0) \setminus (F_{b_j} \cup F_{u_1} \cup \ldots \cup F_{u_m})$, which means that $y \in (F_0 \cap F_{b_i}) \setminus (F_{u_1} \cup \ldots \cup F_{u_m})$ and $y \notin (F_0 \cap F_{b_j}) \setminus (F_{u_1} \cup \ldots \cup F_{u_m})$, proving 2.

Suppose $\Omega = \langle F_{\omega_0}, F_{\omega_1}, \ldots, F_{\omega_p} \rangle$ is the subcollection of Δ consisting of those facets that are not contained in A with

$$0 = \omega_0 < \omega_1 < \ldots < \omega_p.$$

Because of the strict good leaf order Ω is a connected forest and hence a tree.

We claim that $C_{u_1,\ldots,u_m}$ is the localization of the tree Ω at the ideal generated by $\overline{A}$. This follows from two observations. One is that if at the ith step when building $C_{u_1,\ldots,u_m}$ there are facets $F_\alpha, F_\beta \in \Delta$ not containing $F_{u_1} \cup \ldots \cup F_{u_i}$, then F_α, F_β do not contain A and therefore are also facets of Ω. Moreover if $F_\alpha \setminus (F_{u_1} \cup \ldots \cup F_{u_i}) \subseteq F_\beta \setminus (F_{u_1} \cup \ldots \cup F_{u_i})$, then $F_\alpha \setminus A \subseteq F_\beta \setminus A$ and therefore we can conclude that $C_{u_1,\ldots,u_m}$ is a localization Ω and $\{b_1 \ldots b_t\} \subseteq \{\omega_0, \ldots, \omega_p\}$.

So $C_{u_1,\ldots,u_m}$ must be a forest, and since it is connected by 2, it must be a simplicial tree. This settles 3.

By the discussion above we can assume $\omega_p = b_t$ and we will still have $C_{u_1,\ldots,u_m}$ is a localization of Ω. Also note that $F_0 = F_{\omega_0}$ is a good leaf of Ω with a strict good leaf order induced by that on Δ.

To prove 4 we use induction on p. If $p = 1$ or 2, then $C_{u_1,\ldots,u_m}$ will have one or two facets, and in each case $F_{b_t} \setminus A$ clearly must have a free vertex. If $p = 3$, then F_{ω_2} is a leaf of $\Omega_{\omega_2} = \langle F_{\omega_0}, F_{\omega_1}, F_{\omega_2} \rangle$ with a joint F_{w_i} for some $i < 2$. If $F_{\omega_i} \setminus A \in C_{u_1,\ldots,u_m}$, then it acts as a joint of $F_{\omega_2} \setminus A$ so $F_{\omega_2} \setminus A$ is a leaf and must therefore have a free vertex. If $F_{\omega_i} \setminus A \notin C_{u_1,\ldots,u_m}$, then $C_{u_1,\ldots,u_m}$ has at most two facets including $F_{\omega_2} \setminus A$ each of which must have a free vertex. This settles the base cases for induction.

Now suppose $p \geqslant 4$ and $F_{b_t} \setminus A$ has no free vertex in $C_{u_1,\ldots,u_m}$.

By the induction hypothesis, if we consider $\Gamma = \Omega \setminus \langle F_{\omega_{p-1}} \rangle$, then F_{ω_p} will have a free vertex x in $\Gamma_{\overline{A}}$. If x is not a free vertex in $\Gamma_{\overline{A}}$, then $x \in F_{\omega_{p-1}} \setminus A \in \Gamma_{\overline{A}}$. In other words, $x \in F_{\omega_p} \cap F_{\omega_{p-1}}$ and $x \notin F_{\omega_i}$ for any other $i \leqslant p$.

Similarly, if we remove $F_{\omega_{p-2}}$ from Ω we will find a vertex $y \in F_{\omega_p} \cap F_{\omega_{p-2}}$ and $y \notin F_i$ for any other $i \leqslant p$.

By Lemma 3.2, we must then have $F_{\omega_{p-2}} \cap F_{\omega_{p-1}} \subseteq F_{\omega_p}$. Intersecting both sides with F_0 we obtain

$$F_{\omega_p} \cap F_0 \subseteq F_{\omega_{p-1}} \cap F_0 = F_{\omega_{p-2}} \cap F_{\omega_{p-1}} \cap F_0 \subseteq F_{\omega_p} \cap F_0$$

which means that $F_{\omega_p} \cap F_0 = F_{\omega_{p-1}} \cap F_0$; a contradiction. This proves 4 and we are done. □

Proposition 4.11 replaces Proposition 4.10 as a more general version. Back to (7), we start computing Betti numbers of $\mathscr{F}(\Delta)$ for a given tree Δ with good leaf F_0 and strict good leaf order

$$F_0 \cap F_1 \supsetneq F_0 \cap F_2 \supsetneq \cdots \supsetneq F_0 \cap F_q.$$

The formula

$$\beta_{i,j}(\mathscr{F}(\Delta)) = \beta_{i,j}(\mathscr{F}(\langle F_0\rangle)) + \sum_{u=1}^{q} \beta_{i-1,j-|F_u|}(\mathscr{F}((\Delta_{u-1})_{\overline{F_u}}))$$

becomes recursive, since in each step after localization we again have a simplicial tree with a strict induced order on the facets where the last facet remaining is a splitting facet.

To close, we apply the formula to examine some low Betti numbers.

Let $i = 0$. By (7) and (6) we have

$$\beta_{0,j}(\mathscr{F}(\Delta)) = \sum_{u=0}^{q} \delta_{j,|F_u|}.$$

Let $i \geqslant 1$. Because of (7) and (8) we can write

$$\beta_{i,j}(\mathscr{F}(\Delta)) = \sum_{u=1}^{q} \beta_{i-1,j-|F_u|}(\mathscr{F}((\Delta_{u-1})_{\overline{F_u}})) \tag{11}$$

From Proposition 4.11 and (11) we can see that we need the generators of each Δ_u in order to produce a formula for the first graded Betti numbers. To this end, we start from $\Delta_u = \langle F_0, \ldots, F_u\rangle$ so that

$$\begin{aligned}(\Delta_{u-1})_{\overline{F_u}} &= \langle F_i \setminus F_u \mid 0 \leqslant i < u \text{ and } (F_j \setminus F_u) \nsubseteq (F_i \setminus F_u) \text{ for } j \neq i\rangle \\ &= \langle F_i \setminus F_u \mid 0 \leqslant i < u \text{ and } \frac{\operatorname{lcm}(F_j, F_u)}{F_u} \not\mid \frac{\operatorname{lcm}(F_i, F_u)}{F_u} \text{ for } j \neq i\rangle \\ &= \langle F_i \setminus F_u \mid 0 \leqslant i < u \text{ and } \operatorname{lcm}(F_j, F_u) \not\mid \operatorname{lcm}(F_i, F_u) \text{ for } j \neq i\rangle\end{aligned}$$

So we can make our "delta-function" to have the lcm condition built into it. We define

$$\delta_{a,(b,c)} = \begin{cases} 1 & a = |F_b|,\ \operatorname{lcm}(F_d, F_c) \not\mid \operatorname{lcm}(F_b, F_c) \text{ for } 0 \leqslant d < c \\ 0 & \text{otherwise}\end{cases}$$

So (11) becomes

$$\begin{aligned}\beta_{1,j}(\mathscr{F}(\Delta)) &= \sum_{u=1}^{q} \beta_{0,j-|F_u|}(\mathscr{F}((\Delta_{u-1})_{\overline{F_u}})) \\ &= \sum_{u=1}^{q} \sum_{F \text{ facet of } (\Delta_{u-1})_{\overline{F_u}}} \delta_{j-|F_u|,|F|} \\ &= \sum_{u=1}^{q} \sum_{v=0}^{u-1} \delta_{j-|F_u|,(v,u)}\end{aligned}$$

By building appropriate delta functions, one can continue in this manner to build further Betti numbers based on the lcms of the facets.

Acknowledgments The author acknowledges the financial support of NSERC and the hospitality of MSRI in Berkeley, CA where part of this work was completed.

References

1. Caboara, M., Faridi, S., Selinger, P.: Simplicial cycles and the computation of simplicial trees. J. Symb. Comput. **42**(1–2), 74–88 (2007)
2. Eliahou, S., Kervaire, M.: Minimal resolutions of some monomial ideals. J. Algebra **129**(1), 1–25 (1990)
3. Faridi, S.: The facet ideal of a simplicial complex. Manuscripta Math. **109**, 159–174 (2002)
4. Faridi, S.: Cohen-Macaulay properties of square-free monomial ideals. J. Comb. Theory Ser. A **109**(2), 299–329 (2005)
5. Fatabbi, G.: On the resolution of ideals of fat points. J. Algebra **242**(1), 92–108 (2001)
6. Fröberg, R.: A study of graded extremal rings and of monomial rings. Math. Scand. **51**, 22–34 (1982)
7. Gasharov, V., Peeva, I., Welker, V.: The lcm-lattice in monomial resolutions. Math. Res. Lett. **6**(5–6), 521–532 (1999)
8. Grayson, D.R., Stillman, M.E.: Macaulay2, a software system for research in algebraic geometry. Available at http://www.math.uiuc.edu/Macaulay2/Citing/
9. Hà, H.T., Van Tuyl, A.: Splittable ideals and the resolutions of monomial ideals. J. Algebra **309**, 405–425 (2007)
10. Herzog, J., Hibi, T., Trung, N.V., Zheng, X.: Standard graded vertex cover algebras, cycles and leaves. Trans. Am. Math. Soc. **360**(12), 6231–6249 (2008)
11. Villarreal, R.: Monomial Algebras. Monographs and Textbooks in Pure and Applied Mathematics, vol. 238. Dekker, New York (2001)
12. Zheng, X.: Homological properties of monomial ideals associated to quasi-trees and lattices. Ph.D. thesis, Universität Duisburg-Essen (August 2004)

A Survey of Stanley–Reisner Theory

Christopher A. Francisco, Jeffrey Mermin, and Jay Schweig

Dedicated to Tony Geramita in honor of his outstanding research and mentoring

Abstract We survey the Stanley–Reisner correspondence in combinatorial commutative algebra, describing fundamental applications involving Alexander duality, associated primes, f- and h-vectors, and Betti numbers of monomial ideals.

1 Introduction

Stanley–Reisner theory provides the central link between combinatorics and commutative algebra. Pioneered in the 1970s, the correspondence between simplicial complexes and squarefree monomial ideals has been responsible for substantial progress in both fields. Among the most celebrated results are Reisner's criterion for Cohen–Macaulayness, Stanley's proof of the Upper Bound Conjecture for simplicial spheres, and Hochster's formula for computing multigraded Betti numbers of squarefree monomial ideals via simplicial homology. Moreover, techniques such as deformation and polarization can allow one to take a problem about homogeneous ideals and turn it into a question about squarefree monomial ideals.

This paper is written to give a self-contained introduction to Stanley–Reisner theory, directing it especially at relatively new graduate students in commutative algebra and combinatorics. Our goal is to provide enough background to enable readers to progress to more detailed treatments in one of the many excellent references (for example, [11, 13, 15, 17]).

C.A. Francisco (✉) • J. Mermin • J. Schweig
Department of Mathematics, Oklahoma State University, 401 Mathematical Sciences, Stillwater, OK 74078, USA
e-mail: chris@math.okstate.edu; mermin@math.okstate.edu; jay.schweig@okstate.edu

S.M. Cooper and S. Sather-Wagstaff (eds.), *Connections Between Algebra, Combinatorics, and Geometry*, Springer Proceedings in Mathematics & Statistics 76,
DOI 10.1007/978-1-4939-0626-0_5,

In the following section, we define some important terms and explain the basics of the Stanley–Reisner correspondence. We introduce some running examples in Sect. 3. In Sect. 4, we discuss the important role Alexander duality plays in studying squarefree monomial ideals and simplicial complexes and its interaction with associated primes. We develop this topic further in Sect. 5, paying particular attention to shellability and the Cohen–Macaulay property. Section 6 is devoted to connections between f- and h-vectors on the combinatorial side and Hilbert series on the algebraic side. Finally, in Sect. 7, we explore Hochster's formula and the use of simplicial homology to compute multigraded Betti numbers of monomial ideals.

2 Background and Notation

We begin by making explicit some familiar ideas and notation from combinatorics and algebra. Most readers should be comfortable skipping to Sect. 2.3, which defines the Stanley–Reisner correspondence.

2.1 Combinatorics

Fix $n > 0$, and let $X = \{x_1, x_2, \dots, x_n\}$. An *abstract simplicial complex* Δ on vertex set X is a collection of subsets of X (that is, $\Delta \subseteq 2^X$) such that $A \in \Delta$ whenever $A \subseteq B \in \Delta$. The members of Δ are called *simplices* or *faces*, and a simplex of Δ not properly contained in another simplex of Δ is called a *facet*. If all the facets of Δ have the same cardinality, then Δ is called *pure*.

If Δ contains no sets, it is called the *void* complex. Any simplicial complex other than the void complex contains the empty set as a face. This may seem like a minor point, but in the future it will be necessary to differentiate between the void complex, $\emptyset \subseteq 2^X$, and the complex whose only simplex is the empty simplex, $\{\emptyset\} \subseteq 2^X$.

Now fix some ℓ. If $k \leq \ell$, the convex hull of $k + 1$ points in general position (that is, no three are on a single line, no four are in one plane, etc.) in $\mathbb{R}^\ell$ is a *geometric k-simplex*. In general, a collection Δ of geometric simplices in $\mathbb{R}^\ell$ is a *geometric simplicial complex* if $\sigma \cap \tau$ is a geometric simplex in Δ for any $\sigma, \tau \in \Delta$. It is a basic fact from combinatorial topology that every abstract simplicial complex has a geometric realization, and thus we often use the term "simplicial complex" to simultaneously refer to both these points of view. In Sect. 3, each figure is a geometric realization of the corresponding simplicial complex.

The *dimension* of a simplex $A \in \Delta$ is one less than the cardinality of A. This coincides with our usual geometric notion of a simplex, as it requires $k + 1$ points in general position in Euclidean space to have a k-dimensional convex hull.

If $A \notin \Delta$ but $B \in \Delta$ for any $B \subsetneq A$, then A is called a *minimal non-face* of Δ. For example, the minimal non-faces of the bow-tie complex that will be introduced in Example 3.3 are *ad*, *ae*, *bd*, and *be*.

2.2 Algebra

Let k be a field, and let $S = k[X] = k[x_1, x_2, \ldots, x_n]$.

A *monomial* of S is an element m which factors (uniquely, up to order) as a product of the variables in X; it is *squarefree* if no variable appears more than once in this factorization. The *degree* of m is the number of variables in its factorization; if m is squarefree, its degree is equal to the number of variables dividing m. Monomials are partially ordered by divisibility, and this ordering refines degree: If m divides m', then $\deg m \leq \deg m'$, with equality only if $m = m'$.

A *monomial ideal* $I \subset S$ is an ideal which, when viewed as a k-vector space, has a (unique) basis consisting of monomials. Equivalently, a monomial ideal is one with a generating set consisting of monomials. Of course, I is generated by the set of all its monomials, but this generating set is highly redundant. We can remove the redundancy by restricting our attention to those monomials of I which are minimal under divisibility. This yields a generating set which is minimal in the sense that any monomial generating set of I contains it; this is called $\mathrm{gens}(I)$, or the *unique minimal monomial generating set* for I, and its elements are simply referred to as *generators* of I. A *squarefree monomial ideal* is a monomial ideal whose monomial generators are all squarefree.

2.3 The Stanley–Reisner Correspondence

The Stanley–Reisner correspondence arises from two important observations connecting the information in simplices to that in squarefree monomials. The first observation is that these are in natural bijection.

Definition 2.1. Let $A \subset X$. Then the *monomial supported on A* is the squarefree monomial $m_A = \prod_{x_i \in A} x_i$. Conversely, if m is a squarefree monomial, then its *support* is $\mathrm{supp}\, m = \{x_i : x_i \text{ divides } m\}$.

Notation 2.2. *Throughout the paper, we will abuse notation without comment by writing squarefree monomials in place of subsets of X. That is, we will refer to faces of a simplicial complex as m instead of as* $\mathrm{supp}\, m$, *and in the examples we will dispense with the set brackets and the commas.*

The second observation of Stanley–Reisner theory is that simplicial complexes and monomial ideals have opposite behavior with respect to the partial orders of inclusion and divisibility. That is, simplicial complexes are closed under "shrinking," and monomial ideals are closed under "growing":

Proposition 2.3. *Let m be a squarefree monomial. If I is a squarefree monomial ideal and $m \in I$, then $m' \in I$ whenever m divides m'. On the other hand, if Δ is a simplicial complex and $m \in \Delta$, then $m' \in \Delta$ whenever $m' \subset m$.*

The corollary to this observation motivates the definitions at the heart of Stanley–Reisner theory.

Corollary 2.4. *If $I \subset S$ is a squarefree monomial ideal, then the set of squarefree monomials not contained in I forms a simplicial complex.*

Definition 2.5. For a squarefree monomial ideal I, the *Stanley–Reisner complex of* I is the simplicial complex consisting of the squarefree monomials not in I,

$$\Delta_I = \{m \subset X : m \notin I\}.$$

For a simplicial complex Δ, the *Stanley–Reisner ideal of* Δ is the squarefree monomial ideal generated by the non-faces of Δ,

$$I_\Delta = (m \subset X : m \notin \Delta).$$

The *face ring* or *Stanley–Reisner ring* of Δ is the quotient by the Stanley–Reisner ideal, $R_\Delta = S/I_\Delta$.

Observe that the minimal monomial generators of I_Δ are the minimal non-faces of Δ.

The following is immediate.

Proposition 2.6. *If I is a squarefree monomial ideal, then $I_{\Delta_I} = I$. If Δ is a simplicial complex, then $\Delta_{I_\Delta} = \Delta$.*

3 Running Examples

In the examples, we set $X = \{a, b, c, \dots\}$ instead of $\{x_1, x_2, x_3, \dots\}$. Throughout the paper, we will refer to the examples in this section.

Example 3.1. Let K be the complex in Fig. 1. The facets of K are abc, abe, ace, bcd, bde, and cde. The minimal nonfaces of K are ad and bce, so $I_K = (ad, bce)$. Both K and I_K are very well-behaved objects; for example, the geometric realization of K is homeomorphic to a sphere, and I_K is a complete intersection.

Example 3.2. Let Q be the standard triangulation of the real projective plane, as in Fig. 2. The facets of Q are abd, abf, acd, ace, aef, bce, bcf, bde, cdf, and def, and I_Q is generated by the ten minimal nonfaces, all of which have cardinality three:

$$I_Q = (abc, abe, acf, ade, adf, bcd, bdf, bef, cde, cef).$$

The triangulation of the projective plane is globally strange (for example, its homology is characteristic-dependent) but locally well behaved.

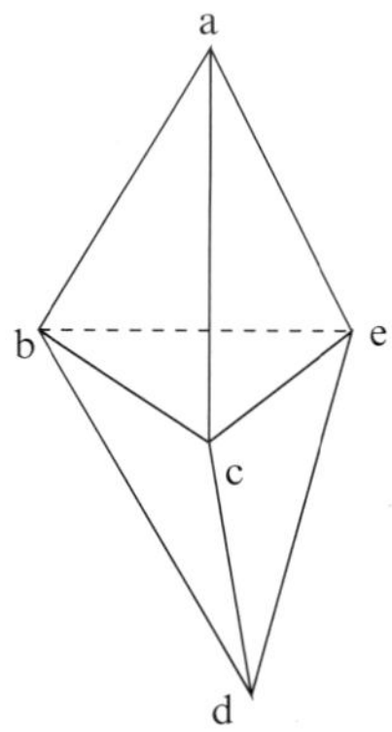

Fig. 1 A simplicial complex homeomorphic to $\mathbb{S}^2$

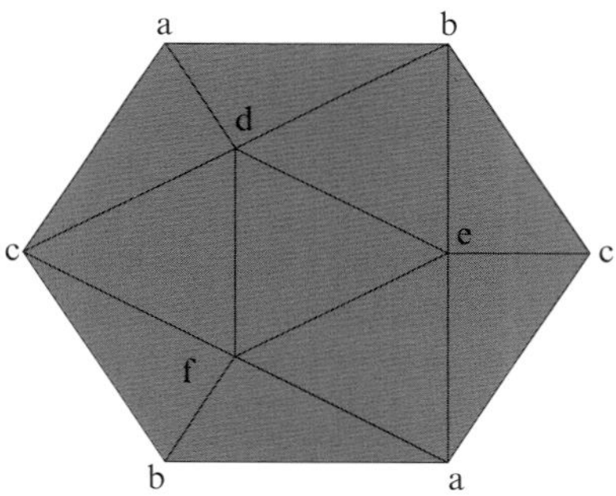

Fig. 2 The standard minimal triangulation of the projective plane (note the identifications on the boundary)

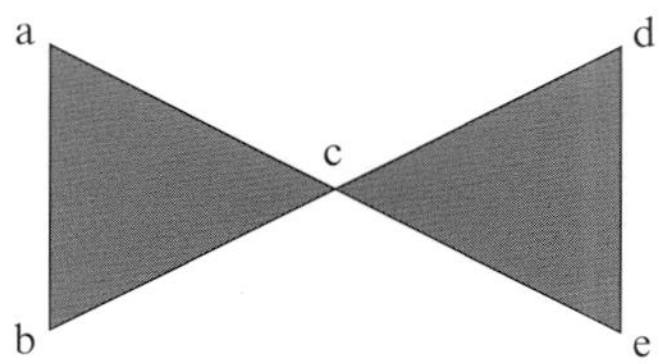

Fig. 3 The bow-tie complex B

Example 3.3. Let B be the "bow-tie" complex shown in Fig. 3. The facets of B are abc and cde, and the minimal non-faces of B are ad, ae, bd, and be, meaning $I_B = (ad, ae, bd, be)$.

4 Alexander Duality and Associated Primes

The notion of Alexander duality comes from algebraic topology, where for a sufficiently "nice" subspace Γ of the n-dimensional sphere $\mathbb{S}^n$ there is an isomorphism between $\tilde{H}_i(\Gamma)$ and $\tilde{H}^{n-i-1}(\Gamma^c)$, where Γ^c is the complement of Γ in $\mathbb{S}^n$. The combinatorial flavor of Alexander duality, which we discuss here, produces a dual complex $\Delta^\vee$ from a simplicial complex Δ and relates this complex to the prime ideals associated to the Stanley–Reisner ideal I_Δ.

Definition 4.1. If Δ is a simplicial complex, the *Alexander dual* of Δ, denoted by $\Delta^\vee$, is the simplicial complex with faces $\{X \setminus m : m \notin \Delta\}$. That is, faces of $\Delta^\vee$ are complements of non-faces of Δ.

Note that the facets of $\Delta^\vee$ are thus complements of *minimal* nonfaces of Δ.

Example 4.2. The complex K of Fig. 1 has minimal non-faces ad and bce, and thus $K^\vee$ has facets bce and ad. Note that $K^\vee$ is the disjoint union of a line segment and solid triangle, meaning $K^\vee$ is homotopy equivalent to $\mathbb{S}^0$, whereas K is homotopy equivalent to $\mathbb{S}^2$.

Example 4.3. If Q is the triangulation of the projective plane shown in Fig. 2, then Q is self-dual. Indeed, the minimal non-faces of Q are abc, abe, acf, ade, adf, bcd, bdf, bef, cde, and cef, and so facets of $Q^\vee$ are the complements of these non-faces: def, cdf, bde, bcf, bce, aef, ace, acd, abf, and abd. Note that these are exactly the facets of Q, and thus $Q^\vee = Q$.

Example 4.4. The complex B of Fig. 3 has minimal non-faces ad, ae, bd, and be, and thus the facets of $B^\vee$ are bce, bcd, ace, and acd. The complex $B^\vee$ is shown below in Fig. 4.

Observe that c is a cone point in both B and $B^\vee$. This illustrates the general fact that a cone point of a complex is also a cone point of the dual.

Remark 4.5. The name "Alexander dual" is justified by the fact that $\Delta^\vee$ is homotopic to the complement of Δ in the $(n-2)$-sphere. See [1, Theorem 2.1] or [11, Remark 5.7].

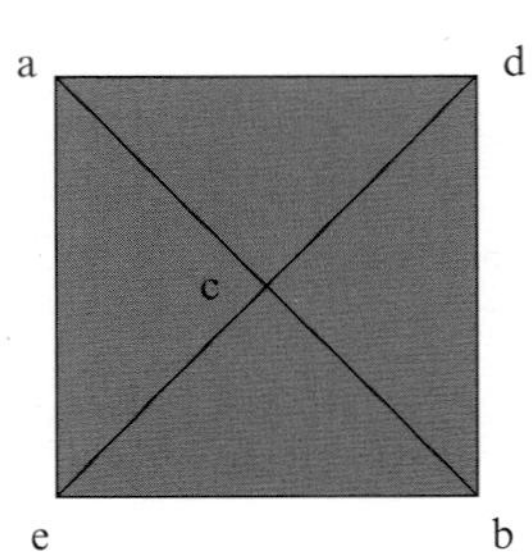

Fig. 4 The dual of the complex B

The translation of Alexander duality to squarefree monomial ideals involves the set of associated primes, which is a natural generalization of the set of prime factors of an integer.

Fact 4.6. Let I be a squarefree monomial ideal. Then I decomposes as an intersection of prime monomial ideals,

$$I = \bigcap_{P \supset I} P.$$

Here P ranges over the primes which contain I and are monomial ideals. Of course, this intersection is highly redundant. For example, the homogeneous maximal ideal appears in the intersection but can be deleted unless I is itself the maximal ideal. It is natural to remove the redundancy by restricting our attention to the monomial primes which are minimal among those containing I; these are called the *associated primes* of I.

Remark 4.7. The discussion above fails badly if I is not a squarefree monomial ideal. Monomial ideals cannot in general be written as intersections of primes, or even of powers of primes. Instead, we can find an *irredundant primary decomposition*: that is, we can write an arbitrary ideal as an irredundant intersection of primary ideals. This decomposition is usually not unique, but the set of primes which occur as radicals of the primary ideals is, and we call these the *associated primes*. The minimal primes containing an ideal are associated, but it is possible for non-minimal primes to be associated as well. Associated primes can be detected as the annihilators of elements as in the last condition of Theorem 4.8. For more details, see [6, Chap. 3].

Theorem 4.8. *Let I be a squarefree monomial ideal. Then the following are equivalent for a monomial prime ideal P.*

1. *P contains I and is minimal among ideals that do.*
2. *I may be written as an irredundant intersection of primary ideals, $I = \bigcap Q_j$, and P is the radical of one of the Q_j in this intersection.*
3. *There is a monomial $m \notin I$ with the property that $(I : m) = P$; that is, $mx \in I$ if and only if $x \in P$.*

Definition 4.9. Let I be a squarefree monomial ideal. If P is a monomial prime ideal satisfying the equivalent conditions of Theorem 4.8, then we say that P is an *associated prime* of I. The set of all associated primes of I is written $\mathrm{Ass}(I)$.

To understand the associated primes of a monomial ideal, we need to study monomial primes. The chief observation is that every monomial prime is generated by a subset of the variables. This yields a bijection—between monomial primes and squarefree monomials—that is the basis of Alexander duality.

Notation 4.10. *If $A \subset V$ is a subset of the variables, then write P_A for the prime ideal generated by the elements of A, $P_A = (x_i : x_i \in A)$. If m is a monomial, write P_m for $P_{\operatorname{supp} m}$.*

This notation allows us to describe the facets of Δ in terms of the associated primes of I.

Proposition 4.11. *Let I be a squarefree monomial ideal, and let $\mathbf{X} = x_1 \dots x_n$ be the product of the variables. Let m be a squarefree monomial. Then P_m is associated to I if and only if $\frac{\mathbf{X}}{m}$ is a facet of Δ_I. More generally, P_m contains I if and only if $\frac{\mathbf{X}}{m} \in \Delta_I$.*

Proof. Observe that P_m contains I if and only if $\operatorname{supp} m$ shares a variable with $\operatorname{supp} \mu$ for every monomial $\mu \in I$. Equivalently, the complement of $\operatorname{supp} m$, $\frac{\mathbf{X}}{m}$, does not contain $\operatorname{supp} \mu$. This means that $\frac{\mathbf{X}}{m}$ is not divisible by any μ, so is not contained in I. In other words, $\frac{\mathbf{X}}{m} \in \Delta_I$. This proves the second claim.

For the first claim, observe that facets are maximal, associated primes are minimal, and the operation taking m to $\frac{\mathbf{X}}{m}$ is order-reversing. (Alternatively, $\frac{\mathbf{X}}{m}$ is a facet if and only if $x_i \frac{\mathbf{X}}{m} \in I$ for all x_i dividing m.)

These observations inspire the definition of the (algebraic) Alexander dual of I as (abusing notation) the ideal generated by the associated primes of I.

Definition 4.12. Let I be a squarefree monomial ideal. Then the *Alexander dual* of I is

$$I^\vee = \big(m : P_m \in \operatorname{Ass}(I)\big).$$

We justify this name by showing that Alexander duality commutes with the Stanley–Reisner correspondence.

Theorem 4.13. *If I is a squarefree monomial ideal, then $\Delta_{(I^\vee)} = (\Delta_I)^\vee$. If Δ is a simplicial complex, then $I_{(\Delta^\vee)} = (I_\Delta)^\vee$. Thus, we can write $I_\Delta^\vee$ or $\Delta_I^\vee$ without confusion.*

Proof. The faces of $\Delta_{(I^\vee)}$ are the monomials m such that P_m does not contain I, i.e., the monomials whose complements are contained in I. The faces of $(\Delta_I)^\vee$ are the complements of the non-faces of Δ_I, i.e., the complements of the monomials in I.

It immediately follows that algebraic Alexander duality is a duality operation, and that it gives us a somewhat more efficient way to compute the associated primes of a squarefree monomial ideal.

Corollary 4.14. *Let I be a squarefree monomial ideal. Then $(I^\vee)^\vee = I$.*

Corollary 4.15. *Let I be a squarefree monomial ideal. Then* $I^\vee = \bigcap_{m \in \mathrm{gens}(I)} P_m$, *and the associated primes of I correspond to the generators of $I^\vee$. That is,* $\mathrm{Ass}(I) = \{P_\mu : \mu \in \mathrm{gens}(I^\vee)\}$.

If we know the Stanley–Reisner ideal of a complex (or vice-versa) we can use it to compute the dual with less work.

Corollary 4.16. *Let* $\mathbf{X} = x_1 \dots x_n$ *be the product of the variables.*

1. *Let Δ be a simplicial complex. Then the facets of $\Delta^\vee$ are the monomials $\frac{\mathbf{X}}{m}$, where m ranges over the generators of I_Δ.*
2. *Let I be a squarefree monomial ideal. Then the generators of $I^\vee$ are the monomials $\frac{\mathbf{X}}{f}$, where f ranges over the facets of Δ_I.*

Corollary 4.17. *Let I be a squarefree monomial ideal. Then I is equidimensional (i.e., all its associated primes have the same height) if and only if $\Delta_I^\vee$ is pure (i.e., all its facets have the same dimension).*

Example 4.18. Let K be the complex in Example 3.1. Then $I_K = (ad, bce)$, so $I_K^\vee = (a,d) \cap (b,c,e) = (ab, ac, ae, bd, cd, de)$. Note that the Stanley–Reisner complex corresponding to $I_K^\vee$ has facets ad and bce. This is consistent with the computation in Example 4.2.

Example 4.19. Let Q be the triangulation of the projective plane from Example 3.2. As we saw in Example 4.3, Q is self-dual, and therefore I_Q is also self-dual.

Example 4.20. Let B be the bow-tie complex in Example 3.3. It has two facets, abc and cde. Therefore $I_B^\vee$ is generated by $\frac{\mathbf{X}}{abc} = de$ and $\frac{\mathbf{X}}{cde} = ab$; i.e., $I_B^\vee = (ab, de)$.

Alexander duality plays a vital role in linking the notions of Cohen–Macaulayness of simplicial complexes and linearity in free resolutions of squarefree monomial ideals, as we discuss in the next section.

5 Shellability, Cohen–Macaulayness, and Linear Resolutions

Shellable complexes occur frequently throughout combinatorics. The shellability condition is particularly helpful because shellable simplicial complexes are homotopic to bouquets of spheres and, in particular, are sequentially Cohen–Macaulay over any field.

Definition 5.1. For a simplex F, recall that its *boundary* ∂F is the union of its codimension-one faces. An ordering $F_1, F_2, \dots, F_t$ of the facets of a simplicial complex Δ is a *shelling* if, for each j with $1 < j \leq t$, the intersection

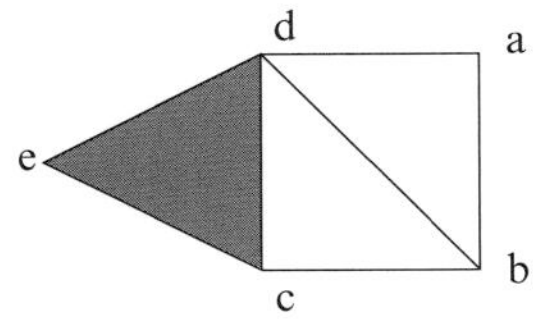

Fig. 5 A shellable simplicial complex

$$\left(\bigcup_{i=1}^{j-1} F_i\right) \cap F_j$$

is a nonempty union of facets of ∂F_j. If there exists a shelling of Δ, then Δ is called *shellable*.

See [2, 3] for more details, particularly in the nonpure case, and motivation.

Example 5.2. The complex in Fig. 5 is shellable. (For example, one shelling is cde, ad, ab, bc, bd.) Note, however, that any sequence of facets of this complex in which the sole two-dimensional facet is not first cannot be a shelling.

The complex K from Example 3.1 is shellable, as we discuss in detail in the next section. The other examples from Sect. 3 are not shellable.

We sketch the proof that shellable complexes are bouquets of spheres. Given a shelling $F_1, \ldots, F_t$ of a simplicial complex Δ, call F_j a *full-restriction facet* if

$$\partial F_j \subseteq \bigcup_{k=1}^{j-1} F_k.$$

Theorem 5.3. *Let Δ be a simplicial complex with shelling $F_1, F_2, \ldots, F_t$. Then Δ is homotopy equivalent to a wedge of spheres of the dimensions of the full restriction facets,*

$$\Delta \simeq \bigvee_{F_j} \mathbb{S}^{\dim F_j},$$

where the wedge is taken over the full restriction facets F_j.

The complex of Fig. 5 is homotopy equivalent to a wedge of two one-dimensional spheres.

Proof (of Theorem 5.3). Let Δ' denote the complex obtained from Δ by removing each full-restriction facet F_j (but keeping ∂F_j in Δ'). Then the given shelling order, after removing the full-restriction facets, is a shelling of Δ'.

A straightforward induction on the number of facets of Δ' shows that it is contractible. If F_j is an i-dimensional full-restriction facet of Δ, then contracting Δ' identifies the boundary of F_j, creating an i-dimensional sphere. The same holds for all full-restriction facets of Δ, proving the result.

Shellability of a simplicial complex implies an important algebraic property known as sequential Cohen–Macaulayness.

Definition 5.4. Let M be a finitely generated graded module. We say that M is *Cohen–Macaulay* if its depth is the same as its dimension; equivalently, if its projective dimension is the same as its codimension. The module M is *sequentially Cohen–Macaulay* if there exists a filtration

$$0 = M_0 \subset M_1 \subset \cdots M_r = M$$

such that each M_i/M_{i-1} is Cohen–Macaulay, and the Krull dimensions of the quotients are increasing:

$$\dim(M_1/M_0) < \dim(M_2/M_1) < \cdots < \dim(M_r/M_{r-1}).$$

See, e.g., [4, 6, 14] for detailed discussions of Cohen–Macaulayness. For more details on sequential Cohen–Macaulayness, see [15], and see [8, 15] for equivalent definitions in terms of Ext modules.

We define a simplicial complex Δ to be (sequentially) Cohen–Macaulay if S/I_Δ is (sequentially) Cohen–Macaulay.

Theorem 5.5. *If Δ is shellable, then Δ is sequentially Cohen–Macaulay. If Δ is also pure, then Δ is Cohen–Macaulay.*

The result that a pure shellable complex is Cohen–Macaulay is [4, Theorem 5.1.13], [11, Theorem 13.45], or [15, Chap. 3, Theorem 2.5]. For the first statement, build the filtration from the Stanley–Reisner ideals of Δ_i, where Δ_i is the subcomplex of Δ generated by the facets of dimension at most i.

The characterization of when a simplicial complex is Cohen–Macaulay is known as *Reisner's criterion*. Recall that the *link* of a face F in a complex Δ, for which we write $\text{link}_\Delta(F)$, is a simplicial complex whose faces are given as follows:

$$\text{link}_\Delta(F) = \{G \in \Delta : F \cup G \in \Delta, F \cap G = \emptyset\}.$$

Theorem 5.6 (Reisner's Criterion). *A simplicial complex Δ is Cohen–Macaulay over k if and only if for any face F of Δ, $\dim_k(\tilde{H}_i(\text{link}_\Delta(F);k)) = 0$ for all $i < \dim(\text{link}_\Delta(F))$.*

That is, Δ is Cohen–Macaulay over k if and only if the homology of each face's link vanishes below its top dimension.

The complex B of Example 3.3 is not Cohen–Macaulay over any field, as $\text{link}_B(c)$ has facets ab and de, meaning $\dim_k(\tilde{H}_0(\text{link}_B(c), k)) = 1$ for any k, whereas $\dim(\text{link}_B(c)) = 1$.

The complex of Fig. 5 is two-dimensional, yet its first homology group is nontrivial. As the link of the empty face is the entire complex, this shows that this complex is not Cohen–Macaulay.

Finally, consider the complex K of Example 3.1. The link of every vertex of K is a 1-sphere (for example, $\text{link}_K(a)$ has facets $bc, ce,$ and eb, while $\text{link}_K(c)$ has facets $ab, bd, de,$ and ea), and the link of every edge of K is a 0-sphere (for example, $\text{link}_K(ab)$ has facets c and e). As $\text{link}_K(\emptyset) = K$, which is a 2-sphere, it follows that no link of a face of K has nontrivial homology below its top dimension. Thus, K is Cohen–Macaulay (as is any simplicial sphere).

The following theorem of Eagon and Reiner [5, Theorem 3] connects Cohen–Macaulayness and free resolutions. (For background on free resolutions, see [13].)

Theorem 5.7. *A simplicial complex Δ is Cohen–Macaulay over k if and only if $I_{\Delta^\vee}$ has a linear free resolution over $S = k[x_1, \ldots, x_n]$.*

Example 5.8. Take Δ to be the simplicial complex of Fig. 5 without the face cde. Then $I_\Delta = (ac, ae, be, bcd, abd, cde)$, and $I_{\Delta^\vee} = (abc, abd, abe, ade, bce, cde)$, which has a linear resolution. Therefore Δ is Cohen–Macaulay. One can verify this by, for example, noting that the codimension of I_Δ is three, as is the projective dimension of S/I_Δ.

Note the importance of the field k in Theorem 5.7. Reisner pointed out the minimal triangulation of the projective plane Q is Cohen–Macaulay if and only if $\text{char}\, k \neq 2$. Thus $I_Q^\vee$ has a linear resolution when $\text{char}\, k \neq 2$ but nonlinear resolution when $\text{char}\, k = 2$.

We can loosen the conditions in Theorem 5.7 to get another useful statement. Sequential Cohen–Macaulayness is a natural generalization of Cohen–Macaulayness in the case in which Δ is not pure. We define the appropriate homological analogue:

Definition 5.9. Let I be a homogeneous ideal, and write (I_d) for the ideal generated by the degree-d forms in I. The ideal I is *componentwise linear* if for all d, (I_d) has a linear resolution.

Remark 5.10. The condition in Definition 5.9 is not as computationally difficult to check as it appears. One needs to only check that (I_d) has a linear resolution for degrees d in which I has minimal generators.

We need only check Herzog and Hibi prove the following theorem [7, Theorem 2.1].

Theorem 5.11. *A simplicial complex Δ is sequentially Cohen–Macaulay over k if and only if $I_{\Delta^\vee}$ is componentwise linear over $S = k[x_1, \ldots, x_n]$.*

One consequence of Theorem 5.11 is a partial converse of Theorem 5.5.

Proposition 5.12. *Fix a field k. Then S/I_Δ is Cohen–Macaulay over k if and only if Δ is pure and S/I_Δ is sequentially Cohen–Macaulay over k.*

Proof. One direction is clear. For the other, assume that Δ is pure and that S/I_Δ is sequentially Cohen–Macaulay over k. By Theorem 5.11, because S/I_Δ is sequentially Cohen–Macaulay over k, $I_{\Delta^\vee}$ is componentwise linear. Moreover, the fact that Δ is pure means that $I_{\Delta^\vee}$ is generated in a single degree. Thus $I_{\Delta^\vee}$ has a linear resolution, and by Theorem 5.7, Δ is Cohen–Macaulay.

One can also detect shellability with similar algebraic methods. Suppose I is a monomial ideal, minimally generated by monomials $m_1, \dots, m_r$, where $\deg m_i \leq \deg m_{i+1}$ for all i. We say that I has *linear quotients* if for each $2 \leq i \leq r$, the ideal $(m_1, \dots, m_{i-1}) : (m_i)$ is generated by a subset of the variables. Herzog and Takayama note in [9] that the statements that I_Δ has linear quotients and $\Delta^\vee$ is nonpure shellable are "almost tautologically equivalent." See also Proposition 6.13. Moreover, if I_Δ has linear quotients and is generated in a single degree, then I_Δ has a linear resolution, and therefore $\Delta^\vee$ is Cohen–Macaulay.

Remark 5.13. We conclude with a brief discussion of the use of Alexander duality in the study of edge and cover ideals of graphs. Given a graph on vertices $\{x_1, \dots, x_n\}$, the edge ideal I_G of G is generated by all monomials $x_i x_j$ such that $\{x_i, x_j\}$ is an edge of G. The cover ideal J_G is generated by monomials m such that each edge of G contains at least one element of $\operatorname{supp} m$, meaning that m is a vertex cover of G. Therefore $J_G = I_G^\vee$, and one can investigate properties of the graph G by studying either the edge or the cover ideal of G, often passing back and forth between the two. See [12] for a survey of recent work on edge and cover ideals.

6 Hilbert Functions and f-Vectors

Probably the most important numerical invariant of a graded ideal is its Hilbert function, which associates with each degree d the dimension of the ideal's degree-d piece. This measures the size of the ideal and contains a lot of other important information, such as its dimension and multiplicity.

Definition 6.1. Let M be a graded module. Then the *Hilbert function* of M is

$$\begin{aligned} \mathrm{HF}_M : \mathbf{Z} &\to \mathbf{Z} \\ d &\mapsto \dim_k(M_d). \end{aligned}$$

In order to use generating function techniques, we also define the *Hilbert series* as the generating function on the Hilbert function, $\mathrm{HS}_M(t) = \sum_d \mathrm{HF}_M(d) t^d$.

Given an ideal I, we study both the Hilbert function of I, HF_I, and the Hilbert function of the quotient of S by I, $\mathrm{HF}_{S/I}$. Observe that $\mathrm{HF}_I + \mathrm{HF}_{S/I} = \mathrm{HF}_S$.

We begin by computing the Hilbert function of S.

Proposition 6.2. *Let $S = k[x_1, \dots, x_n]$. Then $\mathrm{HS}_S(t) = \frac{1}{(1-t)^n}$.*

Proof. We induct on n. If $n = 1$, then $S = k[x]$ has basis $\{1, x, x^2, \dots\}$ and Hilbert series $t^0 + t^1 + t^2 + \cdots = \frac{1}{1-t}$.

In general, let $\mathscr{B}$ be a graded k-basis for $k[x_1, \dots, x_{n-1}]$. Then S decomposes (as a vector space) as

$$S = \bigoplus_{b \in \mathscr{B}} b \cdot k[x_n].$$

If b has degree d, the Hilbert series of the corresponding summand $b \cdot k[x_n]$ is $t^d + t^{d+1} + t^{d+2} + \cdots = \frac{t^d}{1-t}$. We have

$$\begin{aligned}
\mathrm{HS}_S(t) &= \sum_{b \in \mathscr{B}} \mathrm{HS}_{b \cdot k[x_n]}(t) \\
&= \sum_{b \in \mathscr{B}} \frac{t^{\deg b}}{1-t} \\
&= \frac{1}{1-t} \sum_{b \in \mathscr{B}} t^{\deg b} \\
&= \left(\frac{1}{1-t}\right) \mathrm{HS}_{k[x_1,\dots,x_{n-1}]}(t) \\
&= \left(\frac{1}{1-t}\right) \frac{1}{(1-t)^{n-1}}.
\end{aligned}$$

It turns out that the Hilbert series is always a rational function.

Fact 6.3. Let M be a finitely generated S-module. Then

$$\mathrm{HS}_M(t) = \frac{p(t)}{(1-t)^n}$$

for some polynomial $p(t)$. If we write this in lowest terms, it becomes

$$\mathrm{HS}_M(t) = \frac{h(t)}{(1-t)^d},$$

where d is the Krull dimension of M. Here, $h(t)$ is called the *h-polynomial* of M and $h(1)$ is the *multiplicity* of M.

If I is a monomial ideal, then the Hilbert function of I counts the monomials appearing in I, and the Hilbert function of S/I counts the monomials not appearing in I. If I is a squarefree monomial ideal, this is still true, but it is much easier to count the smaller number of squarefree monomials in I. To this end, we define another object, the *squarefree Hilbert function.*

Definition 6.4. Let I be a squarefree monomial ideal. Then the *squarefree Hilbert function* of I is the function

$$\begin{aligned}
\mathrm{HF}_I^{\mathrm{sf}} &: \mathbb{Z} \to \mathbb{Z} \\
d &\mapsto \#\{m \in I : m \text{ is a squarefree monomial of degree } d\}.
\end{aligned}$$

We define $\mathrm{HF}^{\mathrm{sf}}_{S/I}$ similarly, and define the squarefree Hilbert series $\mathrm{HS}^{\mathrm{sf}}_I$ and $\mathrm{HS}^{\mathrm{sf}}_{S/I}$ to be the generating functions on the squarefree Hilbert functions. Note that the squarefree Hilbert series is actually a polynomial, since there are no squarefree monomials of degree greater than n.

In order to understand the relationship between the Hilbert function and squarefree Hilbert function of a squarefree monomial ideal, we need a little more machinery.

Definition 6.5. Let $m = \prod x_i^{e_i}$ be a monomial. Then the *squarefree part* of m is the result of "deleting the exponents" from m,

$$\mathrm{sf}(m) = \prod_{e_i \geq 1} x_i = \prod_{x_i \text{ divides } m} x_i .$$

The idea is that a monomial's presence in or absence from a squarefree ideal depends only on its squarefree part.

Lemma 6.6. *Let I be a squarefree monomial ideal and m a monomial. Then $m \in I$ if and only if* $\mathrm{sf}(m) \in I$.

This observation allows us to compute so-called Stanley decompositions of I and S/I.

Corollary 6.7. *Let I be a squarefree monomial ideal. Then, viewed as vector spaces, I and S/I have the decompositions*

$$I = \bigoplus_{m \in I} m \cdot k[\operatorname{supp} m] \qquad \text{and} \qquad S/I = \bigoplus_{m \notin I} m \cdot k[\operatorname{supp} m],$$

both sums being taken over the set of squarefree monomials.

These decompositions will enable us to compute the Hilbert functions in terms of the corresponding squarefree Hilbert functions.

Lemma 6.8. *Let m be a squarefree monomial. Then the Hilbert series of $m \cdot k[\operatorname{supp} m]$ is $\left(\frac{t}{1-t}\right)^{\deg m}$.*

Proof. By Proposition 6.2, the Hilbert series of $k[\operatorname{supp} m]$ is $\frac{1}{(1-t)^{\deg(m)}}$. Multiplying by m increases the degree of everything by deg m and thus multiplies the Hilbert series by t^{degm}.

Theorem 6.9. *Let I be a squarefree monomial ideal. Then $\mathrm{HS}_I(t) = \mathrm{HS}^{\mathrm{sf}}_I\left(\frac{t}{1-t}\right)$ and $\mathrm{HS}_{S/I}(t) = \mathrm{HS}^{\mathrm{sf}}_{S/I}\left(\frac{t}{1-t}\right)$.*

Proof. By Corollary 6.7 and Lemma 6.8 we have

$$\mathrm{HS}_I(t) = \sum_{m \in I,\ \text{squarefree}} \mathrm{HS}_{m \cdot k[\operatorname{supp} m]}(t)$$

$$= \sum_{m \in I,\ \text{squarefree}} \left(\frac{t}{1-t}\right)^{\deg m}$$

$$= \mathrm{HS}_I^{\mathrm{sf}}\left(\frac{t}{1-t}\right).$$

The computation for the quotient is identical.

The Stanley–Reisner analogue of the squarefree Hilbert Function, which counts squarefree monomials, is the f-vector, which counts faces of a simplicial complex.

Definition 6.10. The *f-vector* of a $(d-1)$-dimensional complex Δ is the sequence $f(\Delta) = (f_{-1}, f_0, \ldots, f_{d-1})$ where f_i is the number of i-dimensional faces of Δ (and $f_{-1} = 1$ whenever Δ is not the void complex). The f-polynomial is the generating function of the f-vector,

$$f_\Delta(t) = f_{-1}t^d + f_0 t^{d-1} + \cdots + f_{d-2}t + f_{d-1}.$$

Note that the coefficient of t^i is the number of *co*dimension-i faces (reversing the convention of the Hilbert series).

The f-vector is perhaps the most natural combinatorial invariant of simplicial complexes. In a sense, study of the f-vector dates back to Euler: The *reduced Euler characteristic* of Δ, which is a topological invariant, is given by

$$\tilde{\chi}(\Delta) = \sum_{i=-1}^{d-1} (-1)^i f_i.$$

The *h-vector* of a complex Δ is the result of an invertible transformation applied to the f-vector of Δ.

Definition 6.11. Given a $(d-1)$-dimensional simplicial complex Δ, define the *h-polynomial* of Δ, written $h_\Delta(t)$, to be the polynomial given by

$$h_\Delta(t) = f_\Delta(t-1).$$

The *h-vector* of Δ is the sequence $h(\Delta) = (h_0, h_1, \ldots, h_d)$ of coefficients of $h_\Delta(t)$:

$$h_\Delta(t) = h_0 t^d + h_1 t^{d-1} + \cdots + h_{d-1}t + h_d.$$

We will see in Theorem 6.15 that the h-vector also appears in the numerator of the Hilbert series of S/I_Δ (expressed as a rational function in simplest terms).

Remark 6.12. The h-vector of a complex Δ is typically studied only when Δ is pure.

Although the f-vector and h-vector contain the same information, properties of some simplicial complexes are often easier to express in terms of the h-vector. One

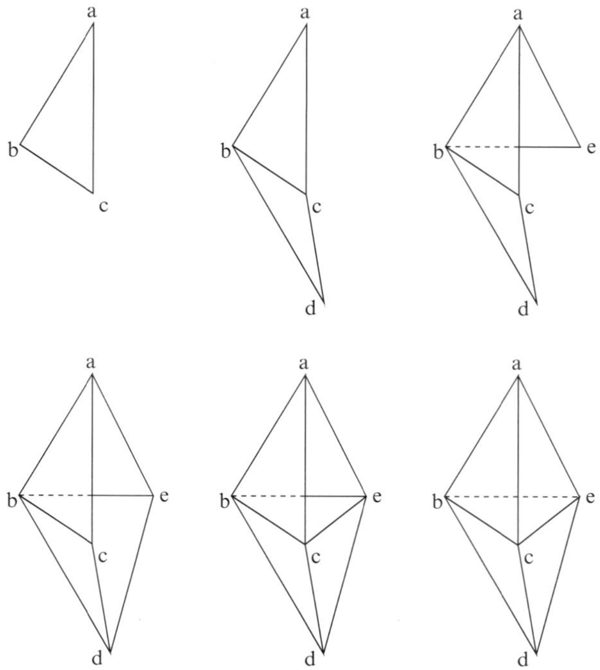

Fig. 6 Shelling the pure simplicial complex K from Example 1

example of this are the *Dehn–Sommerville relations*, which state that the h-vector of a simplicial polytope boundary is palindromic.

Note that the entries of the h-vector are not guaranteed to be nonnegative. Indeed, the complex B in Example 3.3 has facets abc and cde, meaning it has f-vector $(1, 5, 6, 2)$ and h-vector $(1, 2, -1, 0)$. However, when Δ is a pure shellable complex its h-vector counts something, and thus has nonnegative entries. First, we examine a shelling in greater detail.

A shelling of a complex Δ can be thought of as a recipe for building Δ one facet at a time. In the shelling shown in Fig. 6 of the complex K from Example 1, we begin with the void complex (the complex with no faces). Adding the facet abc then adds the faces $\emptyset, a, b, c, ab, ac, bc$, and abc to the complex. With respect to inclusion, this set of faces has a unique minimal face: $\emptyset$. Next, we add the facet bcd, which adds the faces d, bd, cd, and bcd to the complex. Again, this set has a unique minimal element: d. Continuing on, the minimal new faces obtained by adding the facets abe, bde, cde, and ace are, respectively, e, de, ce, and ace.

It turns out that any shelling has this property: with the addition of each facet F_i, the corresponding set of all "new" faces has a unique minimal element with respect to inclusion. Conversely, any ordering of the facets satisfying this property is a shelling. We prove this in the next proposition.

Proposition 6.13. *Let Δ be a simplicial complex and $F_1, F_2, \ldots, F_t$ an ordering of its facets. Then this ordering is a shelling of Δ if and only if the following property holds:*

- *For each j with $1 \leq j \leq t$, the set of faces contained in $\bigcup_{i=1}^{j} F_i$ but not in $\bigcup_{i=1}^{j-1} F_i$ has a unique minimal element with respect to inclusion.*

We often call this unique minimal element the minimal face *associated to F_j.*

Proof. First assume that the property is satisfied, and fix $j > 1$. We need to show that $F_j \cap \left(\bigcup_{i=1}^{j-1} F_i\right)$ is a nonempty union of facets of ∂F_j. Let A be the minimal face associated to F_j, and note that A must be nonempty, as $\bigcup_{i=1}^{j-1} F_i$ already contains the empty set as a face. For a face B of F_j, we have that $B \subseteq F_j \cap \left(\bigcup_{i=1}^{j-1} F_i\right)$ if and only if there is an $a \in A$ with $a \notin B$. As $F_j \setminus a \subseteq \bigcup_{i=1}^{j-1} F_i$ for all $a \in A$, it follows that

$$\bigcup_{a \in A} F_j \setminus a = F_j \cap \left(\bigcup_{i=1}^{j-1} F_i\right),$$

which means the ordering is a shelling.

The converse is proven analogously: Fix $j > 1$, and let $A = \{a \in F_j : F_j \setminus a \subseteq \bigcup_{i=1}^{j-1} F_i\}$. Then a face $B \subseteq F_j$ is contained in $\bigcup_{i=1}^{j-1} F_i$ if and only if it is contained in $F_j \setminus a$ for some $a \in A$, as $F_j \cap \left(\bigcup_{i=1}^{j-1} F_i\right)$ is a union of facets of ∂F_j. Thus, a face of F_j is not contained in $\bigcup_{i=1}^{j-1} F_i$ if and only if it contains A, meaning the property is satisfied.

Looking back at the shelling depicted in Fig. 6, we found that the minimal faces associated to the facets in the shelling were $\emptyset, d, e, de, ce$, and ace. Clearly, a different shelling of K would not necessarily produce the same set of minimal faces. For example, if we use the shelling $abc, ace, abe, bcd, bde, cde$, the associated minimal faces are, in order, $\emptyset, e, be, c, de, cde$.

While these two sets of minimal faces are distinct, they each contain one (-1)-dimensional face (namely, $\emptyset$), two 0-dimensional faces, two 1-dimensional faces, and one 2-dimensional face. The next theorem asserts that this information is recorded by the h-vector of K. Indeed, the f-vector of K is $(1, 5, 9, 6)$, and thus the h-vector of Δ is $(1, 2, 2, 1)$.

Theorem 6.14. *Let Δ be a pure $(d-1)$-dimensional complex, and suppose it has shelling $F_1, F_2, \ldots, F_t$. Then the h-vector $(h_0, h_1, \ldots, h_d)$ is given as follows: For each i,*

$$h_i = |\{j : \text{the minimal face corresponding to } F_j \text{ is } (i-1)\text{-dimensional}\}|.$$

Proof. Consider a facet F_j in a shelling of Δ, and let A be the minimal face associated to F_i. Then every new face obtained by adding F_j to the union $\bigcup_{i=1}^{j-1} F_i$

must contain A, meaning these faces add $(t+1)^{d-|A|}$ to $f_\Delta(t)$, and so

$$h_\Delta(t+1) = f_\Delta(t).$$

The next theorem expresses the relationship between the Hilbert series of a squarefree ideal and the f- and h-vectors of its Stanley–Reisner complex. It is often paraphrased by saying that the relationship between the Hilbert function and the squarefree Hilbert function is the same as the relationship between the f-vector and the h-vector.

Theorem 6.15. *Let Δ be a $(d-1)$-dimensional simplicial complex with f-vector $(f_{-1}, f_0, \dots, f_{d-1})$. Recall that we write R_Δ for the quotient S/I_Δ. Then:*

(1) $HS^{\mathrm{sf}}_{R_\Delta}(t) = \sum_{i=0}^{d} f_{i-1}t^i.$

(2) $HS_{R_\Delta}(t) = \sum_{i=0}^{d} \frac{f_{i-1}t^i}{(1-t)^i} = \frac{1}{(1-t)^d}\sum_{i=0}^{d} h_i t^i.$

Proof. In part (1), for each i, the squarefree monomials in S/I_Δ of degree i are exactly those monomials which correspond to $(i-1)$-dimensional faces of Δ.

For part (2), the first equality follows from part (1) and Theorem 6.9. For the second equality, we first write out the relation $h_\Delta(t) = f_\Delta(t-1)$:

$$\sum_{i=0}^{d} h_i t^{d-i} = \sum_{i=0}^{d} f_{i-1}(t-1)^{d-i}.$$

If we substitute $1/t$ for t, this becomes

$$\sum_{i=0}^{d} \frac{h_i}{t^{d-i}} = \sum_{i=0}^{d} f_{i-1}\left(\frac{1-t}{t}\right)^{d-i},$$

and multiplying through by t^d yields

$$\sum_{i=0}^{d} h_i t^i = \sum_{i=0}^{d} f_{i-1}t^i(1-t)^{d-i}.$$

Now we have

$$\begin{aligned} HS_{R_\Delta}(t) &= \sum_{i=0}^{d} \frac{f_{i-1}t^i}{(1-t)^i} \\ &= \frac{1}{(1-t)^d}\sum_{i=0}^{d} f_{i-1}t^i(1-t)^{d-i} \\ &= \frac{1}{(1-t)^d}\sum_{i=o}^{d} h_i t^i. \end{aligned}$$

We illustrate these ideas using the complex K of Example 4.2. The minimal non-faces of Δ are ae and bcd, and thus $I_\Delta = (ae, bcd)$ and $R_\Delta = k[a,b,c,d,e]/(ae,bcd)$. The f-vector and h-vector of Δ are $(1,5,9,6)$ and $(1,2,2,1)$, respectively. By Theorem 6.15,

$$HS_{R_\Delta}(t) = 1 + \frac{5t}{(1-t)} + \frac{9t^2}{(1-t)^2} + \frac{6t^3}{(1-t)^3} = \frac{1+2t+2t^2+t^3}{(1-t)^3}.$$

7 Hochster's Formula and Betti Numbers

We conclude with some connections of Stanley–Reisner theory to computing Betti numbers of monomial ideals. Many of the results we discuss in this section are for *multigraded* Betti numbers; that is, the degree of the syzygy is a monomial or, equivalently, a vector in $\mathbb{N}^n$.

Notation 7.1. *Throughout the section, let* $\mathbf{b}$ *be a monomial. Abusing notation, we identify* $\mathbf{b}$ *with its exponent vector* $\mathbf{b} \in \mathbb{N}^n$*, and we write* $x_i \in \mathbf{b}$ *to indicate that* x_i *divides* $\mathbf{b}$*. We adopt the common shorthand* $|\mathbf{b}| = \deg \mathbf{b}$.

Hochster's formula, which appeared in [10], has been a central tool in combinatorial commutative algebra for over 30 years. We begin by defining induced subcomplexes of a simplicial complex.

Definition 7.2. Let Δ be a simplicial complex on X, and let $Y \subseteq X$. The *induced subcomplex* $\Delta[Y]$ is the simplicial complex consisting of all faces of Δ whose vertices lie in Y.

Example 7.3. Let B be the bowtie complex from Example 3.3. The induced subcomplex $B[abde]$, shown in Fig. 7, is the union of two disjoint line segments. Note that while B is contractible, $B[abde]$ has nontrivial homology. This illustrates that the collection of induced subcomplexes contains more information about the complex than its homotopy type.

Fig. 7 The induced subcomplex $B[abde]$

Hochster's formula shows that the multigraded Betti numbers of a squarefree monomial ideal I are encoded in the homology of induced subcomplexes of Δ_I. When $\mathbf{b}$ is a multidegree and Δ a simplicial complex, we often write $\Delta[\mathbf{b}]$ to mean the induced subcomplex on the associated subset of X.

Before stating Hochster's formula, we need a quick introduction to (multigraded) Betti numbers. We have done our best to minimize the machinery required in this section. However, the material is necessarily more involved than what comes before. Readers who are interested only in the statements or applications of Hochster's formula should skip ahead to Theorem 7.11.

There are many ways to define Betti numbers, and it is usually most convenient to think of them in terms of a free resolution or as the ranks of certain Tor modules. However, for our purposes, it is easiest to define Betti numbers in terms of the homology of an object called the *(algebraic) upper Koszul complex*, which we construct below.

Notation 7.4. *For each variable x_j, define an object e_j called the* differential *of x_j. Given a monomial g of degree d, write $g = x_{j_1}x_{j_2}\dots x_{j_d}$, and set the* differential of g *equal to*

$$Dg = e_{j_1} \wedge e_{j_2} \wedge \cdots \wedge e_{j_d},$$

where $\wedge$ (pronounced "wedge" and called the exterior product*) is an associative operation satisfying the anticommutativity relation $e_r \wedge e_s = -e_s \wedge e_r$. If k has characteristic other than 2, it follows that $Dg \neq 0$ if and only if g is squarefree. (If* $\operatorname{char} k = 2$*, we take this as an additional axiom.) We endow the differential Dg with multidegree g.*

Definition 7.5. The *Koszul complex* is the algebraic chain complex

$$K_\bullet : 0 \to F_n \to F_{n-1} \to \cdots \to F_1 \to F_0 \to k \to 0,$$

where F_i is the free S-module with basis given by $\{Dg : \deg g = i\}$. We formally set the empty wedge $D(1)$ equal to 1, so $F_0 = S$. The boundary maps are given by the formula $\phi(Dg) = \sum_{x_j \in g} \pm x_j D\left(\frac{g}{x_j}\right)$, where the sign convention is the standard simplicial boundary convention: if $g = \prod x_{j_s}$, then $Dg = \sum (-1)^{s+1} x_{j_s} D\left(\frac{g}{x_{j_s}}\right)$. Note that ϕ preserves multidegree.

Remark 7.6. If we stipulate that $\wedge$ is also distributive over addition, and set $E = \bigoplus F_i$, then E is an anticommutative k-algebra, the *exterior algebra on* $\{x_1, \dots, x_n\}$.

Remark 7.7. The Koszul complex is essentially the usual simplicial chain complex arising from the $(n-1)$-simplex $\{x_1, \dots, x_n\}$, but viewed with coefficients in S. The coefficients in the boundary maps give it multigraded structure. It is a standard exercise in commutative algebra that the Koszul complex is exact and hence a free resolution of k.

Definition 7.8. Let I be a monomial ideal. Then the *(algebraic) upper Koszul complex* of I is the tensor product of I with the Koszul complex,

$$K_{\bullet}(I) : 0 \to IF_n \to IF_{n-1} \to \cdots \to IF_1 \to IF_0 \to I/(x_1, \ldots x_n)I \to 0.$$

The ith Betti number of I is the dimension of the ith homology of this complex,

$$b_i(I) = \dim_k H_i(K_{\bullet}(I)),$$

and the $(i, \mathbf{b})$th multigraded Betti number of I is the dimension of the degree $\mathbf{b}$ part of this homology,

$$b_{i,\mathbf{b}}(I) = \dim_k \left(H_i(K_{\bullet}(I))\right)_{\mathbf{b}}.$$

The module IF_i has k-basis consisting of symbols of the form fDg, where $f \in I$ is a monomial and g is a squarefree monomial of degree i.

The following standard theorems about Betti numbers are considerably easier to derive from one of the many standard treatments (see, for example, [13]) than from the definitions above.

Theorem 7.9. *Let I be a squarefree monomial ideal and $\mathbf{b}$ a multidegree. Then*

(1) $b_{i,\mathbf{b}}(I) = b_{i+1,\mathbf{b}}(S/I)$.
(2) If $\mathbf{b}$ is not squarefree, then $b_{i,\mathbf{b}}(I) = 0$.

The upper Koszul complex $K^{\bullet}(I)$ has k-basis consisting of symbols of the form fDg, for monomials $f \in I$ and squarefree monomials g. The symbol fDg has multidegree fg and homological degree $\deg g$, and its differential is $\phi(fDg) = \sum_{x_j \in g} \pm (fx_j) D(\frac{g}{x_j})$, where the signs alternate according to the standard convention on the order of the x_j.

If we restrict to a squarefree multidegree $\mathbf{b}$, $K^{\bullet}(I)$ becomes a complex $K^{\mathbf{b}}(I)$ of vector spaces, with basis $\mathscr{B} = \left\{B_g = \frac{\mathbf{b}}{g} Dg\right\}_{g \in \mathscr{G}}$, where the index set is

$$\begin{aligned} \mathscr{G} &= \left\{g : g \text{ divides } \mathbf{b} \text{ and } \frac{\mathbf{b}}{g} \in I\right\} \\ &= \left\{g : g \text{ divides } \mathbf{b} \text{ and } \frac{\mathbf{b}}{g} \notin \Delta_I\right\} \\ &= \left\{g : g \text{ divides } \mathbf{b} \text{ and } \frac{\mathbf{b}}{g} \notin (\Delta_I)[\mathbf{b}]\right\} \\ &= ((\Delta_I)[\mathbf{b}])^{\vee}. \end{aligned}$$

Thus, $K^{\mathbf{b}}(I)$ has basis $\{B_g : g \in ((\Delta_I)[\mathbf{b}])^{\vee}\}$, and the differential is $\phi(B_g) = \sum_{x_j \in g} \pm B_{\frac{g}{x_j}}$.

Remark 7.10. Note that $\mathscr{G} = ((\Delta_I)[\mathbf{b}])^\vee$ is a simplicial complex. Bayer, Charalambous, and Popescu call it the *upper Koszul simplicial complex* in multidegree $\mathbf{b}$. See [1, 11]. We will revisit this object later in the section.

Meanwhile, the chain complex associated to $((\Delta_I)[\mathbf{b}])^\vee$ has basis consisting of symbols $\{C_g : g \in ((\Delta_I)[\mathbf{b}])^\vee\}$. The homological degree of C_g is $\dim g = \deg g - 1$, and the differential is $\partial(C_g) = \sum_{x_j \in g} \pm C_{\frac{g}{x_j}}$.

Thus, the map sending B_g to C_g is an isomorphism of chain complexes, shifting homological degree by one. It induces an isomorphism between the ith homology of $K^{\mathbf{b}}(I)$ and the $(i-1)$th homology of $(\Delta_I)[\mathbf{b}]$. This gives us the first form of Hochster's formula:

Theorem 7.11 (Hochster's Formula, Dual Form). *Let I be a squarefree monomial ideal and $\mathbf{b}$ be a squarefree multidegree. Then*

$$b_{i,\mathbf{b}}(I) = \dim_k \tilde{H}_{i-1}\left(((\Delta_I)[\mathbf{b}])^\vee\right)$$

and

$$b_{i,\mathbf{b}}(S/I) = \dim_k \tilde{H}_{i-2}\left(((\Delta_I)[\mathbf{b}])^\vee\right).$$

We can remove the dual from this formula at the price of passing to cohomology. Recall classical Alexander duality (see, e.g., [11, Theorem 5.6]).

Theorem 7.12 (Alexander Duality). *Let Δ be a simplicial complex on ℓ vertices. Then $\tilde{H}_i(\Delta^\vee; k)$ is isomorphic to $\tilde{H}^{\ell-i-3}(\Delta; k)$.*

Applying this to Hochster's formula immediately gives us a second form of Hochster's formula.

Corollary 7.13 (Hochster's Formula, Cohomology Form). *Let I be a squarefree monomial ideal and $\mathbf{b}$ a squarefree multidegree. Then*

$$b_{i,\mathbf{b}}(S/I) = \dim_k \tilde{H}^{|\mathbf{b}|-i-1}(\Delta_I[\mathbf{b}]) \qquad \text{and} \qquad b_{i,\mathbf{b}}(I) = \dim_k \tilde{H}^{|\mathbf{b}|-i-2}(\Delta_I[\mathbf{b}]).$$

Remark 7.14. We can also obtain the cohomological form directly from the upper Koszul complex of S/I. The computation is similar in spirit to the development of the dual form of Hochster's formula, but the details are much messier, and thus we omit it.

Remark 7.15. If k is characteristic zero, then the ranks of corresponding homology and cohomology groups are equal, and we may thus rephrase Corollary 7.13 as:

$$b_{i,\mathbf{b}}(I) = \dim_k \tilde{H}_{|\mathbf{b}|-i-2}(\Delta_I[\mathbf{b}]; k).$$

Example 7.16. Let Δ be the complex from Fig. 5. Then $\Delta[acde]$ is contractible, so $\tilde{H}^i(\Delta[acde], k)$ is trivial for any i, and $b_{i,acde}(I_\Delta) = 0$ for all i (Fig. 8).

However, $\Delta[abcd]$ is homotopy equivalent to the wedge product of two circles (see Fig. 9), and so

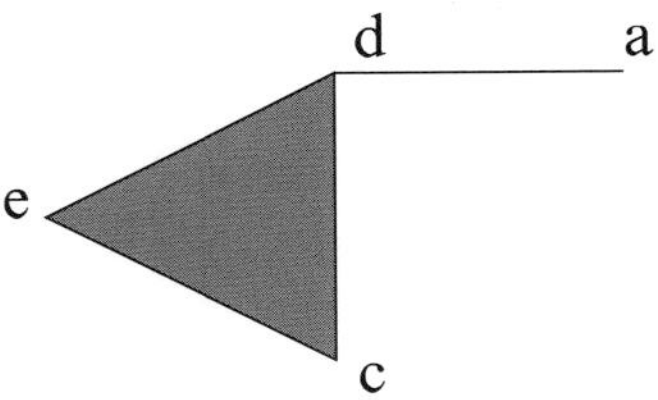

Fig. 8 The induced subcomplex $\Delta[acde]$

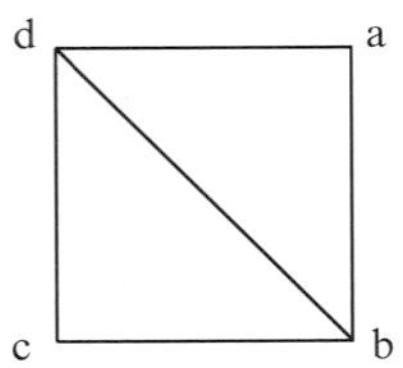

Fig. 9 The induced subcomplex $\Delta[abcd]$

$$b_{1,abcd} = \dim_k \tilde{H}^{|abcd|-1-2}(\Delta[abcd], k) = \dim_k \tilde{H}^1(\Delta[abcd], k) = 2.$$

We turn now to the computation of multigraded Betti numbers of arbitrary monomial ideals via simplicial complexes. When I is not squarefree, there is no Stanley–Reisner complex to work with, and thus Hochster's formula cannot be applied directly. However, the *upper Koszul simplicial complex*, introduced by Bayer et al. [1], continues to make sense. (See also [11].)

Definition 7.17. The *upper Koszul simplicial complex* of a monomial ideal I in multidegree $\mathbf{b}$ is

$$K^{\mathbf{b}}(I) = \{g \in 2^X : \frac{\mathbf{b}}{g} \in I\}$$

(Note: This is the simplicial complex $K_{\mathbf{b}}(I)$ in [1]; we have adopted the notation of [11] to avoid confusion with another simplicial complex defined in [11].)

Bayer, Charalambous, and Popescu prove the following theorem.

Theorem 7.18 ([1]). *Given a monomial ideal I, the multigraded Betti numbers of I are*

$$b_{i,\mathbf{b}}(I) = \dim \tilde{H}_{i-1}(K^{\mathbf{b}}(I); k).$$

Example 7.19. We compute some multigraded Betti numbers of a monomial ideal that is not squarefree to illustrate Theorem 7.18. Let $I = (a^3, b^3, c^4, abc, ac^2, bc^2) \subset S = k[a,b,c]$. We compute $K^{abc^2}(I)$. Note that $abc^2 \in I$, so $\emptyset \in K^{abc^2}(I)$. Moreover, dividing abc^2 by any of a, b, or c yields a monomial in I, meaning $K^{abc^2}(I)$ contains vertices a, b, and c. Because I is generated in degree three and higher, we cannot divide abc^2 by a degree two or higher monomial and stay in I, meaning that the facets of $K^{abc^2}(I)$ are exactly the three isolated vertices a, b, and c. From this, we can compute $b_{i,abc^2}(I)$ for any i. When $i = 0$, we compute dim $\tilde{H}_{-1}(K^{abc^2}(I); k)$, which is zero. When $i = 1$, we are counting the number of connected components of $K^{abc^2}(I)$ minus one, giving us two. Hence $b_{1,abc^2}(I) = 2$. There is no higher homology, meaning there are two minimal first syzygies of multidegree abc^2, and $b_{i,abc^2}(I) = 0$ for $i \neq 1$.

In another direction, Terai proved a beautiful result in [16, Corollary 0.3] showing that the projective dimension of a squarefree monomial ideal and the regularity are dual notions. Suppose M is a finitely graded S-module with minimal free resolution

$$0 \to \bigoplus_j S(-b_{r,j}) \to \cdots \to \bigoplus_j S(-b_{1,j}) \to \bigoplus_j S(-b_{0,j}) \to M \to 0.$$

The *regularity* of M is the maximum value of $b_{i,j} - i$, and it is the label on the bottom row of the Macaulay 2 Betti diagram of the resolution of M.

Theorem 7.20 (Terai). *The regularity of I_Δ equals the projective dimension of $S/I_{\Delta^\vee}$.*

Terai actually proves something a bit stronger, namely that the difference between the regularity of I_Δ and the smallest degree of a generator of I_Δ is the same as the difference between the projective dimension of $S/I_{\Delta^\vee}$ and codim $I_{\Delta^\vee}$. (This is [16, Theorem 2.1], rephrased using the Auslander–Buchsbaum Theorem.) Theorem 5.7 is an immediate consequence because $\Delta^\vee$ is Cohen–Macaulay if and only if pd $S/I_{\Delta^\vee} =$ codim $I_{\Delta^\vee}$, and an ideal has a linear resolution if and only if its regularity equals the smallest degree of a minimal generator.

Bayer, Charalambous, and Popescu generalized these ideas further in [1] with the notion of *extremal Betti numbers*, which are Betti numbers that are nonzero but occupy the upper left corner of a block of Betti numbers that are otherwise zero in a Macaulay 2 Betti diagram. See [1, Sect. 3] for some illustrative examples.

Acknowledgments We thank the referee for his or her careful reading and helpful comments. This work was partially supported by grants from the Simons Foundation (#199124 to Francisco and #202115 to Mermin).

References

1. Bayer, D., Charalambous, H., Popescu, S.: Extremal Betti numbers and applications to monomial ideals. J. Algebra **221**(2), 497–512 (1999)
2. Björner, A., Wachs, M.: Shellable nonpure complexes and posets. I. Trans. Am. Math. Soc. **348**(4), 1299–1327 (1996)

3. Björner, A., Wachs, M.: Shellable nonpure complexes and posets. II. Trans. Am. Math. Soc. **349**(10), 3945–3975 (1997)
4. Bruns, W., Herzog, J.: Cohen-Macaulay Rings. Cambridge Studies in Advanced Mathematics, vol. 39. Cambridge University Press, Cambridge (1993)
5. Eagon, J., Reiner, V.: Resolutions of Stanley-Reisner rings and Alexander duality. J. Pure Appl. Algebra **130**(3), 265–275 (1998)
6. Eisenbud, D.: Commutative Algebra with a View Toward Algebraic Geometry. Springer, New York (1995)
7. Herzog, J., Hibi, T.: Componentwise linear ideals. Nagoya Math. J. **153**, 141–153 (1999)
8. Herzog, J., Sbarra, E.: Sequentially Cohen-Macaulay modules and local cohomology. In: Algebra, Arithmetic and Geometry, Part I, II (Mumbai, 2000). Tata Institute of Fundamental Research Studies in Mathematics, vol. 16, pp. 327–340. Tata Institute of Fundamental Research, Mumbai (2002)
9. Herzog, J., Takayama, Y.: Resolutions by mapping cones. The Roos Festschrift volume, 2. Homology Homotopy Appl. **4**(2, Pt 2), 277–294 (2002)
10. Hochster, M.: Cohen-Macaulay rings, combinatorics, and simplicial complexes. In: Ring Theory, II (Proceedings of Second Conference, University of Oklahoma, Norman, OK, 1975). Lecture Notes in Pure and Applied Mathematics, vol. 26, pp. 171–223. Dekker, New York (1977)
11. Miller, E., Sturmfels, B.: Combinatorial Commutative Algebra. Graduate Texts in Mathematics, vol. 227. Springer, New York (2004)
12. Morey, S., Villarreal, R.H.: Edge ideals: algebraic and combinatorial properties. In: Progress in Commutative Algebra: Ring Theory, Homology, and Decomposition, pp. 85–126. de Gruyter, Berlin (2012)
13. Peeva, I.: Graded Syzygies. Algebra and Applications, vol. 14. Springer, London (2011)
14. Schenck, H.: Computational Algebraic Geometry. London Mathematical Society Student Texts, vol. 58. Cambridge University Press, Cambridge (2003)
15. Stanley, R.P.: Combinatorics and Commutative Algebra. Progress in Mathematics, 2nd edn., vol. 41. Birkhäuser Boston, Boston (1996)
16. Terai, N.: Alexander duality theorem and Stanley-Reisner rings. Free resolutions of coordinate rings of projective varieties and related topics (Kyoto, 1998). Surikaisekikenkyusho Kokyuroku **1078**, 174–184 (1999)
17. Villarreal, R.H.: Monomial Algebras. Monographs and Textbooks in Pure and Applied Mathematics, vol. 238. Dekker, New York (2001)

Numerical Computation of the Hilbert Function and Regularity of a Zero Dimensional Scheme

Zachary A. Griffin, Jonathan D. Hauenstein, Chris Peterson, and Andrew J. Sommese

Abstract Let $R = \mathbb{C}[x_1, \ldots, x_N]$ and let $F = \{f_1, \ldots, f_t\} \subset R$ be a set of generators for an ideal I. Let $Y = \{y_1, \ldots, y_\ell\} \subset \mathbb{C}^N$ be a subset of the set of isolated solutions of the zero locus of F. Let $\mathfrak{m}_{y_i}$ denote the maximal ideal of y_i and let $\mathscr{P}_{y_i}$ denote the $\mathfrak{m}_{y_i}$-primary component of I. Let $J = \cap_{i=1}^{l} \mathscr{P}_{y_i}$ and let $\mathscr{X}$ denote the corresponding zero dimensional subscheme supported on Y. This article presents a numerical algorithm for computing the Hilbert function and the regularity of $\mathscr{X}$. In addition, the algorithm produces a monomial basis for R/J. The input for the algorithm is the polynomial system F and a numerical approximation of each element in Y.

1 Introduction

Homotopy continuation can be used to determine a numerical approximation for each isolated solution of the zero locus of a multivariate polynomial system. By utilizing a deflation method for isolated solutions, such as described in [4, 7–9],

Z.A. Griffin
Harold Vance Department of Petroleum Engineering, Texas A&M University, Mailstop 3116, College Station, TX 77843, USA
e-mail: zacgriffin21@neo.tamu.edu

J.D. Hauenstein
Department of Mathematics, North Carolina State University, Raleigh, NC 27695-8205, USA
e-mail: hauenstein@ncsu.edu

C. Peterson (✉)
Department of Mathematics, Colorado State University, Fort Collins, CO 80523, USA
e-mail: peterson@math.colostate.edu

A.J. Sommese
Department of Applied and Computational Mathematics and Statistics, University of Notre Dame, Notre Dame, IN 46556, USA
e-mail: sommese@nd.edu

S.M. Cooper and S. Sather-Wagstaff (eds.), *Connections Between Algebra, Combinatorics, and Geometry*, Springer Proceedings in Mathematics & Statistics 76, DOI 10.1007/978-1-4939-0626-0_6,

each isolated solution can be computed to arbitrary accuracy using a quadratically convergent Newton's method. Given a numerical approximation of each isolated point in the support of a zero-scheme, this article presents algorithms to compute the Hilbert function, standard monomials, and index of regularity of the zero-scheme. The underlying theory supporting the approach is based on Macaulay's inverse systems [10].

The first section of the paper provides basic background information. Section 3 presents underlying theory and algorithms to compute standard monomials, Hilbert functions, and the index of regularity. Section 4 discusses implementation details. Section 5 consists of two examples applying the algorithm. The final section consists of concluding comments, and a brief discussion of an extension of the algorithm to subschemes of zero-schemes together with practical limitations.

2 Background

This section consists of background information on dual bases, standard monomials, Hilbert functions, and the index of regularity. The section concludes with an example to illustrate these ideas.

2.1 Dual Bases

Following the notation of [1,4], for $\alpha = (\alpha_1, \ldots, \alpha_N) \in (\mathbb{Z}_{\geq 0})^N$, define

$$|\alpha| = \alpha_1 + \cdots + \alpha_N,\ \alpha! = \alpha_1!\alpha_2! \cdots \alpha_N!, \text{ and } \ \partial_\alpha = \frac{1}{\alpha!}\frac{\partial^{|\alpha|}}{\partial x^\alpha}.$$

Let $R = \mathbb{C}[x_1, \ldots, x_N]$, let $y \in \mathbb{C}^N$, and let $\alpha \in (\mathbb{Z}_{\geq 0})^N$. Viewing R as a complex vector space, the linear functional, $\partial_\alpha[y] : R \to \mathbb{C}$, is defined by

$$\partial_\alpha[y](g) = (\partial_\alpha g)(y).$$

When it is clear from context, $\partial_\alpha[y]$ may be written as ∂_α.

Let $y \in \mathbb{C}^N$ and let $S_y = \{\partial_\alpha[y] \mid \alpha \in (\mathbb{Z}_{\geq 0})^N\}$. For each $y \in \mathbb{C}^N$, one can associate the (infinite dimensional) complex vector space, D_y, consisting of all finite $\mathbb{C}$-linear combinations of elements from S_y. It is clear that D_y is a subspace of the vector space, $Hom_{\mathbb{C}}(R, \mathbb{C})$, of linear functionals on R. If $I \subset \mathbb{C}[x_1, \ldots, x_N]$ is an ideal, the *dual space* of I at y is the set of all elements of D_y that vanish on I, namely

$$D_y[I] = \{\partial \in D_y \mid \partial(g) = 0 \text{ for all } g \in I\}.$$

For a set of points $Y = \{y_1, \ldots, y_\ell\}$, define D_Y to be the vector space of all finite $\mathbb{C}$-linear combinations of elements from $S_{y_1} \cup \cdots \cup S_{y_\ell}$ and define the dual space of I at Y by

$$D_Y[I] = \{\partial \in D_Y \mid \partial(g) = 0 \text{ for all } g \in I\}.$$

A *dual basis* for a dual space is a subset of elements in the dual space that forms a $\mathbb{C}$-basis for the space.

Let $\mathscr{V}(I) = \{z \in \mathbb{C}^N \mid g(z) = 0 \text{ for all } g \in I\}$, let $y \in \mathscr{V}(I)$, and let $\mathfrak{m}_y$ denote the maximal ideal of y. If y is an isolated element of $\mathscr{V}(I)$, then $\mathfrak{m}_y$ is a non-embedded associated prime of I. In other words, in an irredundant primary decomposition of I, $\mathfrak{m}_y$ is a minimal associated prime. As a minimal associated prime, $\mathfrak{m}_y$ has a well-defined multiplicity, $\text{mult}_I(y)$, in I. In fact, if $\mathscr{P}_y$ denotes the $\mathfrak{m}_y$-primary component in an irredundant primary decomposition of I, then y being an isolated point implies that $\text{mult}_I(y)$ is the same as the degree of the scheme associated with $\mathscr{P}_y$ which, in turn, is the same as $\dim_{\mathbb{C}} R/\mathscr{P}_y$. The following theorem of Macaulay makes an explicit connection between $\text{mult}_I(y)$ and the dimension of $D_y[I]$ as a complex vector space.

Theorem 2.1. *Let $I \subset \mathbb{C}[x_1, \ldots, x_N]$ be an ideal and let $y \in \mathscr{V}(I)$. Then*

i) $\dim_{\mathbb{C}} D_y[I] < \infty$ *if and only if y is an isolated point in $\mathscr{V}(I)$.*
ii) If y is an isolated point in $\mathscr{V}(I)$, then $\dim_{\mathbb{C}} D_y[I] = \text{mult}_I(y)$.

When y is an isolated point in $\mathscr{V}(I)$, numerical algorithms can be used to compute $\dim_{\mathbb{C}} D_y[I]$ and hence $\text{mult}_I(y)$ [4, 13]. Additionally, the local dimension test of [1] is based on Theorem 2.1.

The following proposition about linear independence in $Hom_{\mathbb{C}}(R, \mathbb{C})$ leads to a stronger formulation of Macaulay's theorem.

Proposition 2.2. *Let Y_1 and Y_2 be finite sets of points in $\mathbb{C}^N$. Then $Y_1 \cap Y_2 = \emptyset$ if and only if $D_{Y_1} \cap D_{Y_2} = 0$.*

Proof. It is clear that $D_{Y_1 \cap Y_2} \subset D_{Y_1} \cap D_{Y_2}$. Therefore, $Y_1 \cap Y_2 \neq \emptyset$ implies $D_{Y_1 \cap Y_2} \neq 0$ yielding $D_{Y_1} \cap D_{Y_2} \neq 0$.

Now suppose that $Y_1 \cap Y_2 = \emptyset$. The statement that $D_{Y_1} \cap D_{Y_2} = 0$ is equivalent to the statement that, for any nonzero elements $\partial_1 \in D_{Y_1}$ and $\partial_2 \in D_{Y_2}$, there exists an element $g \in R$ such that $\partial_1(g) \neq \partial_2(g)$. We will construct a $g \in R$ such that $\partial_1(g) = 0$ but $\partial_2(g) \neq 0$. Let $Y_1 = \{p_1, \ldots, p_r\}$ and $Y_2 = \{q_1, \ldots, q_s\}$. There exists a number B such that

$$\partial_1 = \sum_{\{\alpha \mid |\alpha| \leqslant B\}} \sum_{i=1}^{r} b_{\alpha,i} \partial_\alpha[p_i] \quad \text{and} \quad \partial_2 = \sum_{\{\alpha \mid |\alpha| \leqslant B\}} \sum_{i=1}^{s} c_{\alpha,i} \partial_\alpha[q_i].$$

For any point p, we let $\mathfrak{m}_p$ denote its maximal ideal. By reindexing the elements in Y_2, if necessary, we can assume there exists an α such that $c_{\alpha,s} \neq 0$. Now choose an element $g \in (\cap_{i=1}^{r} \mathfrak{m}_{p_i}^{B+1}) \bigcap (\cap_{i=1}^{s-1} \mathfrak{m}_{q_i}^{B+1})$ such that $\partial_2(g) \neq 0$ (due to the contribution from q_s). By construction, $\partial_1(g) = 0$.

Corollary 2.3. *Let $I \subset \mathbb{C}[x_1, \ldots, x_N]$ be an ideal and $Y = \{y_1, \ldots, y_\ell\} \subset \mathscr{V}(I)$. Then*

i) $\dim_{\mathbb{C}} D_Y[I] < \infty$ *if and only if every element of Y is an isolated point in $\mathscr{V}(I)$.*
ii) If every element of Y is an isolated point in $\mathscr{V}(I)$, then $\dim_{\mathbb{C}} D_Y[I] = \sum_{i=1}^{\ell} \text{mult}_I(y_i)$.

Proof. Part (i) is a corollary of Theorem 2.1. To prove part (ii), note that $\dim_{\mathbb{C}} D_Y[I] \leqslant \sum_{i=1}^{\ell} \text{mult}_I(y_i)$ by Theorem 2.1. The equality of $\dim_{\mathbb{C}} D_Y[I]$ and $\sum_{i=1}^{\ell} \text{mult}_I(y_i)$ follows from Proposition 2.2 by inducting on the number of points in Y and noting that $D_{Y'}[I]$ is a subspace of $D_{Y'}$ for any $Y' \subseteq Y$.

A *graded monomial order* is a monomial ordering that satisfies

- $|\alpha| > |\beta| \implies x^\alpha > x^\beta$
- $x^\alpha > x^\beta \implies x^\gamma x^\alpha > x^\gamma x^\beta$

A *reverse graded monomial order* is a monomial ordering that satisfies

- $|\alpha| < |\beta| \implies x^\alpha > x^\beta$
- $x^\alpha > x^\beta \implies x^\gamma x^\alpha > x^\gamma x^\beta$

Let $>$ be a reverse graded monomial order. For any $k \geq 0$, we can create an ordered list, with respect to $>$, of the $q = \binom{N+k}{k}$ monomials of degree $\leqslant k$, namely

$$1 = x^{\alpha_1} > \cdots > x^{\alpha_q}.$$

By Corollary 2.3, the dual space of the ideal of a zero-scheme, $\mathscr{S}$, is a finite-dimensional vector space over $\mathbb{C}$ whose dimension is equal to the degree of $\mathscr{S}$. It will be convenient to represent the dual space as the row space of a matrix. To that end, let $I \subset \mathbb{C}[x_1, \ldots, x_N]$ be an ideal, let $Y = \{y_1, \ldots, y_\ell\} \subset \mathscr{V}(I)$ be a set of isolated solutions, and let $m = \dim_{\mathbb{C}} D_Y[I]$. Given a basis $B = \{\partial_1, \ldots, \partial_m\}$ for $D_Y[I]$, let A_k be the $m \times q$ matrix where

$$(A_k)_{i,j} = \partial_i(x^{\alpha_j}). \tag{1}$$

The linear independence of B implies that $\text{rank}(A_k) = m$ for $k \gg 0$.

This matrix will be used in Sect. 3 to compute the Hilbert function, index of regularity, and standard monomials. As a function of k, the number of monomials of degree $\leqslant k$ displays rapid growth. Due to this rapid growth in the number of monomials and the notorious ill-conditioning of a monomial basis, numerical computations with the matrix A_k may be difficult. These issues are discussed in Sect. 4.

2.2 Standard Monomials

If g is a nonzero polynomial, let $\text{in}_>(g)$ denote the *initial monomial* of g with respect to a graded monomial order $>$. That is, if $g = \sum_{j=1}^{k} a_j x^{\alpha_j}$, where each $a_j \neq 0$ and $x^{\alpha_1} > x^{\alpha_2} > \cdots > x^{\alpha_k}$, then $\text{in}_>(g) = x^{\alpha_1}$.

If I is an ideal, then $\text{in}_>(I) = \{\text{in}_>(f) | f \in I\}$ is the set of initial monomials of I. A monomial x^α is a *standard monomial* of I with respect to $>$ if $x^\alpha \notin \text{in}_>(I)$.

If $y \in \mathscr{V}(I)$, then monomial orders can naturally be extended to $D_y[I]$. In particular, set $\partial_\alpha > \partial_\beta$ if $x^\alpha > x^\beta$ and

$$\text{in}_>(D_y[I]) = \{\text{in}_>(\partial) \mid \partial \in D_y[I]\}.$$

If y is isolated with $\text{mult}_I(y) = m$ and if $>$ is a reverse graded monomial order then, by the first part of Theorem 3.1 of [9], we have $|\text{in}_>(D_y[I])| = \dim_{\mathbb{C}}(D_y[I]) = m$.

If $B = \{\partial_1, \ldots, \partial_m\}$ is a basis for $D_y[I]$, we say that B is a *reduced dual basis* with respect to $>$ if no two basis elements have the same initial terms, that is, $\text{in}_>(\partial_i) \neq \text{in}_>(\partial_j)$ for $i \neq j$. It is easy to see that given a monomial ordering $>$, any dual basis can be made into a reduced dual basis with respect to $>$. In particular, if B is a reduced dual basis, then $\text{in}_>(B) = \{\text{in}_>(\partial_i) \mid i = 1, \ldots, m\}$ consists of m differential functionals and $\text{in}_>(D_y[I]) = \text{in}_>(B)$.

The standard monomials relate to the initial terms of a dual space. The *opposite monomial ordering* of $>$, denoted $\succ$, is defined by

$$x^\alpha \succ x^\beta \text{ if and only if } x^\beta > x^\alpha.$$

The second part of Theorem 3.1 of [9] provides the following relationship.

Proposition 2.4 (Theorem 3.1 of [9]). *Let I be an ideal such that $\mathscr{V}(I) = \{y\}$. If $>$ and $\succ$ are opposite monomial orderings, then the set of the initial terms of $D_y[I]$ with respect to $\succ$ is the set of standard monomials for I with respect to $>$. That is,*

$$\text{in}_\succ(D_y[I]) = \{\partial_\alpha[y] \mid x^\alpha \notin \text{in}_>(I)\}.$$

2.3 Hilbert Functions

The following information contains only the necessary information regarding Hilbert functions and the index of regularity for this article. This information, along with expanded details, can be found in [3].

Recall that a graded monomial order is a monomial order, $>$, that respects monomial multiplication and satisfies $|\alpha| > |\beta| \implies x^\alpha > x^\beta$. Its opposite ordering, $\succ$, is a reverse graded monomial ordering (i.e., a monomial order, $\succ$, that respects monomial multiplication and satisfies $|\beta| > |\alpha| \implies x^\alpha \succ x^\beta$). Unless otherwise stated, for the rest of this article, when dealing with polynomials, we will only consider graded monomial orderings and when dealing with differentials, we will only consider reverse graded monomial orderings.

The degree of a polynomial with respect to a monomial ordering is the total degree of its initial monomial. Over the set of graded monomial orderings, the degree of a polynomial is invariant. That is, if $>_1$ and $>_2$ are any two graded monomial orderings, then for any polynomial f, $\deg(\text{in}_{>_1}(f)) = \deg(\text{in}_{>_2}(f))$.

Let $R = \mathbb{C}[x_1, \ldots, x_N]$. For $t \in \mathbb{Z}_{\geq 0}$, let $R_{\leqslant t}$ denote the finite-dimensional $\mathbb{C}$-vector space of polynomials of degree $\leqslant t$. For an ideal $I \subset R$, define

$$I_{\leqslant t} = I \cap R_{\leqslant t}.$$

The *affine Hilbert function* of I is $H_I(t) = \dim_{\mathbb{C}} I_{\leqslant t}$ and the affine Hilbert function of R/I is

$$H_{R/I}(t) = \dim_{\mathbb{C}} \mathbb{C}[x_1, \ldots, x_N]_{\leqslant t} - \dim_{\mathbb{C}} I_{\leqslant t}.$$

The following proposition of Macaulay [10] provides a relationship between standard monomials and the Hilbert function for a graded monomial ordering.

Proposition 2.5. *Let $I \subset R$ be an ideal and let $>$ be a graded monomial ordering. Then for all t, $H_{R/I}(t) = H_{R/\langle \mathrm{in}_>(I)\rangle}(t)$. In particular, $H_{R/I}(t)$ is the number of standard monomials of I that are of degree $\leqslant t$.*

For any ideal I, $H_{R/I}(t)$ agrees with a polynomial function in t for $t \gg 0$. The *(affine) Hilbert polynomial* of R/I is the polynomial $HP_{R/I}$ where, for $t \gg 0$, $HP_{R/I}(t) = H_{R/I}(t)$. The *index of regularity* of R/I is the minimum $t_0 \geq 0$ such that $H_{R/I}(t) = HP_{R/I}(t)$ for all $t \geq t_0$. The degree of $HP_{R/I}$ is equal to the dimension of $\mathscr{V}(I)$.

Suppose that $\mathscr{V}(I)$ is zero dimensional and t_0 is the index of regularity of R/I. Since the dimension of $\mathscr{V}(I)$ is zero, the affine Hilbert Polynomial of R/I is a constant and $HP_{R/I}(t) = H_{R/I}(t_0)$ for all $t \geq 0$. Let $h_t = H_{R/I}(t)$. Then

$$1 = h_0 < h_1 < \cdots < h_{t_0} = h_{t_0+1} = h_{t_0+2} = \cdots.$$

In particular, the index of regularity of R/I is the minimum $t_0 \geq 0$ such that $h_{t_0} = h_{t_0+1}$. Since h_t is the number of standard monomials of I that are of degree at most t, the index of regularity of R/I is the maximum degree of the standard monomials. When R/I is a finite-dimensional vector space, we can characterize its Hilbert function by its values $H_{R/I} = \{h_0, h_1, h_2, \ldots\}$ remembering that $h_{t_0} = h_{t_0+1} \implies h_{t_0+1} = h_{t_0+2}$.

2.4 Basic Example

Consider the ideal $I = \langle x_1 - x_2^2, x_1^2 \rangle$. Then $\mathscr{V}(I)$ consists of the single point $\mathbf{0} = (0,0)$. The set

$$B = \{\partial_{(0,0)}[\mathbf{0}], \partial_{(0,1)}[\mathbf{0}], \partial_{(1,0)}[\mathbf{0}] + \partial_{(0,2)}[\mathbf{0}], \partial_{(1,1)}[\mathbf{0}] + \partial_{(0,3)}[\mathbf{0}]\}$$

is a basis for $D_0[I]$. Written more explicitly,

$$B = \{\partial_1[\mathbf{0}], \partial_{x_2}[\mathbf{0}], \partial_{x_1}[\mathbf{0}] + \partial_{x_2^2}[\mathbf{0}], \partial_{x_1 x_2}[\mathbf{0}] + \partial_{x_2^3}[\mathbf{0}]\}.$$

Let $>$ be the graded lexicographic monomial ordering defined by $x_2 > x_1$ and $\succ$ be the opposite monomial ordering. That is, for monomials of degree $\leqslant 3$, we have

$$x_2^3 > x_2^2x_1 > x_2x_1^2 > x_1^3 > x_2^2 > x_2x_1 > x_1^2 > x_2 > x_1 > 1 \text{ and}$$
$$1 \succ x_1 \succ x_2 \succ x_1^2 \succ x_1x_2 \succ x_2^2 \succ x_1^3 \succ x_1^2x_2 \succ x_1x_2^2 \succ x_2^3.$$

It is easy to verify that $S = \{1, x_1, x_2, x_1x_2\}$ is the set of standard monomials of I with respect to $>$. In particular, $H_{R/I} = \{1, 3, 4, 4, \dots\}$ and the index of regularity is 2. Since B is a reduced dual basis with respect to $\succ$, we have

$$\text{in}_{\succ}(D_0[I]) = \text{in}_{\succ}(B) = \{\partial_1[\mathbf{0}], \partial_{x_1}[\mathbf{0}], \partial_{x_2}[\mathbf{0}], \partial_{x_1x_2}[\mathbf{0}]\}$$

which corresponds to S, as required by Proposition 2.4.

Consider the array

	1	x_1	x_2	x_1^2	x_1x_2	x_2^2	x_1^3	$x_1^2x_2$	$x_1x_2^2$	x_2^3
$\partial_1[\mathbf{0}]$	1	0	0	0	0	0	0	0	0	0
$\partial_{x_2}[\mathbf{0}]$	0	0	1	0	0	0	0	0	0	0
$\partial_{x_1}[\mathbf{0}] + \partial_{x_2^2}[\mathbf{0}]$	0	1	0	0	0	1	0	0	0	0
$\partial_{x_1x_2}[\mathbf{0}] + \partial_{x_2^3}[\mathbf{0}]$	0	0	0	0	1	0	0	0	0	1

The matrices $A_0(\mathbf{0}), A_1(\mathbf{0}), A_2(\mathbf{0})$, and $A_3(\mathbf{0})$ correspond to the first column, the first 3 columns, the first 6 columns, and all 10 columns of this array, respectively, which have rank $1, 3, 4$, and 4, respectively. Moreover, the pivot columns of A_k correspond to the monomials in S of degree at most k. Section 3 shows that this is not a coincidence.

3 The Algorithm

Let $I \subset \mathbb{C}[x_1, \dots, x_N]$ be an ideal and $Y = \{y_1, \dots, y_\ell\} \subset \mathscr{V}(I)$ be a collection of isolated points. Let $\mathfrak{m}_{y_i}$ be the maximal ideal of y_i. Let $\mathscr{P}_{y_i}$ be the $\mathfrak{m}_{y_i}$-primary component of I and define

$$J = \bigcap_{i=1}^{\ell} \mathscr{P}_{y_i}. \tag{2}$$

Clearly, $D_Y[I] = D_Y[J]$. The remainder of this section describes an algorithm and the underlying theory for computing the Hilbert function, standard monomials, and index of regularity for J. The input for the algorithm is a numerical approximation for each y_i and generators for the ideal I.

Theorem 3.2 is a key result which supports the computations performed in the algorithm **isolatedHilbertFunction**. Before stating and proving Theorem 3.2, we will need the following lemma.

Lemma 3.1. *Let* $R = \mathbb{C}[x_1, \ldots, x_N]$. *Let* $I \subset R$ *be an ideal and let* $Y = \{y_1, \ldots, y_\ell\} \subset \mathscr{V}(I)$ *be a collection of isolated points in the variety defined by* I. *Let* J *be the ideal consisting of the intersection of the* $\mathfrak{m}_{y_i}$*-primary components of* I *(as in Eq.* (2)*). Let* $g \in R$. *If* $\partial(g) = 0$ *for all* $\partial \in D_Y[I] = D_Y[J]$, *then* $g \in J$.

Proof. If $g \notin J$, then there exists an $i \in \{1, \ldots, \ell\}$ such that $g \notin \mathscr{P}_{y_i}$. Let $>$ be any graded monomial ordering and $\succ$ be its opposite monomial ordering. The standard monomials of $\mathscr{P}_{y_i}$ are a basis for the $\mathbb{C}$-vector space $R/\mathscr{P}_{y_i}$. Since $g \notin I_i$, the image h of g in $R/\mathscr{P}_{y_i}$ is nonzero and can be expressed in terms of the standard monomials of $\mathscr{P}_{y_i}$. In this way, we can write $g = h + F$ where $F \in \mathscr{P}_{y_i}$ and every monomial of h is a standard monomial of $\mathscr{P}_{y_i}$. In particular, we can assume $\text{in}_>(h) \notin \text{in}_>(\mathscr{P}_{y_i})$.

Let $x^\alpha = \text{in}_>(h)$. By Proposition 2.4, there exists a $\partial \in D_{y_i}[\mathscr{P}_{y_i}]$ such that $\text{in}_\succ(\partial) = \partial_\alpha[y_i]$. Since $g = h + F$, $F \in \mathscr{P}_{y_i}$, and $\partial \in D_{y_i}[\mathscr{P}_{y_i}]$, we can conclude that $\partial(g) = \partial(h)$. Let c_α be the coefficient of x^α in h and let d_α be the coefficient of $\partial_\alpha[y_i]$ in ∂. Since $>$ and $\succ$ are opposite orderings, $\partial(h) = c_\alpha d_\alpha \neq 0$. We conclude that $\partial(g) \neq 0$.

Since $y_i \subset Y$ and $J \subset \mathscr{P}_{y_i}$, we have that $D_{y_i}[\mathscr{P}_{y_i}] \subset D_Y[J]$. Therefore, $\partial \in D_Y[J]$ with $\partial(g) \neq 0$.

We are now ready to prove the main theorem regarding the standard monomials, Hilbert function, and index of regularity of J.

Theorem 3.2. *Let* $R = \mathbb{C}[x_1, \ldots, x_N]$, *let* $I \subset R$ *be an ideal, let* $Y = \{y_1, \ldots, y_\ell\} \subset \mathscr{V}(I)$ *where each* y_i *is isolated, and let* J *be the ideal defined by Eq.* (2)*. Let* $>$ *be a graded monomial ordering and* $\succ$ *be its opposite monomial ordering. Let* B *be a basis of* $D_Y[I]$. *For each* $k \geq 0$, *let* A_k *be the matrix defined by Eq.* (1) *with respect to* B *and* $\succ$. *Then, the pivot columns of* A_k *correspond to the standard monomials of* J, *with respect to* $>$, *of degree at most* k. *In particular,* $H_{R/J}(k) = \text{rank}(A_k)$ *and the index of regularity of* R/J *is the minimum* $k_0 \geqslant 0$ *such that* $\text{rank}(A_{k_0}(0)) = \text{rank}(A_{k_0+1}(0))$.

Proof. Fix $k \geqslant 0$ and order the $q = \binom{N+k}{k}$ monomials of degree $\leqslant k$ with respect to $\succ$, that is,

$$1 = x^{\alpha_1} \succ x^{\alpha_2} \succ \cdots \succ x^{\alpha_q}.$$

Suppose that the jth column of A_k is a pivot column. This means that there exists a $\partial \in D_Y[I] = D_Y[J]$ such that $\partial(x^\beta) = 0$ when $x^\beta \succ x^{\alpha_j}$ and $\partial(x^{\alpha_j}) \neq 0$. For any $g \in J$ with $\text{in}_>(g) = x^{\alpha_j}$, since $>$ and $\succ$ are opposite orderings, $\partial(g)$ can be computed using the leading terms of ∂ and g. In particular, $\partial(g) \neq 0$ which is not possible. Thus, x^{α_j} is a standard monomial of J with respect to $>$.

Suppose that x^{α_j} is a standard monomial of J with respect to $>$. If the jth column of A_k is not a pivot column, then this column is the linear span of the first $j-1$ columns. That is, there exists $c_1, \ldots, c_{j-1} \in \mathbb{C}$ such that

$$(A_k)_j = \sum_{p=1}^{j-1} c_p (A_k)_p$$

where $(A_k)_r$ is the rth column of A_k. Define $g = x^{\alpha_j} - \sum_{p=1}^{j-1} c_p x^{\alpha_p}$. By construction, $\partial(g) = 0$ for all $\partial \in D_Y[J]$. Lemma 3.1 yields $g \in J$ which is not possible since $\text{in}_>(g) = x^{\alpha_j}$.

The last statement of the theorem follows directly from the first statement and Proposition 2.5.

The following algorithm to compute the standard monomials, Hilbert function, and index of regularity is justified by Theorem 3.2. Some approaches for computing a numerical dual basis include [4, 5, 13].

Procedure $(S, reg, H) =$ **isolatedHilbertFunction**$(f, Y, >)$
Input A finite set of polynomials $f \subset \mathbb{C}[x_1, \ldots, x_N]$, a set $Y = \{y_1, \ldots, y_\ell\}$ of numerical approximations of distinct isolated solutions of f, and a graded monomial ordering $>$.
Output The set of standard monomials S with respect to $>$ for the ideal $J = \bigcap_{i=1}^{l} \mathscr{P}_{y_i}$ (where $\mathscr{P}_{y_i}$ is the $\mathfrak{m}_{y_i}$-primary component of $I = (f)$), the index of regularity *reg*, and the vector H corresponding to the Hilbert function of R/J up to order *reg*.
Begin

1. Construct the opposite monomial ordering $\succ$ from $>$.
2. Construct a basis B_i for each $D_{y_i}[I]$.
3. Set $B := \bigcup_i B_i$, a basis of $D_Y[I] = D_Y[J]$, and $P := |B|$.
4. Set $H_0 := 1$ and $k := 0$.
5. do
   a. Set $k := k + 1$.
   b. Construct A_k from B and $\succ$.
   c. Set $H_k := \text{rank}(A_k)$.
   while $H_k < P$.
6. Set $reg := k$ and $S :=$ set of monomials corresponding to the pivot columns of A_k.

Remark 3.3. The Hilbert function, index of regularity, and standard monomials for $\sqrt{J}$ can be computed using **isolatedHilbertFunction** by only utilizing the identity differential functionals at each y_i rather than a basis for $D_{y_i}[I]$. This simplification leads to a correspondence between the rows of A_k and the degree k Veronese embedding of the points y_i.

4 Implementation Details

Since the numerical computation of a dual basis at an isolated point was discussed in detail in [1,4,13], we will focus on the computations regarding the matrix A_k defined in Eq. (1). The two main concerns regarding A_k is the ill-conditioning arising from the use of a monomial basis and the growth in the number of columns, which are addressed in Sects. 4.1 and 4.2, respectively.

The ideas presented in the following sections are demonstrated using the polynomial system arising from the nine-point path synthesis problem [11]. We set up the polynomial system $\mathscr{F}_9$ using the nine points of Problem 3 listed in Table 2 of [11] (see also [6, Sect. 5.2]) which has 8,652 isolated nonsingular solutions. We took $\mathscr{Y}_9$ to be 8,652 points in $\mathbb{C}^{12}$, each being an approximation within 10^{-75} of a distinct isolated nonsingular solution.

4.1 Numerical Concerns

There are two main numerical concerns with **isolatedHilbertFunction**, namely the accuracy of the input data and the ill-conditioning of using a monomial basis in A_k. We note that if the set of points Y used in **isolatedHilbertFunction** are exact solutions, then all of the computations can be performed exactly. When the points are numerical approximations, care must be taken to maintain numerical integrity. Since each y_i is an approximation of an isolated solution, using an approximate dual basis with a simple modification of the deflation algorithm of [9], one can create a polynomial system which has a nonsingular isolated solution corresponding to y_i and a basis for $D_{y_i}[I]$. By using Newton's method on this deflated system, the point y_i and a basis for $D_{y_i}[I]$ can be computed to arbitrary accuracy. Rerunning **isolatedHilbertFunction** using the more accurate solutions and higher precision increases the security of the numerical methods.

Due to the ill-conditioning arising from using a monomial basis, the precision one must use to properly determine the rank of A_k may be prohibitively large. For example, the third column of Table 1 presents the largest singular value (i.e., two-norm) of the matrix A_k for the polynomial system $\mathscr{F}_9$ and points $\mathscr{Y}_9$. In particular, reliable rank determination, especially for $k \geqslant 3$, would require the use of higher precision.

We can overcome much of the ill-conditioning by using a preconditioner, namely scaling each row to have unit norm. The last four columns of Table 1 show that after preconditioning, there is a more pronounced indication of the numerical rank.

Table 1 Data for the nine-point path synthesis problem

k	Size of A_k	$\sigma_{\max}(A_k)$	After preconditioning			
			$\sigma_{\max}$	Rank	σ_{rank}	$\sigma_{\text{rank}+1}$
1	8652×13	$7.8 \cdot 10^4$	60.4	13	10.1	–
2	8652×91	$2.9 \cdot 10^9$	46.6	87	0.58	$2.2 \cdot 10^{-14}$
3	8652×455	$1.1 \cdot 10^{14}$	38.7	403	0.052	$1.3 \cdot 10^{-14}$
4	8652×1820	$3.9 \cdot 10^{18}$	33.4	1454	0.025	$1.2 \cdot 10^{-14}$
5	8652×6188	$1.3 \cdot 10^{23}$	29.6	4342	$5.3 \cdot 10^{-5}$	$2.4 \cdot 10^{-14}$
6	8652×18564	$4.1 \cdot 10^{27}$	26.8	8652	$7.2 \cdot 10^{-9}$	–
7	8652×50338	$1.3 \cdot 10^{32}$	24.8	8652	$1.4 \cdot 10^{-7}$	–

4.2 Efficiency Improvements

The number of columns of the matrix A_k is the dimension of $\mathbb{C}[x_1, \ldots, x_N]_{\leqslant k}$, namely $\binom{N+k}{k}$, which grows rapidly as k increases. For $k \geqslant 2$, we can use information already computed from A_{k-1} to control the growth on the number of columns of A_k.

Let B_k be the columns of A_k which correspond to the monomials of degree k, that is, we can write

$$A_k = \left[\, A_{k-1} \; B_k \,\right].$$

Since the pivot columns of A_{k-1} have already been computed, the first reduction is the removal of the nonpivot columns of A_{k-1}. Instead of simply removing the nonpivot columns from A_{k-1}, for improved numerics, one can replace A_{k-1} with a matrix, say $\hat{A}_{k-1}$, whose columns form an orthonormal basis of the linear span of the columns of A_{k-1}, which is the linear span of the pivot columns of A_{k-1}.

The second reduction is based on algebra. In particular, if x^α is not a standard monomial, then, for any $\beta \in (\mathbb{Z}_{\geq 0})^N$, $x^{\alpha+\beta}$ is also not a standard monomial. Let S_j denote the set of standard monomials of degree at most j. If $x^\alpha \in S_k$, then for every $\beta \in (\mathbb{Z}_{\geq 0})^N$ such that $\beta_i \leqslant \alpha_i$, we must have $x^\beta \in S_{|\beta|} \subset S_k$. Therefore, given S_{k-1}, this provides a closedness condition that can be used to generate a set, M_k, consisting of monomials of degree exactly k that forms a superset of $S_k \setminus S_{k-1}$, namely

$$M_k = \{x^\alpha \mid |\alpha| = k \text{ such that } x^\beta \in S_{k-1} \text{ whenever } |\beta| \leqslant k-1 \text{ and } \beta_i \leqslant \alpha_i\}.$$

Let $\hat{B}_k$ consist of the columns of B_k which correspond to the monomials in M_k ordered with respect to $\succ$. We know that the pivot columns of A_k are the pivot columns of A_{k-1} and the columns of $\hat{B}_k$ which are pivot columns of the matrix

$$\left[\, \hat{A}_{k-1} \; \hat{B}_k \,\right].$$

For example, consider the four nonpivot columns of A_2 for the polynomial system $\mathscr{F}_9$ and points $\mathscr{Y}_9$. These columns correspond to the monomials $a\hat{x}$, $\hat{a}x$, $b\hat{y}$, and $\hat{b}y$ which arise from the four quadratic polynomials in $\mathscr{F}_9$, namely

$$n - a\hat{x}, \quad \hat{n} - \hat{a}x, \quad m - b\hat{y}, \quad \text{and} \quad \hat{m} - \hat{b}y.$$

Using only these four monomials, we can immediately remove 13% (48 out of 364) of the columns of B_3, 22% (306 out of 1,365) of the columns of B_4, 32% (1,384 out of 4,368) of the columns of B_5, 40% (4,996 out of 12,376) of the columns of B_6, and 48% (15,336 out of 31,824) of the columns of B_7.

Since the columns of $\hat{A}_{k-1}$ are orthonormal, the third reduction constructs a matrix Q_{k-1} such that $\left[\, \hat{A}_{k-1} \; Q_{k-1} \,\right]$ is unitary. We then have

$$\text{rank}\, A_k = \text{rank}\, \hat{A}_{k-1} + \text{rank}\, Q_{k-1}^* \cdot \hat{B}_k = \text{rank}\, A_{k-1} + \text{rank}\, Q_{k-1}^* \cdot \hat{B}_k$$

where Q_{k-1}^* is the Hermitian adjoint, i.e., conjugate transpose, of Q_{k-1}. Due to the added cost of computing Q_{k-1}, this reduction is only beneficial when the number of columns of Q_{k-1}, which is the difference between the number of rows and the rank of $\hat{A}_{k-1}$, is small.

5 Examples

The following examples discussed in this section were run on an Opteron 250 processor running 64-bit Linux. We used the graded lexicographic monomial ordering $>$ defined by $x_N > x_{N-1} > \cdots > x_1$ so that its opposite monomial ordering $\succ$ has

$$1 \succ x_1 \succ x_2 \succ \cdots \succ x_N \succ x_1^2 \succ x_1x_2 \succ \ldots.$$

Though exact solutions are known for the following examples (either by hand or using symbolic software), these examples utilized numerical approximations of the solutions generated by Bertini [2] with the matrix computations performed using Matlab.

5.1 A Collection of Isolated Solutions

Consider $I = \langle x_1^2 + x_2 + x_3 - 1, x_1 + x_2^2 + x_3 - 1, x_1 + x_2 + x_3^2 - 1 \rangle \subset \mathbb{C}[x_1, x_2, x_3]$ from [12]. This ideal has the following five isolated solutions:

$$y_1 = (1, 0, 0)\,, \; y_2 = (0, 1, 0)\,, \; y_3 = (0, 0, 1)\,,$$

$$y_4 = \left(-1 - \sqrt{2}, -1 - \sqrt{2}, -1 - \sqrt{2}\right), \quad \text{and} \quad y_5 = \left(\sqrt{2} - 1, \sqrt{2} - 1, \sqrt{2} - 1\right).$$

Table 2 Computing the Hilbert function, index of regularity, and standard monomials for I at $\hat{Y}_i$

	Hilbert function	Index of reg.	Standard monomials
$\hat{Y}_1$	$1, 2, 2, \ldots$	1	$1, x_1$
$\hat{Y}_2$	$1, 3, 4, 4, \ldots$	2	$1, x_1, x_3, x_1x_3$
$\hat{Y}_3$	$1, 4, 6, 6, \ldots$	2	$1, x_1, x_2, x_3, x_1x_2, x_1x_3$
$\hat{Y}_4$	$1, 4, 7, 7, \ldots$	2	$1, x_1, x_2, x_3, x_1x_2, x_1x_3, x_2x_3$
$\hat{Y}_5$	$1, 4, 7, 8, 8, \ldots$	3	$1, x_1, x_2, x_3, x_1x_2, x_1x_3, x_2x_3, x_1x_2x_3$

Table 3 Computing the Hilbert function and index of regularity for the ideal and its radical for $\mathscr{I}$ at $\hat{A}$

	Hilbert function	Index of reg.
Ideal	$1, 4, 10, 15, 16, 16, \ldots$	4
Radical ideal	$1, 4, 5, 5 \ldots$	2

Solutions y_1, y_2, and y_3 have multiplicity 2 while y_4 and y_5 have multiplicity 1. The points $\hat{y}_i$ that were used in the algorithm were approximations to the y_i accurate to ten digits. Table 2 displays the Hilbert function, index of regularity, and standard monomials using **isolatedHilbertFunction** with $\hat{Y}_i = \{\hat{y}_1, \ldots, \hat{y}_i\}$, $i = 1, \ldots, 5$. All of these computations, including the computation of $\hat{Y}_i$, took less than a second.

5.2 *A System with Isolated and Positive-Dimensional Solution Components*

Consider

$$\mathscr{I} = \langle (x_2 - x_1^2)(x_1 - 2)^2, (x_1x_2 - x_3)(x_2 - 2)^2, (x_2^2 - x_1x_3)(x_3 - 2) \rangle \subset \mathbb{C}[x_1, x_2, x_3].$$

This ideal has a one-dimensional solution component, namely the twisted cubic $C = \{(t, t^2, t^3) \mid t \in \mathbb{C}\}$, and the following five isolated solutions:

$$a_1 = (2, 1, 2)\,,\; a_2 = \left(\sqrt{2}, 2, 2\right),\; a_3 = \left(-\sqrt{2}, 2, 2\right),$$

$$a_4 = (2, 0, 0)\,,\;\text{ and }\; a_5 = (2, 2, 2)\,.$$

Solutions $a_1, \ldots, a_4$ have multiplicity 2 and a_5 has multiplicity 8. Let $\hat{A} = \{\hat{a}_1, \ldots, \hat{a}_5\}$ where $\hat{a}_i$ is a ten-digit approximation to a_i. Table 3 displays the computed Hilbert function and index of regularity for both J and the radical of J. All of these computations, including the computation of $\hat{A}$, took less than a second.

Table 4 Computing the Hilbert function and index of regularity for I_n at $\hat{Y}_n$

n	Hilbert function	Index of reg.	Time
3	$1, 4, 9, 12, 12, \ldots$	3	<1 s
4	$1, 5, 15, 31, 45, 48, 48, \ldots$	5	13 s
5	$1, 6, 20, 45, 70, 80, 80, \ldots$	5	1 min 10 s
6	$1, 7, 27, 71, 131, 177, 192, 192, \ldots$	6	32 min 48 s

Table 5 Computing the Hilbert function and index of regularity for $\sqrt{I_n}$

n	Hilbert function	Index of reg.	Time
3	$1, 4, 10, 16, 18, 18, \ldots$	4	1 s
4	$1, 5, 15, 28, 36, 37, 37, \ldots$	5	4 s
5	$1, 6, 21, 51, 96, 141, 162, 167, 168, 168, \ldots$	8	11 s
6	$1, 7, 28, 78, 168, 294, 425, 506, 536, 542, 543, 543, \ldots$	10	1 min 52 s

5.3 A Sequence of Systems

For each $n \geq 3$, consider a generalization of the *mth191* system of Bernd Sturmfels presented in [4, 8], namely $I_n = \langle f_1, \ldots, f_n \rangle \subset \mathbb{C}[x_1, \ldots, x_n]$ where

$$f_i = x_i^3 - x_i^2 + \sum_{j=1}^{n} x_j^2 - 1.$$

Let $Y_n = \{e_1, \ldots, e_n\} \subset \mathbb{C}^n$ be the set of standard basis vectors of $\mathbb{C}^n$ and $\hat{Y}_n = \{\hat{e}_1, \ldots, \hat{e}_n\}$ where each $\hat{e}_i$ is a 12-digit approximation to e_i. For $n = 3, 4, 5, 6$, the multiplicity of each e_i is 4, 12, 16, and 32, respectively. Table 4 displays the Hilbert function and index of regularity using **isolatedHilbertFunction** for the ideal I_n at $\hat{Y}_n$ for $3 \leqslant n \leqslant 6$. We note that less than 1 % of the time was spent actually performing rank computations. In particular, for $n = 6$, approximately 22 % of the time was spent computing dual bases using an implementation of [4] and approximately 77 % of the time was spent by Matlab constructing the matrices A_k from the points and dual bases.

Table 5 displays the Hilbert function and index of regularity using **isolatedHilbertFunction** for $\sqrt{I_n}$ for $3 \leqslant n \leqslant 6$ using the points in $\mathscr{V}(I_n)$ computed by Bertini to 12 digits. For $n = 6$, Bertini took 27 s to compute the points with **isolatedHilbertFunction** taking 85 s. Of this 85 s, roughly 60 s was used to construct the matrices A_k and roughly 25 s was used to compute ranks.

6 Concluding Remarks

As before, let $Y = \{y_1, \ldots, y_\ell\} \subset \mathbb{C}^N$, let $\mathfrak{m}_{y_i}$ denote the maximal ideal of y_i, and let $\mathscr{P}_{y_i}$ denote an $\mathfrak{m}_{y_i}$-primary ideal. Let $J = \cap_{i=1}^{l} \mathscr{P}_{y_i}$, let L be an ideal such that $\mathscr{V}(L) \cap \mathscr{V}(J) = \emptyset$, and let $I = J \cap L$. From the algorithm, **isolatedHilbertFunction**, one can compute the Hilbert function of R/J from a set of generators for I and a set of numerical approximations $\hat{Y} = \{\hat{y}_1, \ldots, \hat{y}_l\} \subset \mathbb{C}^N$ to the elements in Y. Several algorithms follow, in a natural manner, from **isolatedHilbertFunction** (and modifications).

One of the more immediate applications is a direct algorithm to determine if a set of points satisfies the *Uniform Position Property*. A set of points $Y = \{y_1, \ldots, y_\ell\} \subset \mathbb{C}^N$ has the Uniform Position Property if the Hilbert function of any subset Y' of Y only depends on the cardinality of Y'. This property can be checked by iterating **isolatedHilbertFunction** over the power set of Y. While the power set of Y has cardinality 2^l, the ability to parallelize the computation suggests it is feasible on sets consisting of fewer than 30 points.

Examples were presented involving the Hilbert function of R/J and of $R/\sqrt{J}$. The approach for computing the Hilbert function of $R/\sqrt{J}$ can be generalized as follows. Let $v = (v_1, \ldots, v_l) \in (\mathbb{Z}_{\geq 0})^l$ and let $J(v) = \cap_{i=1}^{l}(\mathscr{P}_{y_i}, \mathfrak{m}_{y_i}^{v_i})$. The Hilbert function of $R/\sqrt{J}$ was computed by utilizing only the identity differential functionals at each y_i rather than a basis for $D_{y_i}[I]$. This corresponds to computing $R/J(v)$ when $v = (1, 1, \ldots, 1)$. In general, by utilizing a subset, dependent on v_i, of the differential functionals at each y_i, the Hilbert function of $R/J(v)$ can be computed for any v. When all the entries of v are sufficiently large, then $J(v) = J$. When some of the entries of v are equal to zero, then the corresponding points are excluded. The scheme defined by $J(v)$ is a subscheme of the scheme defined by J. The ability to make such a broad range of direct computations on subschemes of zero-schemes is due to the ability to numerically decompose varieties over $\mathbb{C}$, a feature which is often difficult in an exact setting.

Finally, it is important to note the limits of such computations. As part of **isolatedHilbertFunction**, reliable rank computations must be made on the matrices A_k. These matrices can become very large and have numerical entries. The limit of reliable rank computations on such matrices is not completely clear. Such a limit must depend on the minimum of the number of rows and columns as well as the relationship between the nonzero singular values and singular values which should be zero. Since the number of rows of A_k is constant, we have a constant bound on the minimum of the number of rows and columns. For example, Table 1 shows that successful computations can be made on a matrix with approximately 8,600 rows and 50,000 columns. Computations involving a matrix where the minimum of the number of rows and columns is 30,000 would require a tremendous amount of patience and memory. Matrices where the minimum is more than 100,000 are currently out of reach.

Acknowledgments The second author was partially supported by NSF grant DMS-1262428. The third author was partially supported by NSF grant DMS-1228308. The fourth was partially supported by the Duncan Chair of the University of Notre Dame and DARPA/AFRL G-2457-2.

References

1. Bates, D.J., Hauenstein, J.D., Peterson, C., Sommese, A.J.: A numerical local dimension test for points on the solution set of a system of polynomial equations. SIAM J. Numer. Anal. **47**(5), 3608–3623 (2009)
2. Bates, D.J., Hauenstein, J.D., Sommese, A.J., Wampler, C.W.: Bertini: software for numerical algebraic geometry. Available at bertini.nd.edu (2006)
3. Cox, D., Little, J., O'Shea, D.: Ideals, Varieties, and Algorithms. Undergraduate Texts in Mathematics, 2nd edn. Springer, New York (1996)
4. Dayton, B.H., Zeng, Z.: Computing the multiplicity structure in solving polynomial systems. In: ISSAC'05, pp. 116–123. ACM, New York (2005)
5. Hao, W., Sommese, A.J. and Zeng, Z.: Algorithm 931: An algorithm and software for computing multiplicity structures at zeros of nonlinear systems, Transactions of Mathematical Software, 40 (2013), Article 5: appears online at dx.doi.org/10.1145/2513109.2513114
6. Hauenstein, J.D., Sottile, F.: Algorithm 921: alphaCertified: certifying solutions to polynomial systems. Trans. Math. Softw. **38**(4), 28 (2012)
7. Hauenstein, J.D., Wampler, C.W.: Isosingular sets and deflation. Found. Comput. Math. (2013). doi:10.1007/s10208-013-9147-y
8. Leykin, A., Verschelde, J., Zhao, A.: Newton's method with deflation for isolated singularities of polynomial systems. Theor. Comp. Sci. **359**, 111–122 (2006)
9. Leykin, A., Verschelde, J., Zhao, A.: Higher-order deflation for polynomial systems with isolated singular solutions. In: Algorithms in Algebraic Geometry. The IMA Volumes in Mathematics and Its Applications, vol. 146, pp. 79–97. Springer, New York (2008)
10. Macaulay, F.S.: The Algebraic Theory of Modular Systems. Cambridge University Press, Cambridge (1916)
11. Morgan, A.P., Sommese, A.J., Wampler, C.W.: Complete solution of the nine-point path synthesis problem for four-bar linkages. ASME J. Mech. Des. **114**(1), 153–159 (1992)
12. Ojika, T.: Modified deflation algorithm for the solution of singular problems. J. Math. Anal. Appl. **123**, 199–221 (1987)
13. Zeng, Z.: The closedness subspace method for computing the multiplicity structure of a polynomial system. Contemp. Math. **496**, 347–362 (2009)

Regularity of Squarefree Monomial Ideals

Huy Tài Hà

Dedicated to Tony Geramita, a great teacher, colleague and friend.

Abstract We survey a number of recent studies of the Castelnuovo–Mumford regularity of squarefree monomial ideals. Our focus is on bounds and exact values for the regularity in terms of combinatorial data from associated simplicial complexes and/or hypergraphs.

1 Introduction

Castelnuovo–Mumford regularity (or simply regularity) is an important invariant in commutative algebra and algebraic geometry that governs the computational complexity of ideals, modules, and sheaves. Computing or finding bounds for the regularity is a difficult problem. Many simply stated questions and conjectures have been verified only in very special cases. For instance, the Eisenbud–Goto conjecture, which states that the regularity of a projective variety is bounded by the difference between its degree and codimension, has been proved only for arithmetic Cohen–Macaulay varieties, curves, and surfaces.

The class of squarefree monomial ideals is a classical object in commutative algebra, with strong connections to topology and combinatorics, which continues to inspire much of current research. During the last few decades, advances in computer technology and speed of computation have drawn many researchers'

H.T. Hà (✉)
Department of Mathematics, Tulane University, 6823 St. Charles Ave.,
New Orleans, LA 70118, USA
e-mail: tha@tulane.edu

S.M. Cooper and S. Sather-Wagstaff (eds.), *Connections Between Algebra, Combinatorics, and Geometry*, Springer Proceedings in Mathematics & Statistics 76,
DOI 10.1007/978-1-4939-0626-0_7,

attention toward problems and questions involving this class of ideals. Investigating the regularity of squarefree monomial ideals has evolved to be a highly active research topic in combinatorial commutative algebra.

In this paper, we survey a number of recent studies on the regularity of squarefree monomial ideals. Even for this class of ideals, the list of results on regularity is too large to exhaust. Our focus will be on studies that find bounds and/or compute the regularity of squarefree monomial ideals in terms of combinatorial data from associated simplicial complexes and hypergraphs. Our aim is to provide readers with an adequate overall picture of the problems, results, and techniques and to demonstrate similarities and differences between these studies. At the same time, we hope to motivate interested researchers, and especially graduate students, to start working in this area. We shall showcase results in a timeline to exhibit (if possible) trends in developing new and/or more general results from previous ones.

The paper is structured as follows. In the next section we collect basic notation and terminology from commutative algebra and combinatorics that will be used in the paper. This section provides readers who are new to the research area the necessary background to start. More advanced readers can skip this section and go directly to Sect. 3. In Sect. 3, we outline a number of inductive results that appear to be common techniques in many recent works. Some of these results arise from simple short exact sequences in commutative algebra, while others may require heavier topological machineries to prove. Section 4 is devoted to studies that provide bounds for the regularity of squarefree monomial ideals. In Sect. 5, we discuss studies that explicitly compute the regularity of squarefree monomial ideals and identify classes of ideals with small regularity. In Sects. 3–5, we do not generally prove results being surveyed. Rather, occasionally for initial theorems of various inductive techniques we shall sketch the proofs to demonstrate the methods used. The last section provides a number of open problems and questions that we would like to see answered.

2 Preliminaries

We begin by recalling familiar notation and terminology from commutative algebra and combinatorics. We follow standard texts [3, 29, 38, 44] in these fields.

We shall investigate the connection between commutative algebra and combinatorics via the notion of Stanley–Reisner ideals and edge ideals. To illustrate a general framework, we will not include studies that are based on constructions that do not necessarily give one-to-one correspondences between squarefree monomial ideals and combinatorial structures; for instance, *path ideals* of directed trees (cf. [7]).

Throughout the paper, K will denote any infinite field and $R = K[x_1, \ldots, x_n]$ will be a polynomial ring over K. For obvious reasons, we shall identify the variables $x_1, \ldots, x_n$ with the vertices of simplicial complexes and hypergraphs being discussed. By abusing notation, we also often identify a subset V of the vertices $X = \{x_1, \ldots, x_n\}$ with the squarefree monomial $x^V = \prod_{x \in V} x$ in the polynomial ring R. In examples, we use $a, b, c, \ldots$ as variables instead of the x_is.

2.1 Simplicial Complexes

A *simplicial complex* Δ over the vertex set $X = \{x_1, \ldots, x_n\}$ is a collection of subsets of X such that if $F \in \Delta$ and $G \subseteq F$ then $G \in \Delta$. Elements of Δ are called *faces*. Maximal faces (with respect to inclusion) are called *facets*. For $F \in \Delta$, the *dimension* of F is defined to be $\dim F = |F| - 1$. The *dimension* of Δ is $\dim \Delta = \max\{\dim F \mid F \in \Delta\}$. The complex is called *pure* if all of its facets are of the same dimension.

Let Δ be a simplicial complex and let $Y \subseteq X$ be a subset of its vertices. The *induced subcomplex* of Δ on Y, denoted by $\Delta[Y]$, is the simplicial complex with vertex set Y and faces $\{F \in \Delta \mid F \subseteq Y\}$.

Definition 2.1. Let Δ be a simplicial complex over the vertex set X, and let $\sigma \in \Delta$.

1. The *deletion* of σ in Δ, denoted by $\mathrm{del}_\Delta(\sigma)$, is the simplicial complex obtained by removing σ and all faces containing σ from Δ.
2. The *link* of σ in Δ, denoted by $\mathrm{link}_\Delta(\sigma)$, is the simplicial complex whose faces are

$$\{F \in \Delta \mid F \cap \sigma = \emptyset, \sigma \cup F \in \Delta\}.$$

Definition 2.2. A simplicial complex Δ is recursively defined to be *vertex decomposable* if either

1. Δ is a simplex; or
2. there is a vertex v in Δ such that both $\mathrm{link}_\Delta(v)$ and $\mathrm{del}_\Delta(v)$ are vertex decomposable, and all facets of $\mathrm{del}_\Delta(v)$ are facets of Δ.

A vertex satisfying condition (2) is called a *shedding vertex*, and the recursive choice of shedding vertices are called a *shedding order* of Δ.

Recall that a simplicial complex Δ is said to be *shellable* if there exists a linear order of its facets $F_1, F_2, \ldots, F_t$ such that for all $k = 2, \ldots, t$, the subcomplex $\left(\bigcup_{i=1}^{k-1} \overline{F_i}\right) \bigcap \overline{F_k}$ is pure and of dimension $(\dim F_k - 1)$. Here $\overline{F}$ represents the simplex over the vertices of F. It is a celebrated fact that *pure* shellable complexes give rise to *Cohen–Macaulay Stanley–Reisner rings*. For more details on Cohen–Macaulay rings and modules, we refer the reader to [8]. The notion of Stanley–Reisner rings will be discussed later in the paper. Recall also that a ring or module is *sequentially Cohen–Macaulay* if it has a filtration in which the factors are Cohen–Macaulay and their dimensions are increasing. This property corresponds to (*nonpure*) shellability in general.

Vertex decomposability can be thought of as a combinatorial criterion for shellability and sequentially Cohen–Macaulayness. In particular, for a simplicial complex Δ,

$$\Delta \text{ vertex decomposable } \Rightarrow \Delta \text{ shellable } \Rightarrow \Delta \text{ sequentially Cohen–Macaulay.}$$

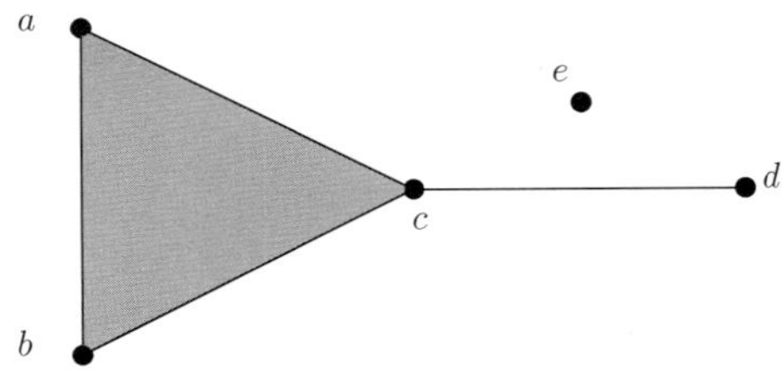

Fig. 1 A vertex decomposable simplicial complex

Example 2.3. The simplicial complex Δ in Fig. 1 is a nonpure simplicial complex of dimension 2. It has three facets; the facet $\{a, b, c\}$ is of dimension 2, the facet $\{c, d\}$ is of dimension 1, and the facet $\{e\}$ is of dimension 0. The complex Δ is vertex decomposable with $\{e, d\}$ as a shedding order.

2.2 Hypergraphs

A hypergraph $H = (X, \mathscr{E})$ over the vertex set $X = \{x_1, \ldots, x_n\}$ consists of X and a collection $\mathscr{E}$ of nonempty subsets of X; these subsets are called the *edges* of H. A hypergraph H is *simple* if there is no nontrivial containment between any pair of its edges. Simple hypergraphs are also referred to as *clutters* or *Sperner systems*. All hypergraphs considered in this paper will be simple.

When working with a hypergraph H, we shall use $X(H)$ and $\mathscr{E}(H)$ to denote its vertex and edge sets, respectively. We shall denote by $\text{is}(H)$ the set of (*isolated*) vertices that do not belong to any edge in H, and let H^{red} be the hypergraph obtained by removing vertices in $\text{is}(H)$ from the vertex set of H. An edge $\{v\}$ consisting of a single vertex is often referred to as an *isolated loop* (this is not to be confused with an isolated vertex).

Let $Y \subseteq X$ be a subset of the vertices in H. The *induced subhypergraph* of H on Y, denoted by $H[Y]$, is the hypergraph with vertex set Y and edge set $\{E \in \mathscr{E} \mid E \subseteq Y\}$. The *contraction* of H to Y is the hypergraph with vertex set Y and edges being minimal nonempty elements of $\{E \cap Y \mid E \in \mathscr{E}\}$.

Definition 2.4. Let H be a simple hypergraph.

1. A collection C of edges in H is called a *matching* if the edges in C are pairwise disjoint. The maximum size of a matching in H is called its *matching number*.
2. A collection C of edges in H is called an *induced matching* if C is a matching, and C consists of all edges of the induced subhypergraph $H[\cup_{E \in C} E]$ of H. The maximum size of an induced matching in H is called its *induced matching number*.

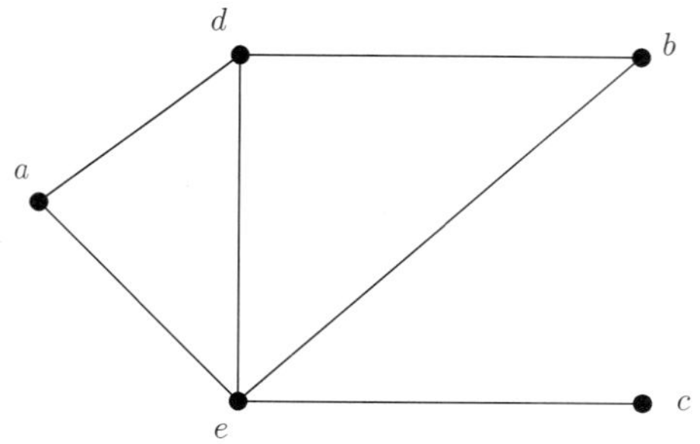

Fig. 2 A simple graph whose independence complex is in Fig. 1

Example 2.5. Figure 1 can be viewed as a hypergraph over the vertex set $\{a,b,c,d,e\}$ with edges $\{a,b,c\},\{c,d\}$ and $\{e\}$. The collection $\{\{a,b,c\},\{e\}\}$ forms an induced matching in this hypergraph.

A *graph* is a hypergraph in which all edges are of cardinality 2. The *complement* of a graph G, denoted by G^c, is the graph with the same vertex set and an edge E is in G^c if and only if E is not in G.

Definition 2.6. A graph G is called *chordal* if it has no induced cycles of length ≥ 4.

A hypergraph H is d*-uniform* if all its edges have cardinality d. For an edge E in H, let $N(E) = \{x \in X \mid \exists F \subseteq E \text{ s.t. } F \cup \{x\} \in \mathscr{E}\}$ be the set of *neighbors* of E, and let $N[E] = N(E) \cup E$.

Definition 2.7. Let $H = (X, \mathscr{E})$ be a simple hypergraph. Let E be an edge and let Y be a subset of the vertices in H.

1. Define $H \setminus E$ to be the hypergraph obtained by deleting E from the edge set of H. This is often referred to as the *deletion* of E from H.
2. Define $H \setminus Y$ to be the hypergraph obtained from H by deleting the vertices in Y and all edges containing any of those vertices.
3. Define H_E to be the contraction of $H \setminus N(E)$ to $X \setminus N[E]$.

Note that when H is a graph and E is an edge, then H_E is just the induced subgraph of H over the vertex set $X \setminus N[E]$.

Definition 2.8. Let $H = (X, \mathscr{E})$ be a simple hypergraph.

1. A collection of vertices V in H is called an *independent set* if there is no edge $E \in \mathscr{E}$ such that $E \subseteq V$.
2. The *independence complex* of H, denoted by $\Delta(H)$ is the simplicial complex whose faces are independent sets in H.

Example 2.9. The simplicial complex Δ in Fig. 1 is the independence complex of the graph in Fig. 2.

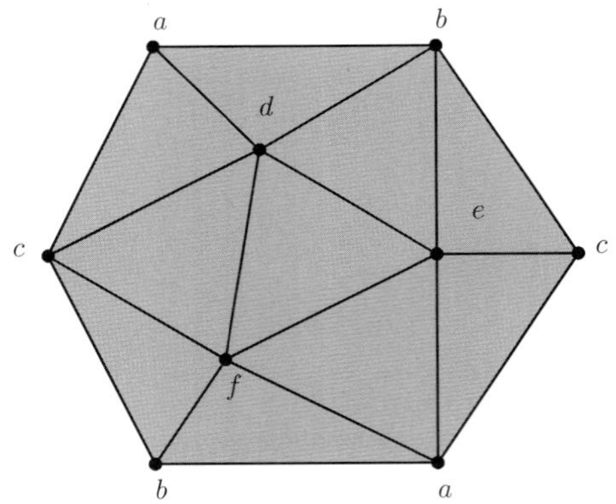

Fig. 3 A minimal triangulation of the real projective plane

Remark 2.10. We call a hypergraph H vertex decomposable (shellable, sequentially Cohen–Macaulay) if its independence complex $\Delta(H)$ is vertex decomposable (shellable, sequentially Cohen–Macaulay).

2.3 Stanley–Reisner Ideals and Edge Ideals

The Stanley–Reisner ideal and edge ideal constructions are well-studied correspondences between commutative algebra and combinatorics. Those constructions arise by identifying minimal generators of a squarefree monomial ideal with the minimal nonfaces of a simplicial complex or the edges of a simple hypergraph.

Stanley–Reisner ideals were developed in the early 1980s (cf. [44]) and have led to many important homological results (cf. [8, 43]).

Definition 2.11. Let Δ be a simplicial complex on X. The *Stanley–Reisner ideal* of Δ is defined to be

$$I_\Delta = \left(x^F \mid F \subseteq X \text{ is not a face of } \Delta\right).$$

Example 2.12. Let Δ be the simplicial complex in Fig. 1 and let $R = K[a,b,c,d,e]$. Then the minimal nonfaces of Δ are $\{a,d\}, \{a,e\}, \{b,d\}, \{b,e\}, \{c,e\}$, and $\{d,e\}$. Thus,

$$I_\Delta = (ad, ae, bd, be, ce, de).$$

Example 2.13. The simplicial complex Δ in Fig. 3 represents a minimal triangulation of the real projective plane. Its Stanley–Reisner ideal is

$$I_\Delta = (abc, abe, acf, ade, adf, bcd, bdf, bef, cde, cef).$$

The edge ideal construction was introduced in [48] for graphs and later generalized to hypergraphs in [26]. This construction is similar to that of facet ideals in [19].

Definition 2.14. Let H be a simple hypergraph on X. The *edge ideal* of H is defined to be

$$I(H) = \big(x^E \mid E \subseteq X \text{ is an edge in } H\big).$$

The notion of Stanley–Reisner ideal and edge ideal give the following one-to-one correspondences that allow us to pass back and forth from squarefree monomial ideals to simplicial complexes and simple hypergraphs.

$$\left\{\begin{array}{l}\text{simplicial complexes}\\ \text{over } X\end{array}\right\} \leftrightarrow \left\{\begin{array}{l}\text{squarefree monomial}\\ \text{ideals in } R\end{array}\right\} \leftrightarrow \left\{\begin{array}{l}\text{simple hypergraphs}\\ \text{over } X\end{array}\right\}$$

Stanley–Reisner ideals and edge ideals are also closely connected via the notion of independence complex.

Lemma 2.15. *Let H be a simple hypergraph and let $\Delta = \Delta(H)$ be its independence complex. Then*

$$I_\Delta = I(H).$$

Example 2.16. The edge ideal of the graph G in Fig. 2 is the same as the Stanley–Reisner ideal of its independence complex, the simplicial complex in Fig. 1.

2.4 Alexander Duality

The Alexander duality theory for simplicial complexes carries nicely over to squarefree monomial ideals and have proved to be a significant tool in the study of these ideals.

Definition 2.17. Let Δ be a simplicial complex over the vertex set X. The Alexander dual of Δ, denoted by $\Delta^\vee$, is the simplicial complex over X with faces

$$\{X \setminus F \mid F \not\in \Delta\}.$$

Notice that $\Delta^{\vee\vee} = \Delta$. If $I = I_\Delta$, then we shall denote by $I^\vee$ the Stanley–Reisner ideal of the Alexander dual $\Delta^\vee$. Also, if $I = I(H)$, then we shall denote by $H^\vee$ the simple hypergraph corresponding to $I^\vee$.

It is a celebrated result of Terai [45] that the regularity of a squarefree monomial ideal can be related to the projective dimension of its Alexander dual.

Theorem 2.18. *Let $I \subseteq R$ be a squarefree monomial ideal. Then*

$$reg(I) = pd(R/I^{\vee}).$$

Theorem 2.18 basically says that studying the regularity of squarefree monomial ideals is equivalent to studying the projective dimension of squarefree monomial ideals. Many studies in commutative algebra take the later point of view.

2.5 Castelnuovo–Mumford Regularity

The regularity of graded modules over the polynomial ring R can be defined in various ways. Let $\mathfrak{m}$ denote the maximal homogeneous ideal in R.

Definition 2.19. Let M be a finitely generated graded R-module. For $i \geq 0$, let

$$a^i(M) = \begin{cases} \max\left\{l \in \mathbb{Z} \,\middle|\, \left[H^i_{\mathfrak{m}}(M)\right]_l \neq 0\right\} & \text{if } H^i_{\mathfrak{m}}(M) \neq 0 \\ -\infty & \text{otherwise.} \end{cases}$$

The *regularity* of M is defined to be

$$\operatorname{reg}(M) = \max_{i \geq 0}\{a^i(M)\}.$$

Note that $a^i(M) = 0$ for $i > \dim M$, so the regularity of M is well defined. This invariant can also be computed via the minimal free resolution (cf. [9, 15]).

Definition 2.20. Let M be a graded R-module and let

$$0 \to \bigoplus_{j \in \mathbb{Z}} R(-j)^{\beta_{pj}(M)} \to \cdots \to \bigoplus_{j \in \mathbb{Z}} R(-j)^{\beta_{0j}(M)} \to M \to 0$$

be its minimal free resolution. Then the regularity of M is given by

$$\operatorname{reg}(M) = \max\{j - i \mid \beta_{ij}(M) \neq 0\}.$$

By looking at the minimal free resolution, it is easy to see that $\operatorname{reg}(R/I) = \operatorname{reg}(I) - 1$, so we shall work with $\operatorname{reg}(I)$ and $\operatorname{reg}(R/I)$ interchangeably.

The focus of this paper is on squarefree monomial ideals. As we have seen, squarefree monomial ideals are in direct one-to-one correspondence with simplicial complexes, and the regularity of these ideals can also be computed from the reduced homology groups of corresponding simplicial complexes. We summarize this connection in the following lemma.

Lemma 2.21. *For a simplicial complex Δ, the following are equivalent:*

1. $reg(R/I_\Delta) \geq d$.
2. $\tilde{H}_{d-1}(\Delta[S]) \neq 0$, *where $\Delta[S]$ denotes the induced subcomplex on some subset S of vertices.*
3. $\tilde{H}_{d-1}(link_\Delta \sigma) \neq 0$ *for some face σ of Δ.*

Proof. The equivalence of (1) and (2) follows directly from Definition 2.20, together with Hochster's formula for graded Betti numbers (as stated in [38, Corollary 5.12]). The equivalence of (1) and (3) follows directly from the local cohomology characterization of regularity, together with the fact that $H^i_{\mathfrak{m}}(R/I_\Delta)_{-\sigma} \cong \tilde{H}^{i-|\sigma|-1}(\mathrm{link}_\Delta \sigma)$ (see [38, Chap. 13.2]).

Remark 2.22. For simplicity, if $I = I_\Delta$, then we sometimes write $\mathrm{reg}(\Delta)$ for $\mathrm{reg}(I)$, and if $I = I(H)$, then we write $\mathrm{reg}(H)$ for $\mathrm{reg}(I)$.

3 Regularity and Induction

The backbone of most of the studies that we survey is mathematical induction based on combinatorial structures of given simplicial complexes and hypergraphs. The technique is to relate the regularity of a squarefree monomial ideal corresponding to a simplicial complex and/or hypergraph to that of smaller sub-ideals corresponding to subcomplexes and/or subhypergraphs. In this section, we discuss a number of inductive results that lie in the core of most of these studies. It is worth noting that there are also inductive results that go from smaller simplicial complexes and/or hypergraphs to larger ones; for instance, in the work of [6, 40]. Since we eventually are interested in bounds or values for the regularity, those works are beyond the scope of our survey.

We start the section by a few crude inductive bounds for the regularity of a hypergraph (or simplicial complex) in terms of that of subhypergraphs (or subcomplexes) that follow directly from Lemma 2.21.

Lemma 3.1.

1. *Let H be a simple hypergraph. Then $reg(H) \geq reg(H')$ for any induced subhypergraph H' of H.*
2. *Let Δ be a simplicial complex. Then $reg(\Delta) \geq reg(link_\Delta(\sigma))$ for any face σ of Δ.*

For any homogeneous ideal $I \subseteq R$ and any homogeneous element $h \in R$ of degree d, the following short exact sequence is standard in commutative algebra:

$$0 \longrightarrow \frac{R}{I:h}(-d) \xrightarrow{\times h} \frac{R}{I} \longrightarrow \frac{R}{I+h} \longrightarrow 0. \tag{1}$$

By taking the long exact sequence of local cohomology modules associated with (1), we get

$$\operatorname{reg}(I) \le \max\{\operatorname{reg}(I : h) + d, \operatorname{reg}(I, h)\}. \tag{2}$$

Remark 3.2. If h is an indeterminate of R appearing in I, then it was shown in [14, Lemma 2.10] that, in fact, $\operatorname{reg}(I)$ is always equal to either $\operatorname{reg}(I : h) + 1$ or $\operatorname{reg}(I, h)$.

In practice, induction usually starts by deleting a vertex or a face (edge). That is, h is often taken to be the variable corresponding to a vertex or the product of variables corresponding to a face (edge) of the simplicial complex (hypergraph).

Remark 3.3. Let $I = I_\Delta$ be the Stanley–Reisner ideal of a simplicial complex Δ, and let $h = x^\sigma$ for a face σ of dimension $d - 1$ in Δ. Then $I : h = I_{\operatorname{link}_\Delta(\sigma)}$ and $I + h = I_{\operatorname{del}_\Delta(\sigma)} + h$.

For a subset V of the vertices in a hypergraph H, let $H : V$ and $H + V$ denote the hypergraphs corresponding to the squarefree monomial ideals $I(H) : x^V$ and $I(H) + x^V$, respectively. As a consequence of (2), we have the following inductive bounds.

Theorem 3.4.

1. *Let Δ be a simplicial complex and let σ be a face of dimension $d - 1$ in Δ. Then*

$$\mathit{reg}(\Delta) \le \max\{\mathit{reg}(\mathit{link}_\Delta(\sigma)) + d, \mathit{reg}(\mathit{del}_\Delta(\sigma))\}.$$

2. *Let H be a simple hypergraph and let V be a collection of d vertices in H. Then*

$$\mathit{reg}(H) \le \max\{\mathit{reg}(H : V) + d, \mathit{reg}(H + V)\}.$$

Another basic exact sequence in commutative algebra is:

$$0 \longrightarrow \frac{R}{I \cap J} \longrightarrow \frac{R}{I} \oplus \frac{R}{J} \longrightarrow \frac{R}{I + J} \longrightarrow 0.$$

Let E be an edge of a simple hypergraph H. By taking $I = I(H \backslash E)$ and $J = (x^E)$, this sequence gives

$$0 \longrightarrow \frac{R}{(x^E) \cap I(H \setminus E)} \longrightarrow \frac{R}{(x^E)} \oplus \frac{R}{I(H \setminus E)} \longrightarrow \frac{R}{I(H)} \longrightarrow 0.$$

Taking the associated long exact sequence of cohomology modules again, we get

$$\operatorname{reg}(H) \le \max\{|E|, \operatorname{reg}(H \setminus E), \operatorname{reg}\big((x^E) \cap I(H \setminus E)\big) - 1\}. \tag{3}$$

Recall that H_E is the contraction of $H \setminus N(E)$ to the vertices $X \setminus N[E]$. It is easy to see that $(x^E) \cap I(H \setminus E) = x^E(y \mid y \in N(E)) + I(H_E)$. Also, since the variables in $N[E]$ do not appear in H_E, by taking the tensor product of minimal free resolutions, we get

$$\operatorname{reg}\big((x^E) \cap I(H \setminus E)\big) = \operatorname{reg}(I(H_E)) + |E|.$$

Thus, (3) gives the following inductive bound.

Theorem 3.5. *Let H be a simple hypergraph and let E be an edge of cardinality d in H. Then*

$$reg(H) \leq \max\{d, reg(H \setminus E), reg(H_E) + d - 1\}.$$

Induction also works even when we do not necessarily split the edges of H into disjoint subsets. Kalai and Meshulam [31] obtain the following powerful result. This result was later extended to arbitrary (not necessarily squarefree) monomial ideals by Herzog [28].

Theorem 3.6. *Let $I_1, \ldots, I_s$ be squarefree monomial ideals in R. Then*

$$reg\left(R \Big/ \sum_{i=1}^{s} I_i\right) \leq \sum_{i=1}^{s} reg(R/I_i).$$

In particular, for edge ideals of a hypergraph and subhypergraphs, we have the following inductive bound.

Corollary 3.7. *Let H and $H_1, \ldots, H_s$ be simple hypergraphs over the same vertex set X such that $\mathscr{E}(H) = \bigcup_{i=1}^{s} \mathscr{E}(H_i)$. Then*

$$reg(R/I(H)) \leq \sum_{i=1}^{s} reg(R/I(H_i)).$$

4 Combinatorial Bounds for Regularity

In this section, we examine various bounds for the regularity of a squarefree monomial ideal in terms of combinatorial data from associated simplicial complex and hypergraph. These bounds can be proved using inductive results from Sect. 3. There are also works in the literature that relate the regularity of (squarefree) monomial ideals to other algebraic invariants (cf. [1, 11, 22, 30]). These works are beyond the scope of this survey.

For edge ideals of graphs, it turns out that the matching number and induced matching number provide nice upper and lower bounds for the regularity. The following result is due to Katzman [32, Lemma 2.2].

Theorem 4.1. *Let G be a simple graph and let $\nu(G)$ be the maximum size of an induced matching in G. Then*

$$reg(I(G)) \geq \nu(G) + 1.$$

This result is generalized in [26, Theorem 6.5] for properly connected hypergraph, and extended for all simple hypergraph in [41, Corollary 3.9]. In fact, the result is a direct consequence of the inductive bound in Lemma 3.1.

Theorem 4.2. *Let H be a simple hypergraph. Suppose that $\{E_1, \ldots, E_s\}$ forms an induced matching in H. Then*

$$reg(H) \geq \sum_{i=1}^{s} (|E_i| - 1) + 1.$$

Proof. Let H' be the induced subhypergraph of H on the vertex set $\bigcup_{i=1}^{s} E_i$. Since $\{E_1, \ldots, E_s\}$ forms an induced matching in H, these are the only edges in H'. Thus,

$$\operatorname{reg}(H') = \sum_{i=1}^{s} (|E_i| - 1) + 1.$$

Moreover, by Lemma 3.1, $\operatorname{reg}(H) \geq \operatorname{reg}(H')$. The result now follows.

Remark 4.3. It is clear that if H consists of disjoint edges then the bound in Theorem 4.2 becomes an equality. On the other hand, as we shall see later in Example 4.11, the regularity of a hypergraph can be arbitrarily larger than the right-hand side of the bound in Theorem 4.2.

Lower bounds and upper bounds are more interesting when they go together. A natural invariant related to the induced matching number is the (*minimax*) matching number. The following result was proved in [26, Theorem 6.7] and [50, Theorem 11].

Theorem 4.4. *Let G be a simple graph. Let $\beta(G)$ be the minimum size of a maximal matching G. Then*

$$reg(G) \leq \beta(G) + 1.$$

We can also recover the proof of Theorem 4.4 as a direct consequence of an inductive bound, Corollary 3.7. For that we first need a simple lemma.

Lemma 4.5. *Let G be a simple graph and assume that an edge $\{u, v\}$ forms a maximal matching of size 1 in G. Then*

$$reg(R/I(G)) = 1.$$

Proof. Since $\text{reg}(R/I(G)) = \text{reg}(I(G))-1$, it suffices to show that $\text{reg}(I(G)) = 2$. The assertion is trivial if G consists of exactly one edge $\{u, v\}$. We can then induct on the number of edges in G, making use of Theorem 3.4.

Proof (of Theorem 4.4). Let $\beta = \beta(G)$ and let $\{E_1, \ldots, E_\beta\}$ be a maximal matching in G. For each E_i, let G_i be the subgraph of G consisting of E_i and all edges incident to its vertices. Since $\{E_1, \ldots, E_\beta\}$ forms a maximal matching in G, we have $\bigcup_{i=1}^{\beta} \mathscr{E}(G_i) = \mathscr{E}(G)$. The result now follows from Corollary 3.7 and Lemma 4.5.

Remark 4.6. If G consists of disjoint edges, then the bound in Theorem 4.4 becomes an equality. On the other hand, by taking G to be the complement of a chordal graph (in which case, $\text{reg}(I(G)) = 2$), one can make $\beta(G) + 1$ arbitrarily larger than $\text{reg}(I(G))$.

A number of generalizations for Theorem 4.4 have been developed. Woodroofe [50] was the first to observe that giving a matching in a graph G is a simple way to "cover" the edges of G by subgraphs. One can also consider giving a matching in G as a special way to "pack" edges in G. These ideas have been extended to give better bounds for graphs and to obtain bounds for hypergraphs in general.

Let $c(G)$ denote the minimum number of subgraphs in G whose complements are chordal such that every edge in G belong to at least one of those subgraphs. Woodroofe [50, Lemma 1] extends Theorem 4.4 to give the following bound.

Theorem 4.7. *Let G be a simple graph. Then*

$$reg(G) \leq c(G) + 1.$$

Moradi and Kiani [39, Theorem 2.1] improve the bound in Theorem 4.4 for the class of vertex decomposable graphs. Define $\gamma(G)$ to be the maximum number of vertex disjoint paths of length at most 2 in G such that paths of length 1 form an induced matching.

Theorem 4.8. *Let G be a vertex decomposable graph. Then*

$$reg(G) \leq \gamma(G) + 1.$$

The author and Woodroofe [27] extend this result to any graph making use of packing of stars in a graph. A *star* consists of a vertex x as its *center* and edges xy_is incident to x (the vertices y_is are referred to as the *outer* vertices of the star). A star is *nondegenerate* if there are at least two edges incident to its center. A collection P of stars in G is called *center-separated* if for any pair of stars S_1 and S_2 in P, at least two outer vertices and the center of S_1 are not contained in S_2 (and vice-versa). For a center-separated collection of nondegenerate stars P, let ζ_P denote the number of stars in P plus the size of the remaining induced matching after deleting the vertices of P from G. Let $\zeta(G)$ be the maximum value of ζ_P over all center-separated collection of nondegenerate stars P. It is not hard to see that $\zeta(G) \leq \gamma(G)$.

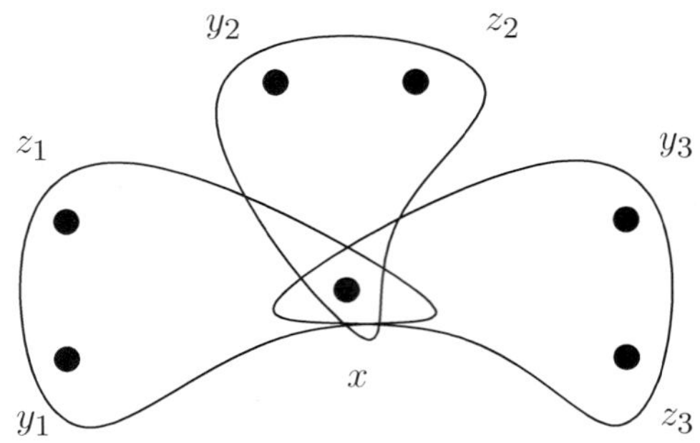

Fig. 4 A 3-uniform hypergraph with regularity greater than $(3-1)\cdot\beta$

Theorem 4.9. *Let G be a simple graph. Then*

$$reg(G) \leq \zeta(G) + 1.$$

Example 4.10. Let G be the graph in Fig. 2. It is easy to see that $\zeta(G) = 1$. For instance, the nondegenerate star centered at b gives a maximal center-separated packing of stars in G, whose removal results in a subgraph of isolated vertices $\{a, c\}$. The bound in Theorem 4.9 gives $\text{reg}(I(G)) \leq 2$. In fact, we have the equality for this example.

For hypergraphs the problem becomes more subtle. In particular, using the matching number to bound the regularity as in Theorem 4.4 is no longer possible.

Example 4.11. For $s > 1$, consider the hypergraph H_s with edges $\{x, y_i, z_i\}$ (for $i = 1, \dots, s$) on the vertex set $\{x, y_1, \dots, y_s, z_1, \dots, z_s\}$. Figure 4 illustrates H_3. We have that the induced matching number and minimax matching number of H_s are both 1. On the other hand, it is straightforward to compute that $\text{reg}(I(H_s)) = s + 2$, which can be taken to be arbitrarily far from $(3-1)\beta(H_s) = 2$.

A generalization of Theorem 4.4 is obtained in [27, Theorem 1.2] replacing the notion of a matching by a 2-collage.

Definition 4.12. Let H be a simple hypergraph. A subset C of the edges in H is called a *2-collage* if for each edge E in H, there exists a vertex $v \in E$ such that $E \setminus \{v\}$ is contained in some edge of C. For a uniform hypergraph, this condition is equivalent to requiring that for any edge E in H there is an edge F in C such that the symmetric difference of E and F has cardinality exactly 2.

Theorem 4.13. *Let H be a simple hypergraph, and let* $\{E_1, \dots, E_s\}$ *be a 2-collage in H. Then*

$$reg(H) \leq \sum_{i=1}^{s} (|E_i| - 1) + 1.$$

Instead of packing a 2-collage in a hypergraph, the concept of *edgewise domination* was introduced by Dao and Schweig [13], as a covering invariant, to generalize Theorem 4.4.

Definition 4.14. A collection C of edges in a hypergraph H is called *edgewise dominant* if every vertex $x \in H^{\text{red}}$ not contained in some edge of C or contained in an isolated loop has a neighbor contained in some edge of C. Define

$$\epsilon(H) = \min\{|C| \mid C \text{ is edgewise dominant}\}.$$

Theorem 4.15. *For any simple hypergraph H, we have*

$$reg(H) \leq |X(H)| - \epsilon(H^{\vee}).$$

To prove Theorem 4.15, the inductive method of Theorem 3.4 was put in a more general perspective by considering a hereditary family of hypergraphs that allows one to go from a given hypergraph H to the deletion and contraction of H at a vertex.

Definition 4.16. Let Φ be a collection of simple hypergraphs. We say that Φ is *hereditary* if for any hypergraph $H \in \Phi$ and any subset V of the vertices of H, H^{red}, $H + V$ and $H : V$ are all in Φ.

The proof of Theorem 4.15 is based on the following more general version of Theorem 3.4 (see [13, Lemma 3.3]).

Theorem 4.17. *Let Φ be a hereditary family of simple hypergraphs, and let $f : \Phi \to \mathbb{N}$ be a function such that $f(H^{red}) = f(H)$ for all $H \in \Phi$, $f(H) \leq |X(H)|$ when H contains no edges, and $f(H) = 0$ when H has only isolated loops. Furthermore, suppose that f satisfies the following condition: for any $H \in \Phi$ with at least one edge of cardinality ≥ 2, there exists a sequence of subsets $A_1, \ldots, A_t$ of the vertices such that, writing H_i for the hypergraph corresponding to $H + \sum_{j=1}^{i} A_j$, the following two properties are satisfied.*

- $|is(H_t)| > 0$ *and* $f(H_t) + |is(H_t)| \geq f(H)$, *and*
- *for each* i, $f(H_{i-1} : A_i) + |is(H_{i-1})| + |A_i| \geq f(C)$.

Then for any $H \in \Phi$, we have

$$reg(H^{\vee}) = pd(H) \leq |X(H^{red})| - f(H).$$

This technique, restricted to graphs [14], gives the following numerical bound for the regularity (see [14, Theorem 4.1]).

Theorem 4.18. *Let G be a graph. Assume that the resolution of $I(G)$ is linear up to the k-th step for some $k \geq 1$, and let d be the maximum degree of a vertex in G. Then*

$$reg(G) \leq \log_{\frac{k+4}{2}}\left(\frac{d}{k+1}\right) + 3.$$

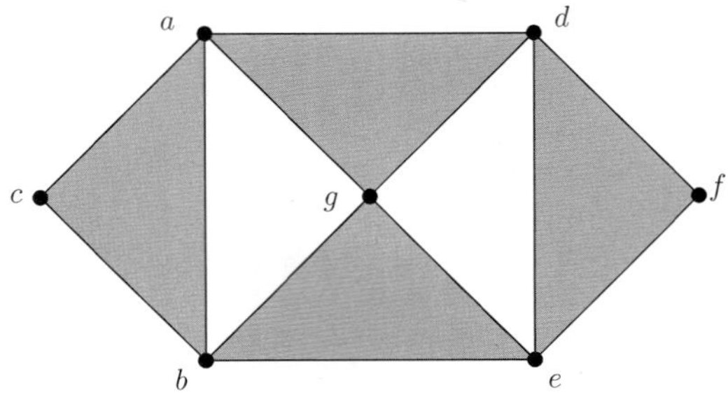

Fig. 5 Comparing bounds in Theorems 4.13, 4.15 and 4.20

Remark 4.19. It is not clear how strong the bound in Theorem 4.18 is in comparison with other known bounds using packing and covering invariants.

Lin and McCullough, in [36], introduced the notion of *labeled hypergraphs* that, in complement to the well-studied constructions of Stanley–Reisner ideals and edge ideals, also gives a one-to-one correspondence to squarefree monomial ideals, and used this notion to study the regularity of squarefree monomial ideals. Note that the *unlabeled* version of their construction coincides with the *dual hypergraph* notion in hypergraph theory (cf. [3]). In the same spirit of employing induction, Lin and McCullough [36, Theorem 4.9] obtain the following bound. We shall rephrase their result in terms of edge ideals of hypergraphs.

An edge E of a simple hypergraph H is said to contain a *free vertex* if there exists $x \in E$ such that x does not belong to any other edge in H.

Theorem 4.20. *Let $H = (X, \mathscr{E})$ be a simple hypergraph, and let $H' = (X, \mathscr{E}')$ be the hypergraph obtained by removing all edges with free vertices from H. Let $\beta(H')$ be the matching number of H'. Then*

$$reg(H) \leq |X| - |\mathscr{E}| + |\mathscr{E}'| - \beta(H') + 1.$$

Remark 4.21. It is not quite clear which bounds among Theorems 4.13, 4.15, and 4.20 are best in general. The following examples were given in [36] to illustrate this.

Example 4.22. Let $I = (abc, def, adg, beg)$ be the edge ideal of the hypergraph in Fig. 5. Then the bounds for $\text{reg}(I)$ in Theorems 4.13 and 4.15, respectively, are 9 and 6. On the other hand, the bound for $\text{reg}(I)$ in Theorem 4.20 is 5.

Example 4.23. Let $I = (abc, abd, acd, bcd)$ be the edge ideal of the hypergraph in Fig. 6. Then the bounds for $\text{reg}(I)$ in Theorems 4.13 and 4.15 are both 3, while the bound using labeled hypergraph in Theorem 4.20 is 4.

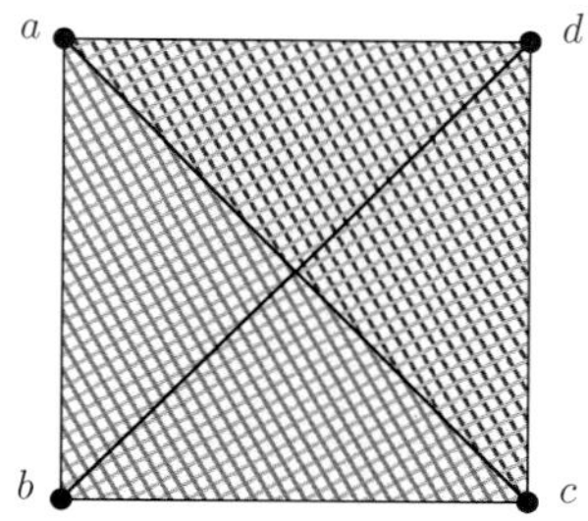

Fig. 6 Comparing bounds in Theorems 4.13, 4.15 and 4.20

5 Small Regularity and Computing Regularity

In this section we survey studies that explicitly compute the regularity for special classes of squarefree monomial ideals and identify combinatorial structures that force the ideals to have small regularity. Ideals of regularity 0 are trivial. A squarefree monomial ideal has regularity 1 if and only if it is generated by a collection of variables. Thus, we shall only consider ideals with regularity at least 2.

The following result was originally stated and proved by Wegner [49] using topological language, and re-stated in terms of monomial ideals by Fröberg [21] (see also [18, 25]). It is in fact a direct consequence of Lemma 2.21.

Theorem 5.1. *Let G be a simple graph. Then reg$(I(G)) = 2$ if and only if G^c is a chordal graph.*

Proof. We briefly sketch out the proof. Let Δ be the independence complex of G. By Lemma 2.21, $\text{reg}(R/I(G)) = 1$ iff no induced subcomplex of Δ has any balls of positive dimension. This is the case iff the 1-skeleton of Δ has no induced cycle of length ≥ 4. Note that the 1-skeleton of Δ is exactly the complement of G. In fact, an induced cycle of length ≥ 4 in G^c gives a homology cycle in Δ.

It is then natural to investigate squarefree monomial ideals with regularity 3. To the best of our knowledge, it is still an open question to characterize these ideals; though partial results have been obtained.

Let P be a given collection of graphs. A graph G is said to be *P-free* if it contains no induced subgraphs that are the same as elements in P. In particular, a graph G is called *claw-free* if it contains no 4 vertices on which the induced subgraph is a star, and a graph G is *C_4-free* if it contains no induced 4-cycles. The following result was proved in [42, Theorem 1.2] (see also [14, Theorem 3.4]).

Theorem 5.2. *Let G be a claw-free simple graph such that G^c contains no C_4. Then*

$$\textit{reg}(I(G)) \leq 3.$$

Example 5.3. A 5-cycle is a claw-free graph whose complement contains no C_4, and $\text{reg}(I(C_5)) = 3$.

For bipartite graphs, the edge ideals having regularity 3 can be characterized. Recall that a graph G is *bipartite* if the vertices $X(G)$ can be partitioned into disjoint subsets $X(G) = Y \cup Z$ such that edges of G connect a vertex in Y and a vertex in Z. The *bipartite complement* of a bipartite graph G, denoted by G^{bc}, is the bipartite graph over the same partition of the vertices, and for $y \in Y, z \in Z$, $\{y, z\} \in G^{\text{bc}}$ if and only if $\{y, z\} \notin G$. The following classification was given in [20, Theorem 3.1].

Theorem 5.4. *Let G be a connected bipartite graph. Then* $reg(G) = 3$ *if and only if* G^c *has induced cycles (of length* ≥ 4*) and* G^{bc} *has no induced cycles of length* ≥ 6.

Let us now turn to classes of ideals for which the regularity can be computed explicitly. In most of the studies that we survey, this is the case when inductive inequalities in Sect. 3 become equalities, and the method is to find combinatorial invariants that behaves well with respect to the induction processes.

The following result was first proved for forests in [51, Theorem 2.18] and [25, Corollary 3.11] and extended to chordal graphs in [26, Theorem 6.8].

Theorem 5.5. *Let G be a chordal graph and let* $\nu(G)$ *be the maximum size of an induced matching in G. Then*

$$reg(G) = \nu(G) + 1.$$

To prove Theorem 5.5, we shall need a lemma (see [26, Lemma 5.7 and Theorem 6.2]) which states that for chordal graphs, one can always find an edge E such that the inductive inequality of Theorem 3.5 becomes an equality, and that this process can be done recursively.

Lemma 5.6. *Let G be a chordal graph. Then there exists an edge E in G such that*

1. $reg(G) = \max\{reg(G_E) + 1, reg(G \setminus E)\}$.
2. *$G \setminus E$ and G_E are both chordal graphs.*

Proof (of Theorem 5.5). We use induction on the number of edges in G. If G consists of only isolated vertices or a single edge, then the assertion is trivial. For the induction step, in light of Lemma 5.6 it suffices to show that for an edge E in G satisfying the conclusion of Lemma 5.6, we have

$$\nu(G) = \max\{\nu(G_E) + 1, \nu(G \setminus E)\}.$$

This is indeed true. It is easy to see that an induced matching in $G \setminus E$ is also an induced matching in G, and an induced matching in G_E together with E would form an induced matching in G. Thus, $\nu(G) \geq \max\{\nu(G_E) + 1, \nu(G \setminus E)\}$. On the other hand, suppose that $C = \{E_1, \ldots, E_c\}$ is an induced matching in G of maximum size. If $E \notin C$, then C is an induced matching in $G \setminus E$. If $E \in C$, then $C \setminus \{E\}$ forms an induced matching in G_E. Therefore, $\nu(G) \leq \max\{\nu(G_E) + 1, \nu(G \setminus E)\}$.

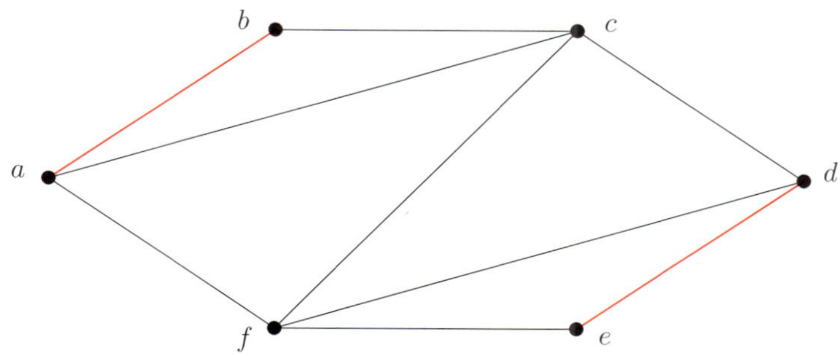

Fig. 7 A chordal graph

Example 5.7. Let G be the chordal graph in Fig. 7 and let $I = I(G)$. Then

$$0 \to R(-5)^2 \oplus R(-6) \to R(-4)^{10} \oplus R(-5)^2 \to R(-3)^{16} \oplus R(-4) \to R(-2)^9 \to I \to 0$$

is the minimal free resolution of I giving $\text{reg}(I) = 3$. On the other hand, it is easy to see that $\nu(G) = 2$ and $\{ab, cd\}$ is a maximal induced matching.

Recall that a collection of vertices V in a graph G is called a *vertex cover* if for any edge E in G, $V \cap E \neq \emptyset$. A *minimal* vertex cover is with respect to inclusion. Note that the smallest size of a vertex cover in G is equal to the height of $I(G)$. A graph G is called *unmixed* if all minimal vertex cover of G has the same cardinality. For an unmixed graph G, it is known (cf. [23]) that $2\text{ht}I(G) \geq |X(G)|$. A graph G is called *very well-covered* (see [37]) if G is unmixed, has no isolated vertices, and $2\text{ht}I(G) = |X(G)|$.

Theorem 5.5 has been extended to a number of classes of graphs. In particular, for the following classes of graph, the regularity can be computed by its induced matching number.

Theorem 5.8. *Let G be a simple graph. Let $\nu(G)$ be the maximum size of an induced matching in G. Then*

$$reg(G) = \nu(G) + 1$$

in the following cases:

1. *G is a sequentially Cohen–Macaulay bipartite graph (see [47]);*
2. *G is an unmixed bipartite graph (see [35]);*
3. *G is a very well-covered graph (see [37]);*
4. *G is a vertex decomposable graph without cycles of length 5 (see [33], the case where G is C_5-free and C_4-free was proved in [5]).*

Remark 5.9. Chordal and sequentially Cohen–Macaulay bipartite graphs are vertex decomposable. Also, bipartite graphs are C_5-free. Therefore, in Theorem 5.8, 1. follows from 4.

Example 5.10. Let G be the induced 5-cycle. Then G is not C_5-free nor very well-covered. Its independence complex is also an induced 5-cycle, which is vertex decomposable. Thus, the class of vertex decomposable graphs is not fully covered by Theorem 5.8.

Moreover, the minimal free resolution of $I = I(G)$ is given by

$$0 \to R(-5) \to R(-3)^5 \to R(-2)^5 \to I \to 0.$$

Therefore, $\text{reg}(I) = 3$. On the other hand, the induced matching number of G is $\nu(G) = 1$. Observe that $\text{reg}(I(G)) \neq \nu(G) + 1$ in this example.

The problem is again more subtle when moving to hypergraphs. To do so, we need to identify classes of hypergraphs where inductive inequalities in Sect. 3 become equalities, and to find combinatorial invariants of hypergraphs that respect induction processes.

It is proved by the author and Woodroofe in [27, Theorem 1.5] that the inductive inequalities of Theorem 3.4 are equalities for all vertex decomposable simplicial complexes and hypergraphs.

Theorem 5.11. *Let H be a simple hypergraph with edge ideal $I = I(H)$. Suppose that H is vertex decomposable and x is the initial vertex in its shedding order. Then*

$$reg(I) = \max\{reg(I : x) + 1, reg(I, x)\}.$$

The following example arises from communication with Chris Francisco. The author would like to thank Chris Francisco for his help.

Example 5.12. Let $I = (acd, bcd, abe, bce, bcf, cdf) \subseteq R = K[a, \ldots, f]$. Then $\text{reg}(I) = 3$, $\text{reg}(I, a) = 3 = \text{reg}(I)$ and $\text{reg}(I : a) + 1 = 4 > \text{reg}(I)$. Thus, in this example, $\text{reg}(I)$ is equal to the smaller value between $\text{reg}(I, a)$ and $\text{reg}(I : a) + 1$.

Example 5.10 shows that even for vertex decomposable graphs, the induced matching number is no longer the right invariant to compute the regularity (i.e., the regularity may be strictly bigger than the induced matching number). On the other hand, for a smaller class of vertex decomposable hypergraphs, those that are triangulated, it was proved in [26] that this invariant still works.

Triangulated hypergraphs were introduced in [26] as a generalization of chordal graphs. Recall that the *distance* between two edges E and F in a d-uniform hypergraph H is the minimum length of a sequence of edges $E_0 = E, E_1, \ldots, E_s = F$ such that $|E_{i-1} \cap E_i| = d - 1$. A d-uniform hypergraph H is called *properly connected* if for any intersecting edges E and F, the distance between E and F is exactly $d - |E \cap F|$. Note that all simple graph are properly connected. Recall also that for a vertex x, the set of *neighbors* of x is given by

$$N(x) = \{y \in X \mid \exists E \in \mathscr{E} \text{ s.t. } \{x, y\} \subseteq E\}.$$

Definition 5.13. A d-uniform properly connected hypergraph $H = (X, \mathscr{E})$ is said to be *triangulated* if for every nonempty subset $Y \subseteq X$, the induced subgraph $H[Y]$ contains a vertex $x \in Y \subseteq X$ such that the induced hypergraph of $H[Y]$ on $N(x)$ is a complete hypergraph consisting of all subsets of d elements of $N(x)$.

Note that triangulated hypergraphs are vertex decomposable. The following theorem was proved in [26, Theorem 6.8].

Theorem 5.14. *Let H be a d-uniform properly connected hypergraph, and assume that H is triangulated. Let $\nu(H)$ denote the induced matching number of H. Then*

$$reg(H) = (d-1)\nu(H) + 1.$$

Using their notion of labeled hypergraphs, Lin and McCullough [36, Proposition 4.1 and Theorem 4.12] also compute the regularity explicitly for special classes of hypergraphs. We shall again rephrase their results in terms of edge ideals of hypergraphs. A hypergraph H is called *saturated* if every edge in H contains free vertices.

Theorem 5.15. *Let $H = (X, \mathscr{E})$ be a simple hypergraph with edge ideal $I = I(H)$. The following are equivalent:*

1. *H is saturated; and*
2. *The Taylor resolution of I is minimal.*

In this case, we have

$$reg(I) = |X| - |\mathscr{E}| + 1.$$

For a vertex x in a hypergraph $H = (X, \mathscr{E})$, let $W_x = \{E \in \mathscr{E} \mid x \in E\}$. Denote by S_E the set $\{W_x \mid x \in E\}$, and let $\mathscr{P} = \bigcup_{E \in \mathscr{E}} S_E$. An element W_x in $\mathscr{P}$ is called *simple* if it does not contain any other element of $\mathscr{P}$ as a proper subset. The hypergraph H is said to have *isolated simple edges* if for any edge E in H not containing free vertices, the set S_E has exactly one simple maximal element.

Theorem 5.16. *Let H be a simple hypergraph with edge ideal $I = I(H)$. Suppose that H has isolated simple edges. Then*

$$reg(I) = |X| - |V| + \sum_{W \in \mathscr{P},\ W \text{ simple}} (|W| - 1) + 1.$$

Example 5.17. Let $I = (ab, bcdef, ac, eg, fg, gh, hi) \subseteq R = K[a, b, \ldots, i]$ be the edge ideal of H. Then the set $\mathscr{P}$ of H is depicted in Fig. 8, where filled points represent edges in H with free vertices and unfilled points represent edges in H without free vertices. Note that, as a hypergraph, this figure represents the dual hypergraph of H, and also the labeled hypergraph corresponding to I in the sense

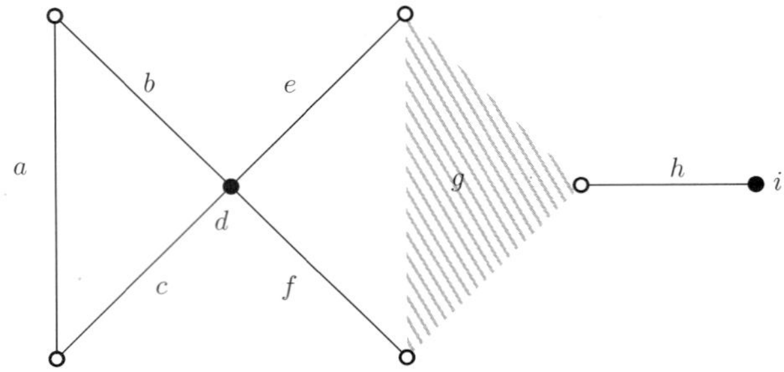

Fig. 8 A labeled hypergraph having isolated simple edges

of [36]. It can be seen that H has isolated simple edges, and those simple sets W_xs in $\mathscr{P}$ are labeled by a and g. Theorem 5.16 gives

$$\text{reg}(I) = 9 - 7 + (2 - 1) + (3 - 1) + 1 = 6.$$

Remark 5.18. It can be seen that the ideal I in Example 5.17 is not sequentially Cohen–Macaulay. Thus, its associated simplicial complex $\Delta(I)$ is not vertex decomposable. This example shows that inductive inequalities in Theorems 3.4 and 5.11 may become equalities for a larger class of squarefree monomial ideals than those corresponding to vertex decomposable simplicial complexes.

6 Open Problems and Questions

In this last section of the paper, we state a number of open problems and conjectures that we would like to see answered. We first observe that the combinatorial bounds in Theorems 4.13, 4.15, and 4.20 in practice are not easy to compute. It is desirable to find bounds that relate to more familiar combinatorial invariants of hypergraphs and simplicial complexes.

Problem 6.1. Find bounds for the regularity of a squarefree monomial ideal in terms of familiar combinatorial invariants and structures of associated hypergraph and simplicial complex.

Let $\chi(H)$ denote the *chromatic number* of a hypergraph H, i.e., the least number of colors needed to color the vertices of H such that no edge (that is not an isolated loop) in H is mono-colored. It was observed in [50, Theorem 2] that for any independent set T in G,

$$\text{reg}(R/I(G)) \leq \chi((G \setminus T)^c). \tag{4}$$

Problem 6.2. Extend (4) to hypergraphs. Investigate the relation between the regularity of a hypergraph and its coloring properties.

As pointed out in Example 5.3, the class of graphs with regularity ≤ 3 is strictly larger than that of claw-free graphs whose complements contain no induced C_4. We would like to see this class of graphs classified.

Problem 6.3. Give a combinatorial characterization for squarefree monomial ideals I such that $\operatorname{reg}(I) = 3$. In particular, classify simple graphs G such that $\operatorname{reg}(I(G)) = 3$.

It was shown in Theorem 5.11 that for vertex decomposable simplicial complexes (and hypergraphs) the inductive inequalities in Sect. 3 become equalities. It is natural to seek for combinatorial invariants that measure the regularity of these complexes (and hypergraphs).

Problem 6.4. Let H be a vertex decomposable hypergraph. Compute $\operatorname{reg}(H)$.

Examples 5.12 and 5.17 show that inductive inequalities in Theorem 3.4 may become equalities for a larger class of simplicial complexes (and hypergraphs) than vertex decomposable ones. In particular, we would like to classify squarefree monomial ideals for which the conclusion of Theorem 5.11 holds.

Problem 6.5. Characterize squarefree monomial ideals I for which there exists a variable x such that

$$\operatorname{reg}(I) = \max\{\operatorname{reg}(I : x) + 1, \operatorname{reg}(I, x)\}.$$

Recall that the inductive bound of Theorem 3.5 comes from the following short exact sequence

$$0 \longrightarrow \frac{R}{(x^E) \cap I(H \setminus E)} \longrightarrow \frac{R}{(x^E)} \oplus \frac{R}{I(H \setminus E)} \longrightarrow \frac{R}{I(H)} \longrightarrow 0. \qquad (5)$$

The bound in Theorem 3.5 would become an equality if the mapping cone construction applied to (5) results in the *minimal* free resolution of $R/I(H)$. We would like to be able to identify all such edges E for which this is true.

Problem 6.6. Characterize all edges E in a simple hypergraph H such that the mapping cone construction applied to the short exact sequence (5) gives the minimal free resolution of $R/I(H)$.

Hà and Van Tuyl [26, Theorem 3.2] partially answered this problem. Problem 6.6, in fact, comes from a larger picture, when a squarefree monomial ideal I is splitted into the sum of two sub-ideals. We would also like to see in general which combinatorial structures would enforce the minimality of the mapping cone construction applied to a similar short exact sequence.

Problem 6.7. Let H and H' be simple hypergraphs on the same vertex set X with disjoint edge sets $\mathscr{E}(H)$ and $\mathscr{E}(H')$. Let $I = I(H)$ and $J = I(H')$. Find combinatorial conditions for the mapping cone construction applied to the short exact sequence

$$0 \longrightarrow \frac{R}{I \cap J} \longrightarrow \frac{R}{I} \oplus \frac{R}{J} \longrightarrow \frac{R}{I+J} \longrightarrow 0$$

to give the minimal free resolution of $R/I + J$.

Along the same line, it is also desirable to see when the inductive inequalities in Theorem 3.6 and Corollary 3.7 become equalities.

Problem 6.8. Let $H, H_1, \ldots, H_s$ be simple hypergraphs over the same vertex set X, and assume that $\mathscr{E}(H) = \bigcup_{i=1}^{s} \mathscr{E}(H_i)$. Find combinatorial conditions for the following equality to hold:

$$\operatorname{reg}(R/I(H)) = \sum_{i=1}^{s} \operatorname{reg}(R/I(H_i)).$$

An important problem in commutative algebra during the last few decades is to study the regularity of powers of ideals. It is known (cf. [2, 12, 34, 46]) that for a homogeneous ideal $I \subseteq R$, $\operatorname{reg}(I^q)$ is asymptotically a linear function, i.e., there exist constants a, b, and q_0 such that for $q \geq q_0$, $\operatorname{reg}(I^q) = aq + b$. While the coefficient a is well-understood, the constants b and q_0 are quite mysterious. Many recent studies are devoted to investigating these constants (cf. [4, 10, 16, 17, 24]). Even for edge ideals of graphs the exact values for these constants are still out of reach.

Problem 6.9. Let I be a squarefree monomial ideal. Find combinatorial interpretations of the linear function $\operatorname{reg}(I^q)$ for $q \gg 0$.

For a very simple family of edge ideals, we would like to see constants b and q_0 evaluated explicitly.

Problem 6.10. Let $I = I(C_k)$ be the edge ideal of a k-cycle. Find the linear form of $\operatorname{reg}(I^q)$ for $q \gg 0$.

Computational experiments lead us to conjecture that for this class of edge ideals, the regularity of I^q becomes a linear function as soon as $q \geq 2$.

Conjecture 6.11. Let $I = I(C_k)$ be the edge ideal of a k-cycle. Then $\operatorname{reg}(I^q)$ is a linear function for $q \geq 2$.

Acknowledgments The author would like to thank an anonymous referee for a careful reading and many helpful comments. The author would also like to thank S.A. Seyed Fakhari for pointing out a mistake in our original definition of H_E in Theorem 3.5.

References

1. Ahmad, S., Anwar, I.: An upper bound for the regularity of ideals of Borel type. Commun. Algebra **36**(2), 670–673 (2008)
2. Bagheri, A., Chardin, M., Hà, H.T.: The eventual shape of Betti tables of powers of ideals. Math. Res. Lett. (to appear). arXiv:1110.0383
3. Berge, C.: Hypergraphs: Combinatorics of Finite Sets. North-Holland, New York (1989)
4. Berlekamp, D.: Regularity defect stabilization of powers of an ideal. Math. Res. Lett. **19**(1), 109–119 (2012)
5. Biyikoglu, T., Civan, Y.: Vertex decomposable graphs, codismantlability, Cohen-Macaulayness and Castelnuovo-Mumford regularity. Preprint. arXiv:1205.5631
6. Biyikoglu, T., Civan, Y.: Bounding Castelnuovo-Mumford regularity of graphs via Lozin's transformation. Preprint. arXiv:1302.3064
7. Bouchat, R.R., Hà, H.T., O'Keefe, A.: Path ideals of rooted trees and their graded Betti numbers. J. Comb. Theory Ser. A **118**(8), 2411–2425 (2011)
8. Bruns, W., Herzog, J.: Cohen-Macaulay Rings. Cambridge Studies in Advanced Mathematics, vol. 39. Cambridge University Press, Cambridge (1993)
9. Chardin, M.: Some results and questions on Castelnuovo-Mumford regularity. In: Syzygiez and Hilbert Functions. Lecture Notes in Pure and Applied Mathematics, vol. 254, pp. 1–40. Chapman & Hall/CRC, London/Boca Raton (2007)
10. Chardin, M.: Powers of ideals and the cohomology of stalks and fibers of morphisms. Preprint. arXiv:1009.1271
11. Cimpoeaş, M.: Regularity for certain classes of monomial ideals. An. Ştiint. Univ. "Ovidius" Constanţa Ser. Mat. **15**(1), 33–46 (2007)
12. Cutkosky, S.D., Herzog, J., Trung, N.V.: Asymptotic behaviour of the Castelnuovo-Mumford regularity. Composito Math. **118**, 243–261 (1999)
13. Dao, H., Schweig, J.: Bounding the projective dimension of a squarefree monomial ideal via domination in clutters. Preprint. arXiv:1301.2665
14. Dao, H., Huneke, C., Schweig, J.: Bounds on the regularity and projective dimension of ideals associated to graphs. Preprint. arXiv:1110.2570
15. Eisenbud, D., Goto, S.: Linear free resolutions and minimal multiplicity. J. Algebra **88**(1), 89–133 (1984)
16. Eisenbud, D., Harris, J.: Powers of ideals and fibers of morphisms. Math. Res. Lett. **17**(2), 267–273 (2010)
17. Eisenbud, D., Ulrich, B.: Notes on regularity stabilization. Proc. Am. Math. Soc. **140**(4), 1221–1232 (2012)
18. Eisenbud, D., Green, M., Hulek, K., Popescu, S.: Restricting linear syzygies: algebra and geometry. Compositio Math. **141**, 1460–1478 (2005)
19. Faridi, S.: The facet ideal of a simplicial complex. Manuscripta Math. **109**(2), 159–174 (2002)
20. Fernández-Ramos, O., Gimenez, P.: Regularity 3 in edge ideals associated to bipartite graphs. Preprint. arXiv:1207.5553
21. Fröberg, R.: On Stanley-Reisner rings. In: Topics in Algebra, vol. 26, Part 2, pp. 57–70 (Warsaw, 1988). Banach Center Publications/PWN, Warsaw (1990)
22. Frühbis-Krüger, A., Terai, N.: Bounds for the regularity of monomial ideals. Pragmatic 1997 (Catania). Matematiche (Catania) **53**(Suppl.), 83–97 (1998/1999)
23. Gitler, I., Valencia, C.E.: Bounds for invariants of edge-rings. Commun. Algebra **33**, 1603–1616 (2005)
24. Hà, H.T.: Asymptotic linearity of regularity and a^*-invariant of powers of ideals. Math. Res. Lett. **18**(1), 1–9 (2011)
25. Hà, H.T., Van Tuyl, A.: Splittable ideals and the resolutions of monomial ideals. J. Algebra **309**(1), 405–425 (2007)
26. Hà, H.T., Van Tuyl, A.: Monomial ideals, edge ideals of hypergraphs, and their minimal graded free resolutions. J. Algebr. Comb. **27**(2), 215–245 (2008)

27. Hà, H.T., Woodroofe, R.: Results on the regularity of squarefree monomial ideals. Preprint. arXiv:1301.6779
28. Herzog, J.: A generalization of the Taylor complex construction. Commun. Algebra **35**(5), 1747–1756 (2007)
29. Herzog, J., Hibi, T.: Monomial Ideals. Graduate Texts in Mathematics, vol. 260. Springer, Berlin (2011)
30. Hoa, L.T., Trung, N.V.: On the Castelnuovo-Mumford regularity and the arithmetic degree of monomial ideals. Math. Z. **229**(3), 519–537 (1998)
31. Kalai, G., Meshulam, R.: Intersections of Leray complexes and regularity of monomial ideals. J. Comb. Theory Ser. A **113**(7), 1586–1592 (2006)
32. Katzman, M.: Characteristic-independence of Betti numbers of graph ideals. J. Comb. Theory Ser. A **113**(3), 435–454 (2006)
33. Khosh-Ahang, F., Moradi, S.: Regularity and projective dimension of edge ideal of C_5-free vertex decomposable graphs. Preprint. arXiv:1305:6122
34. Kodiyalam, V.: Asymptotic behaviour of Castelnuovo-Mumford regularity. Proc. Am. Math. Soc. **128**(2), 407–411 (1999)
35. Kummini, M.: Regularity, depth and arithmetic rank of bipartite edge ideals. J. Algebr. Comb. **30**(4), 429–445 (2009)
36. Lin, K.-N., McCullough, J.: Hypergraphs and the regularity of square-free monomial ideals. Preprint. arXiv:1211.4301
37. Mahmoudi, M., Mousivand, A., Crupi, M., Rinaldo, G., Terai, N., Yassemi, S.: Vertex decomposability and regularity of very well-covered graphs. J. Pure Appl. Algebra **215**(10), 2473–2480 (2011)
38. Miller, E., Sturmfels, B.: Combinatorial Commutative Algebra. Graduate Texts in Mathematics, vol. 227. Springer, Berlin (2004)
39. Moradi, S., Kiani, D.: Bounds for the regularity of edge ideals of vertex decomposable and shellable graphs. Bull. Iran. Math. Soc **36**(2), 267–277 (2010)
40. Morales, M., Pour, A.A.Y., Zaare-Nahandi, R.: Regularity of edge ideal of a graph. J. Pure Appl. Algebra **216**(12), 2714–2719 (2012)
41. Morey, S., Villarreal, R.H.: Edge Ideals: Algebraic and Combinatorial Properties. Progress in Commutative Algebra, vol. 1, pp. 85–126. de Gruyter, Berlin (2012)
42. Nevo, E.: Regularity of edge ideals of C_4-free graphs via the topology of the lcm-lattice. J. Comb. Theory Ser. A **118**(2), 491–501 (2011)
43. Peeva, I.: Graded Syzygies. Algebra and Applications, vol. 14. Springer, London (2011)
44. Stanley, R.: Combinatorics and Commutative Algebra. Progress in Mathematics, vol. 41. Birkhäuser, Boston (1983)
45. Terai, N.: Alexander duality theorem and Stanley-Reisner rings. Free resolution of coordinate rings of projective varieties and related topics (Japanese) (Kyoto, 1998). Surikaisekikenkyusho Kokyuroku No. 1078, pp. 174–184 (1999)
46. Trung, N.V., Wang, H.: On the asymptotic behavior of Castelnuovo-Mumford regularity. J. Pure Appl. Algebra **201**(1–3), 42–48 (2005)
47. Van Tuyl, A.: Sequentially Cohen-Macaulay bipartite graphs: vertex decomposability and regularity. Arch. Math. (Basel) **93**(5), 451–459 (2009)
48. Villarreal, R.H.: Monomial Algebras. Monographs and Textbooks in Pure and Applied Mathematics, vol. 238. Dekker, New York (2001)
49. Wegner, G.: d-collapsing and nerves of families of convex sets. Arch. Math. (Basel) **26**, 317–321 (1975)
50. Woodroofe, R.: Matchings, coverings, and Castelnuovo-Mumford regularity. Preprint. arXiv:1009.2756
51. Zheng, X.: Resolutions of facet ideals. Commun. Algebra **32**, 2301–2324 (2004)

Planar Graphs and the Koszul Algebra Structure for Trivariate Monomial Ideals

Jared L. Painter

This manuscript is dedicated to Tony Geramita for all his works and contributions to the mathematical community

Abstract We describe the Koszul algebra for rings given by quotients of trivariate monomial ideals using planar graphs representing the minimal free resolution. Much is known about the Koszul algebra structure for such ideals, but a complete classification is still unknown. It is known that we can use the minimal free resolution to determine the Koszul algebra structure for trivariate monomial ideals. We will see what new insight we can gain about this classification using planar graphs and provide additional information to previous results. By inspecting the planar graphs we also find some surprising results about which lower Bass numbers we can obtain for trivariate monomial ideals.

1 Introduction

Free resolutions for quotients of monomial ideals have been studied heavily over the past 20 years. Over polynomial rings in three variables it is known that minimal free resolutions for quotients of monomial ideals can be represented by planar graphs [9,10]. Additionally, we know that if I is generic, then the planar graph representing the minimal free resolution is a planar triangulation [4,9]. More detailed information on what the matrices of the minimal free resolution look like is given in [11]. When R is an Artinian local ring the minimal free resolution can also provide information about the lower Bass numbers of R and the Koszul algebra structure of R. For codepth 3 local rings the classification of the Koszul algebra was originally

J.L. Painter (✉)
Houston Baptist University, 7502 Fondren Rd., Houston, TX 77074, USA
e-mail: jpainter@hbu.edu

S.M. Cooper and S. Sather-Wagstaff (eds.), *Connections Between Algebra, Combinatorics, and Geometry*, Springer Proceedings in Mathematics & Statistics 76,
DOI 10.1007/978-1-4939-0626-0_8,

done in characteristic zero by Weyman [13], and in general by Avramov et al. [3]. Recently new restrictions on the parameters of the Koszul algebra have been given by Avramov [2, 3.1]. In [12] we use the Koszul algebra classification provided by Avramov [2] and information on the structure of the minimal free resolution of R provided in [11] to compute the Koszul algebra for trivariate monomial ideals.

Throughout, we will assume that $S = \Bbbk[x, y, z]$ is a trivariate polynomial ring over a field $\Bbbk$, with homogeneous maximal ideal $\mathfrak{m} = \langle x, y, z \rangle$, and $I \subseteq \mathfrak{m}^2$ is an $\mathfrak{m}$-primary monomial ideal of S. We also set $R = S/I$. In this case it is known that if I is minimally generated by n monomials, then the minimal free resolution of R has the form

$$\mathbb{F} := 0 \longrightarrow S^m \xrightarrow{f_3} S^{m+n-1} \xrightarrow{f_2} S^n \xrightarrow{f_1} S \longrightarrow R \longrightarrow 0. \tag{1}$$

We will think of the maps f_i as matrices under the standard basis.

Our primary goal in this manuscript is to see what new insight we can gain about the Koszul algebra, A of R and the lower Bass numbers of R, by inspecting the planar graph representing the minimal free resolution of R. We use classification put forth in [2] and the methods outlined in [12] to describe A. In [12] we learn how to compute some of the parameters of A from the minimal free resolution. These methods directly relate to when we obtain nonzero entries from I in the matrices of the minimal free resolution of R. In [12, 3.6] we learn that the parameter p of A is precisely the number of columns in f_2 which are contained in I. Additionally, in [12, 3.8] it is shown that r is the number of rows in f_3 which are dependent mod I. The number r is also directly related to the first Bass number of R; specifically, the first Bass number of R, $\mu_R^1 = m+n-1-r$, see [11, 2.9]. More information on what it means for a row in f_3 to be dependent mod I is given in the discussion following Definition 3.1.

In Sect. 2 we will provide background on minimal free resolutions for monomial ideals and discuss how we can represent minimal free resolutions with planar graphs. We will use planar graphs in Proposition 2.4 to describe when we can expect nonzero entries from the ideal to appear in the minimal free resolution of R. In Sect. 3 we will provide background on the Koszul algebra structure. Corollary 3.2 will describe how to compute p from the planar graph in general and how to compute r in special cases. For the reader's benefit we will also give a restatement of [2, 2.1 and 3.1] in Theorem 3.3.

We begin Sect. 4 by inspecting the planar graphs for rings given in the Koszul Algebra classification of generic monomial ideals provided in [12, 5.2]. In Theorem 4.2 we will impose conditions on the minimal generators of I to produce quotient rings which are Golod. In Example 4.3 we will look at the planar graph for a family of rings given in [12, 5.3] with Koszul class **B**. Examples of rings in Koszul class **B** were originally found by Brown in [5] and all have type 2. Additional examples of non-Artinian rings with type 1 and 3 have been found by Christensen and Veliche in [7].

We would also like to use planar graphs to explore which values p and q can take on in the Koszul class $\mathbf{H}(p, q)$. We know in general that $0 \leq p \leq n-1$ for an n-generated monomial ideal by Painter [12, 3.7]. In Theorem 5.2 we will show that for generic ideals p can be any integer such that $0 \leq p \leq n-1$, and q can be any integer such that $0 \leq q \leq n-2$, except $q = n-3$. Since $q \neq n-3$ this also implies that $\mu_R^1 \neq m+2$, when I is generic. Upon inspection of the planar graphs leading up Theorem 5.2 we will see that $\mu_R^1 = m+k$, where $1 \leq k < n$, except $k = 2$, when I is generic. This leads us to ask the question; can $\mu_R^1 = m+2$ for non-generic monomial ideals? Currently we know of only one example where $\mu_R^1 = m+2$ when $n = 5$, which we will provide in Example 6.6. We will also demonstrate the diversity of the Koszul class $\mathbf{H}(p, q)$ by showing that p need not be close to q in Example 5.4.

To conclude we will compile several open questions about the Koszul algebra, the lower Bass numbers, and even the structure of the minimal free resolution of R. Conjectures 6.2 and 6.3 are given in [12] and are restated in Sect. 6 along with some additional discussion relating to planar graphs. We would like to keep in mind that many of the questions relating to the Koszul algebra and Bass numbers would be relatively simple to solve, if we knew more about the structure of the minimal free resolution of R.

2 Planar Graphs and Resolutions

We begin by offering some background on minimal free resolutions, and how we represent these resolutions with planar graphs. Much of this can be found in [10, Chap. 3] and [11]. For consistency we will adopt the notation set in [11,12]. We will assume that $I = \langle m_1, \ldots, m_n \rangle$ is a monomial ideal minimally generated by n monomials. We denote the least common multiple of monomials $m_1, \ldots, m_j$ by $m_{1\ldots j}$, specifically $m_{ij} = [m_i, m_j]$. Throughout this paper the monomial m_i will be represented by $x^{a_i} y^{b_i} z^{c_i}$. We say that a monomial m' *strongly divides* a monomial m, denoted $m' \| m$, if m' divides m/x_i for all variables x_i dividing m. If a monomial m' strictly divides a monomial m, we will write $m' |_< m$.

The set of second syzygies for R are generated by,

$$\sigma_{ij} = \frac{m_{ij}}{m_j} e_j - \frac{m_{ij}}{m_i} e_i, \text{ for } 1 \leq i < j \leq n. \tag{2}$$

However, the set $\{\sigma_{ij}\}_{i<j}$ is rarely a minimal generating set for the second syzygies, which is what we are interested in. We will use the *ordered minimal second syzygies*, denoted S_2 as defined in [11, 2.3]. This requires an ordering to be set on the σ_{ij}'s. We use the standard dictionary order on the indices of each σ_{ij}, namely $\sigma_{ij} < \sigma_{kl}$, if either $i < k$ or $j < l$ when $i = k$.

Definition 2.1. Let $Z_2 = \sum_{i<j} S\sigma_{ij} \subseteq S^n$ be the set of all second syzygies of R. If I is a monomial ideal, then $\sigma_{ij} \in S_2$ if and only if both of the following conditions are satisfied.

1. $\sigma_{ij} \in Z_2 - \mathfrak{m}Z_2$.
2. $\sigma_{ij} \neq \sum_{k<l} a_{kl}\sigma_{kl}$, $a_{kl} \in S$, in which $\sigma_{ij} < \sigma_{kl}$ for all k, l such that a_{kl} is a unit.

In general if I is minimally generated by monomials $m_1, \ldots, m_n$, then the matrix f_1 is $[m_1 \ldots m_n]$. The matrix f_2 consists of all column vectors σ_{ij} where $\sigma_{ij} \in S_2$ from Definition 2.1. We order the columns in f_2 using the standard dictionary ordering on the indices of each σ_{ij} set above.

Since we are concerned with a minimal generating set for the second syzygies it would be nice if we knew when $\sigma_{ij} \in S_2$. By Painter [11, 2.5] we know that $\sigma_{ij} \in S_2$, if there is no other minimal generator m_k of I such that $m_k|_{<}m_{ij}$. But this doesn't take care of every case. If there is such an m_k it need not imply that $\sigma_{ij} \notin S_2$. Condition 2 in Definition 2.1 determines which second syzygies are minimal when we have a choice of which second syzygies we can remove from a minimal generating set of Z_2. Specifically, we may have several second syzygies, say σ_{ij}, σ_{ik}, and σ_{jk} where any two of these will generate the other. When this occurs condition 2 of Definition 2.1 says we remove the smallest σ_{ij} of those from which we have a choice.

We will now discuss how we can use planar graphs to represent the minimal free resolution of R. We refer the reader to [9, 10] for more on planar graphs representing minimal free resolutions.

Definition 2.2. A *planar graph* is a graph G with an embedding into a surface homeomorphic to the plane $\mathbb{R}^2$.

Simply speaking a planar graph is a graph drawn in the plane, where edges can only intersect edges at a vertex of the graph. In [10, 3.17] it is shown that the minimal free resolution of R can be represented by some planar graph; moreover, if R is Artinian this graph is almost 3-connected. The almost 3-connected property tells us that each vertex must have at least two edges associated with it, and that each edge will border either one or two faces. We will refer to an edge which borders one face as an outer edge and an edge which borders two faces as an interior edge. It should be noted that in [10] they define planar *maps* the same way we have defined planar *graphs*. In general planar graphs can usually be embedded into the plane in many ways. The notion planar *map* is used because the embedding of the graph into the plane depends on the differentials chosen in the free resolution. Since we have chosen a specific structure for f_2 in Definition 2.1 it is clear how the planar graph of R will be embedded into $\mathbb{R}^2$.

We will now describe how these graphs represent minimal free resolutions. For our ideals the planar graph G representing the minimal free resolution of R will have n vertices, $m + n - 1$ edges, and m faces by (1). The vertices represent the minimal generators of the ideal or the columns of f_1, the edges represent the minimal second syzygies or the columns of f_2, and the faces represent the minimal third syzygies or the columns of f_3. Generally we will choose to refer to the minimal second and third syzygies as the boundaries and cycles, respectively. We notice that since R is Artinian, the graph is always bounded with paths eventually connecting the three pure power generators of I. Since the graph G is almost 3-connected we know that all of the edges on the outer boundary of the graph only correspond to one minimal cycle, and any interior edge will correspond to exactly two minimal cycles. For the matrices of the resolution this means every outer edge relates to a row in f_3 with exactly one nonzero entry, and any interior edge relates to a row in f_3 with exactly two nonzero entries. Thus every row in f_3 will have at most two nonzero entries.

We are mainly interested in what information we can gain about the minimal free resolution by a visual inspection of the planar graph. Specifically, which graphs will yield nonzero entries from I in the matrices of the minimal free resolution of R. The importance of this can be found in Corollary 3.2, where we will learn how to compute the parameters p and r of the Koszul algebra by simply identifying which edges and faces of the planar graph admit nonzero entries from I in the matrices of the minimal free resolution of R.

To draw the planar graph for R we will use labels which help describe the structure of the resolution. We will label each vertex according to its corresponding generator. Each edge will be labeled according to m_{ij}, where m_i and m_j are the two vertices corresponding to this edge. The faces will be labeled as $m_{1\ldots j}$, where $m_1, \ldots, m_j$ are the vertices corresponding to the face. To illustrate this we consider the following example.

Example 2.3. Let $I = \langle x^3, y^3, x^2z, z^3 \rangle$, then the free resolution of R is given by,

$$0 \to S^2 \xrightarrow{\begin{bmatrix} z & 0 \\ -y^3 & 0 \\ x & z^2 \\ 0 & -x^2 \\ 0 & y^3 \end{bmatrix}} S^5 \xrightarrow{\begin{bmatrix} -y^3 & -z & 0 & 0 & 0 \\ x^3 & 0 & -x^2z & -z^3 & 0 \\ 0 & x & y^3 & 0 & -z^2 \\ 0 & 0 & 0 & y^3 & x^2 \end{bmatrix}} S^4 \xrightarrow{\begin{bmatrix} x^3 & y^3 & x^2z & z^3 \end{bmatrix}} \cdots \tag{3}$$

The planar graph representing this resolution will have four vertices, five edges, and two faces. Using our previously described labels we get the following planar graph for R.

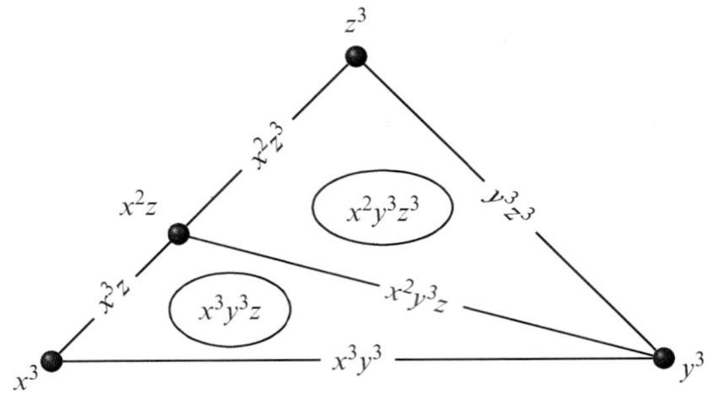

(4)

Usually we choose not to label the edges and faces unless it benefits the understanding of a problem. We use the monomial labels because it directly relates to the computation of the free resolution. For example, consider the face label x^3y^3z for the graph in (4) this corresponds with the first column of f_3 in (3). Note the entries in this column are given by the computations,

$$\frac{x^3y^3z}{x^3y^3} = z \qquad -\frac{x^3y^3z}{x^3z} = -y^3 \qquad \frac{x^3y^3z}{x^2y^3z} = x.$$

We can generalize this type of computation for any cycle in f_3. Specifically, if $m_1, \ldots, m_k$ are vertices corresponding to a face in f_3, then each entry of this cycle in f_3 is given by,

$$\pm \frac{m_{1\ldots k}}{m_{ij}}, \quad \text{where } \sigma_{ij} \text{ is a boundary of the cycle.} \tag{5}$$

Recall we are interested in when entries in the matrices of the minimal free resolution are contained in I. We can use the expression in (5) to explain when we get nonzero pure power entries in f_3 which are also contained in I. Since the columns in the matrix of f_2 consist of the $\sigma_{ij} \in S_2$ we will focus on when nonzero entries of each column of f_2 are contained in I. Moreover, since σ_{ij} is a column vector we refer to the nonzero row entries of σ_{ij} simply as the nonzero entries of σ_{ij}. In Proposition 2.4 we will give a precise description of when nonzero entries from σ_{ij} are contained in I. We should note that a complete answer as to when nonzero entries of f_3 are contained in I is still unclear. Using results from [11] we can describe in general when we get nonzero pure power entries from I in f_3 using planar graphs.

Proposition 2.4. *Let $I = \langle m_1, \ldots, m_n \rangle$ and let G be the planar graph associated with R, then following hold.*

1. *Suppose $\sigma_{ij} \in S_2$, then $gcd(m_i, m_j) = 1$ if and only if each nonzero entry in σ_{ij} is contained in I.*
2. *If there is a face on G with label $x^a y^{b'} z^{c'}$, where x^a is a vertex of the face and $y^{b'} z^{c'}$ is a boundary of the face, then $\pm x^a$ is an entry for this cycle in f_3.*

3. *If there is a face on G with label $x^a y^b z^{c'}$, where x^a and y^b are vertices of the face, and $x^a z^{c'}$ and $y^b z^{c'}$ are boundaries of the face, then both $\pm x^a$ and $\pm y^b$ are entries for a cycle in f_3.*
4. *If there is a cycle in f_3 such that all three pure power generators of I are entries in f_3, then R is Gorenstein and $I = \langle x^a, y^b, z^c \rangle$.*

Proof. We note that part 1 is proved in [11, 2.6], and both 2 and 3 follow immediately from (5). Part 4 follows from the fact that if we had all three pure power entries in one cycle then all three pure power generators must be vertices of this cycle. This implies that there is only one face in the planar graph. If there was another vertex it would have to be on the outer boundary since the graph must be almost 3-connected. Say this vertex is $x^{a'} y^{b'}$. But since there would be no minimal generator dividing $[z^c, x^{a'} y^{b'}] = x^{a'} y^{b'} z^c$ this would imply that this label would be a minimal boundary, which would give us more than one face for the graph. This is a contradiction so $I = \langle x^a, y^b, z^c \rangle$, and R is Gorenstein. □

Revisiting Example 2.3 the three boundaries with labels x^3y^3, x^2y^3z, and y^3z^3 all satisfy part 1 of Proposition 2.4, and both faces satisfy part 2. By inspecting the minimal free resolution in (3) we verify that we have three columns in f_2 which are contained in I, and exactly two nonzero entries from I in f_3. If part 3 of Proposition 2.4 is satisfied, we will have a planar graph of the following form,

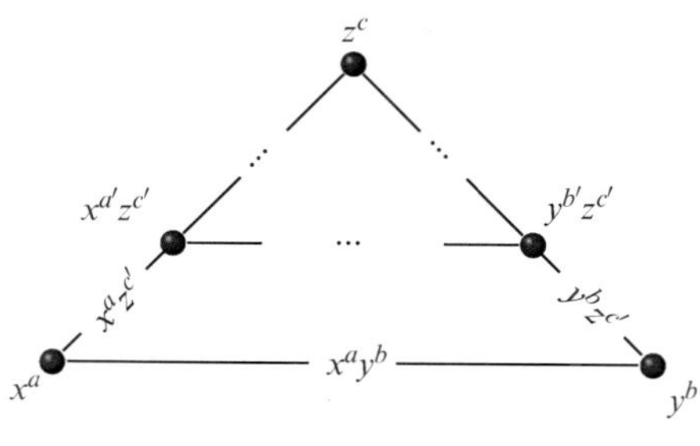

(6)

We note that if we add a vertex of the form $x^\alpha y^\beta$, then $x^a z^{c'}$ and $y^b z^{c'}$ can no longer be boundaries of the same cycle. In Sect. 4 we will see that this case is special in that it represents a particular class for the Koszul Algebra.

Definition 2.5. A monomial ideal $I = \langle m_1, \ldots, m_n \rangle$ is *generic* if whenever two distinct minimal generators m_i and m_j have the same positive degree in some variable, there is another minimal generator m_k such that $m_k \| m_{ij}$.

Part 2 of Proposition 2.4 can be simplified if we restrict to generic monomial ideals. When I is generic the planar graph for R will be a planar triangulation. This means that each column in f_3 will have exactly three nonzero pure power entries [11, 4.1]. Moreover, [11, 3.9] says that if we get nonzero entries from I in f_3, each of these entries must be the same. Thus if I is generic part 2 of Proposition 2.4 says that $\pm x^a$ is an entry of any cycle in f_3 with vertices of the form x^a, $y^{b_1}z^{c_1}$, and $y^{b_2}z^{c_2}$. It is shown in [11, 3.10] that there are at most $n-2$ such entries from I in f_3 when I generic. This was the case in Example 2.3.

3 Koszul Algebra Preliminaries

In this section we introduce the Koszul algebra structure as outlined in [2], so we can learn how to identify the Koszul algebra class of R from the planar graph. For more on the Koszul complex we refer the reader to [6, 8]. Let K be the Koszul complex of $\mathfrak{m}$ over R. Since $\{x, y, z\}$ is not a regular sequence over R the homology of K is nonzero. We define $A = H(K)$ to be the Koszul algebra, which is a graded $\Bbbk$-algebra. The nontrivial graded components of A are given by $A_i = H_i(K) \neq 0$, $1 \leq i \leq 3$. Since the Koszul complex is defined over the exterior algebra, K has a graded commutative structure. Here K is just the deleted resolution of $\mathfrak{m}$ over R, and thus has the following form.

$$0 \longrightarrow R \xrightarrow{\varphi_3=\begin{bmatrix} z \\ -y \\ x \end{bmatrix}} R^3 \xrightarrow{\varphi_2=\begin{bmatrix} -y & -z & 0 \\ x & 0 & -z \\ 0 & x & y \end{bmatrix}} R^3 \xrightarrow{\varphi_1=\begin{bmatrix} x & y & z \end{bmatrix}} R \longrightarrow 0 \qquad (7)$$

We will choose the standard basis elements e_1, e_2, e_3 of R^3 for the graded degree 1 component of K, then $e_1 \wedge e_2$, $e_1 \wedge e_3$, and $e_2 \wedge e_3$ are the ordered generators of the graded degree 2 part. Finally we have that $e_1 \wedge e_2 \wedge e_3$ generates the degree 3 part of K. In [2, 1.2.2] we are given that,

$$\operatorname{rank}_{\Bbbk}(A_1) = n, \qquad \operatorname{rank}_{\Bbbk}(A_2) = m + n - 1, \qquad \text{and} \qquad \operatorname{rank}_{\Bbbk}(A_3) = m. \qquad (8)$$

This tells us exactly the number of generators for each graded component of A, it also indicates that A has a significant relation to the minimal free resolution of R. We will classify the algebra structure of A according to the classification given in [2] which requires the use of several parameters.

Definition 3.1. For $R = S/I$ and $A = H(K)$ we define the following for A.

1. $p = \operatorname{rank}_{\Bbbk}(A_1^2)$.
2. $q = \operatorname{rank}_{\Bbbk}(A_1 \cdot A_2)$.
3. $r = \operatorname{rank}_{\Bbbk}(\delta_2)$, where $\delta_2 : A_2 \longrightarrow \operatorname{Hom}_{\Bbbk}(A_1, A_3)$, where $\delta_2(x)(y) = xy$ for $x \in A_2$ and $y \in A_1$.

In [12, 3.6 and 3.8] it is shown how we can obtain p and r from the minimal free resolution of R. The parameter p is the number of columns in f_2 which are contained in I, and r is the number of rows in f_3 which are dependent mod I. We say that a row r_k in f_3 is dependent mod I if there exist $a_i \in S$ such that

Table 1 Bass series and invariants for Koszul algebra classes

Class	$I_R^R(t)$	p	q	r
T	$\dfrac{m+lt-2t^2-t^3+t^4}{1-t-lt^2-(m-3)t^3-t^5}$	3	0	0
B	$\dfrac{m+(l-2)t-t^2+t^4}{1-t-lt^2-(m-1)t^3+t^4}$	1	1	2
$\mathbf{G}(r)$ $[r \geq 2]$	$\dfrac{m+(l-r)t-(r-1)t^2-t^3+t^4}{1-t-lt^2-nt^3+t^4}$	0	1	r
$\mathbf{H}(0,0)$	$\dfrac{m+lt+t^2-t^3}{1-t-lt^2-mt^3}$	0	0	0
$\mathbf{H}(p,q)$ $[p+q \geq 1]$	$\dfrac{m+(l-q)t-pt^2-t^3+t^4}{1-t-lt^2-(m-p)t^3+qt^4}$	p	q	q

$$r_k - (a_1 r_1 + \cdots + a_{k-1} r_{k-1} + a_{k+1} r_{k+1} + \cdots + a_{m+n-1} r_{m+n-1}) \in IS^m. \quad (9)$$

We will now state how to find p and r from the planar graph representing the minimal free resolution of R. Since the proof of this follows immediately from Proposition 2.4 and [12, 3.6 and 3.8] we will state it as a corollary.

Corollary 3.2. *Let I be an $\mathfrak{m}$-primary monomial ideal, and let G be the planar graph representing the minimal free resolution of R given in* (1)*. Then we have the following description for the parameters p and r of A.*

1. *p is the distinct number of edges on G with vertices x^a and $y^{b'}z^{c'}$, where at least one of the exponents $b', c' > 0$.*
2. *If I is generic, then r is the number of faces in G with vertices x^a, $y^{b_i}z^{c_i}$, and $y^{b_j}z^{c_j}$, where $b_i > b_j \geq 0$ and $0 \leq c_i < c_j$.*
3. *If I is not generic, and the planar graph G of R has that all faces are triangles except for one face satisfying part 3 of Proposition 2.4, then $p = 1$ and $r = 2$.*

We would like to expand Corollary 3.2 further, but it is still unclear as to when a row in f_3 is dependent mod I in general. For the reader's convenience we will restate [12, 2.10], which is a collection of information from [2, 2.1 and 3.1].

Theorem 3.3 (Avramov). *Let R be a local ring of codepth 3 with minimal free resolution as given in* (1) *and let $l = n - 1$. Table 1 lists the Bass series $I_R^R(t)$ along with values for the previously defined invariants for various classes of the Koszul algebra for R.*

We note that R is Golod if and only if it is in $\mathbf{H}(0,0)$ [2, 1.4.3]. Using planar graphs we will be able to give a more precise description of Golod rings admitted from generic monomial ideals than that which was given in [12, 5.2]. We will ignore the class $\mathbf{G}(r)$ as it seems unlikely that we will have any rings given by monomial ideals in this class. This is due to the fact that in the class $\mathbf{G}(r)$, $r \geq 2$ and $p = 0$. There are currently no known examples of monomial ideals where this can happen, and upon inspection it seems that we would be unable to construct a planar graph that would give us $r \geq 2$ and $p = 0$. We will also assume that R is not Gorenstein; otherwise, R would be in the class $\mathbf{C}(3)$, see [12] and [2, 2.1].

4 Identifying the Koszul Structure from Planar Graphs

A complete classification of the Koszul algebra structure is given for generic monomial ideals in [12, 5.2]. We will explore the classification for generic monomial ideals from [12, 5.2] more for planar graphs. The reader should refer to [12, 4.2] to learn how to distinguish between the classes **T** and $\mathbf{H}(3,0)$. For a specific example of a monomial ideal in which R is in $\mathbf{H}(3,0)$ the reader should see [12, 4.3].

Example 4.1. We begin by looking at the planar graph for generic monomial ideals in the Koszul class **T**. If $I = \langle x^a, y^b, z^c, x^{a_1}y^{b_1}z^{c_1}, \ldots, x^{a_\rho}y^{b_\rho}z^{c_\rho}\rangle$ with $a_i, b_i, c_i > 0$ for all $1 \leq i \leq \rho$ and is generic, then R is in **T**. Here the planar graph for R will have the following form.

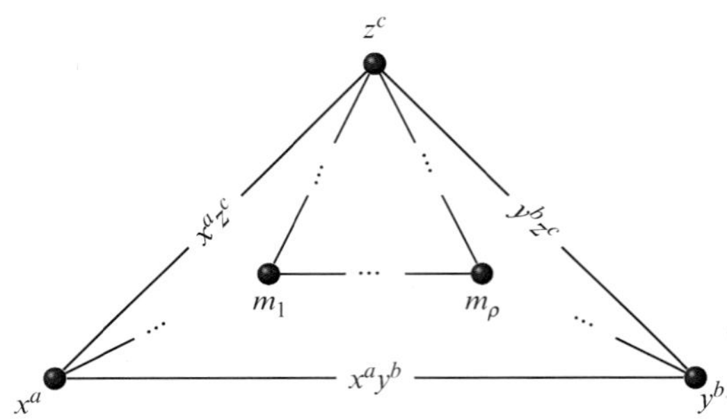

(10)

The key here is that we have the three distinguished edges between the pure power vertices, so $p = 3$. Since I is generic and generated by four or more monomials we can conclude that there is no face of G satisfying part 2 in Corollary 3.2, which implies that $r = 0$. We can then use the description in [1] and [12, 4.2] to conclude that R is in **T**.

The class $\mathbf{H}(p,q)$, where $p + q \geq 1$ can be quite diverse even for generic monomial ideals. We will provide an example to illustrate the diversity of $\mathbf{H}(p,q)$ in Sect. 5. We would like to give some additional insight into the class $\mathbf{H}(0,0)$, since rings in this class are Golod. By Painter [12, 5.1 and 5.2], we know that when I is generic R will be Golod if there is no $\sigma_{ij} \in S_2$ such that $\gcd(m_i, m_j) = 1$. For the graph this means that there are no edges between the pure power vertices, and there are also no edges from any pure power vertex to an opposite outer edge. This implies we must have the existence of at least three vertices of the form $x^{\alpha_1}y^{\beta_1}$, $x^{\alpha_2}z^{\gamma_1}$, and $y^{\beta_2}z^{\gamma_2}$, which would yield the following planar graph.

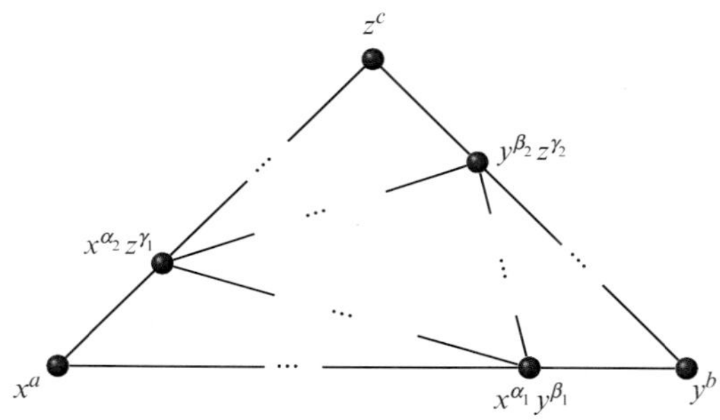

(11)

Using (11) we offer a more detailed description of when R is in $\mathbf{H}(0,0)$ with respect to the generators of the ideal.

Theorem 4.2. *Let $I = \langle x^a, y^b, z^c, m_1, \ldots, m_\rho\rangle$ be a generic monomial ideal minimally generated by $n = \rho + 3$ monomials. If I has three distinguished minimal generators, $x^{\alpha_1}y^{\beta_1}$, $x^{\alpha_2}z^{\gamma_1}$, and $y^{\beta_2}z^{\gamma_2}$, with $\alpha_i, \beta_i, \gamma_i > 0$ for $i = 1, 2$, then R is $\boldsymbol{H}(0,0)$ and is Golod if the following hold.*

1. $[x^{\alpha_1}y^{\beta_1}, x^{\alpha_2}z^{\gamma_1}, y^{\beta_2}z^{\gamma_2}] = m_{1\ldots\rho}$.
2. $min\{\alpha_1, \alpha_2\} = min\{a_i^+\}_{i=1}^{\rho}$, $min\{\beta_1, \beta_2\} = min\{b_i^+\}_{i=1}^{\rho}$, *and* $min\{\gamma_1, \gamma_2\} = min\{c_i^+\}_{i=1}^{\rho}$.
3. *Without loss of generality if $\alpha_1 > \alpha_2$, then $\beta_1 < \beta_2$ and $\gamma_1 > \gamma_2$.*

The notation a_i^+ means that we are only taking the minimum of nonzero exponents on x. Conditions 1 and 2 imply that these three distinguished generators have the largest degrees on the variables out of all of the generators $m_1, \ldots, m_\rho$, and the smallest positive degree on each variable out of all of the minimal generators of I. Condition 3 is saying that the three distinguished generators must have different degrees on each variable, and that each distinguished generator must have the largest degree on one variable and the smallest degree on the other variable for all degrees on the variables in the generators $m_1, \ldots, m_\rho$.

Proof. Since I is generic we know that $p \geq q = r$ by Painter [12, 5.1] and Theorem 3.3. Therefore, it is enough to show that $p = 0$. We will prove this by attempting to construct I so that $p \neq 0$ and get a contradiction. If $p \neq 0$, Corollary 3.2 says that the planar graph must have an edge with corresponding vertices x^a and $y^{b'}z^{c'}$. If I is generated by only six monomials, then the pure powers and the three distinguished generators are the only generators of I. Since $\alpha_1 > \alpha_2$, $\beta_1 < \beta_2$, and $\gamma_1 > \gamma_2$ by condition 3 we have

$$[x^{\alpha_1}y^{\beta_1}, x^{\alpha_2}z^{\gamma_1}] = x^{\alpha_1}y^{\beta_1}z^{\gamma_1}, \quad [x^{\alpha_1}y^{\beta_1}, y^{\beta_2}z^{\gamma_2}] = x^{\alpha_1}y^{\beta_2}z^{\gamma_2},$$

and

$$[x^{\alpha_2}z^{\gamma_1}, y^{\beta_2}z^{\gamma_2}] = x^{\alpha_2}y^{\beta_2}z^{\gamma_1}.$$

Since there are no other generators of I which strictly divide any of the above least common multiples they must each be edge labels in the planar graph. Thus, we have the following planar graph.

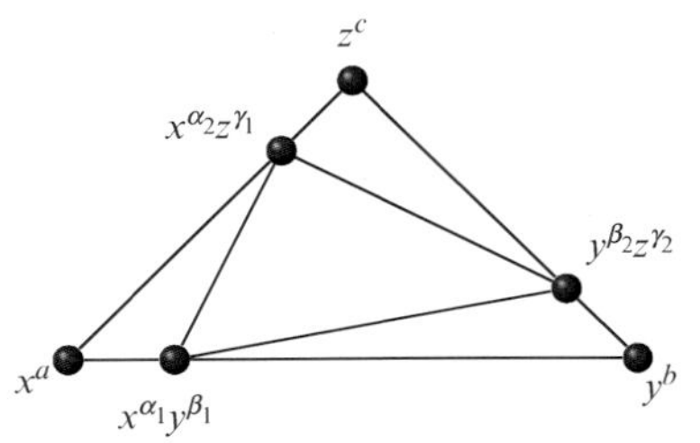

This implies that both p and r are zero by Corollary 3.2.

(12)

Now suppose that I is minimally generated by $n > 6$ monomials, and suppose that the planar graph has an edge with corresponding vertices x^a and $y^{b'}z^{c'}$ such that $y^{b'}z^{c'}$ is not one of the three distinguished generators. Since β_2 is the largest exponent on y, and γ_2 is the smallest positive exponent on z, for the generators $m_1, \ldots, m_\rho$, we know that $b' < \beta_2$ and $c' > \gamma_2$. We want to show that the edge with label $x^a y^{b'} z^{c'}$ is not a minimal boundary. To do this we refer to [11, 2.1 and 2.4]. Notice that since β_1 is the smallest positive degree on the variable y for all generators, $x^{\alpha_1} y^{\beta_1} |_{<} x^a y^{b'} z^{c'}$. If $\beta_1 < b'$, then we are done by Painter [11, 2.1 and 2.4], as this will imply that $x^{\alpha_1} y^{\beta_1} \| x^a y^{b'} z^{c'}$. If $\beta_1 = b'$, then there exists another minimal generator, m_i of I, such that $m_i \| [x^{\alpha_1} y^{\beta_1}, y^{b'} z^{c'}] = x^{\alpha_1} y^{b'} z^{c'}$ since I is generic. But this would imply that $m_i \| x^a y^{b'} z^{c'}$. Therefore, the edge with label $x^a y^{b'} z^{c'}$ cannot be a minimal boundary. □

This result allows us to identify when some generic monomial ideals admit Golod rings by inspecting the generators of I. However, we can have generic ideals where the distinguished generators do not satisfy the conditions in Theorem 4.2 but still admit Golod rings. For example, one could check that the ideal $I = \langle x^5, y^5, z^5, x^3y^2, x^2z^3, y^3z^2, x^4yz \rangle$ admits a Golod ring by computing $p = q = 0$ from the minimal free resolution. We would like to find a more precise description as to which monomial ideals admit Golod rings in general, but it is currently not clear how we would go about this.

We now revisit part 3 from Proposition 2.4 which yields a special class of the Koszul algebra, namely the class **B**. Unfortunately we do not yet have a complete description of all monomial ideals in this class. Rings in this class were first introduced in [5] and were all type 2. Examples of non-Artinian rings in class **B** with type 1 and type 3 are given in [7]. In [12, 5.3] a family of these examples are given which have type $n - 3$, where n is the number of generators of I. This family of examples follows that of the graph in (6) and is currently the most diverse family of these rings with respect to the socle degree.

Example 4.3. Let $I = \langle x^a, y^b, z^c, x^{a_1}z^{c'}, x^{a_2}y^{b_2}z^{c'}, \ldots, x^{a_{\rho-1}}y^{b_{\rho-1}}z^{c'}, y^{b_\rho}z^{c'} \rangle$ with $\rho \geq 2$ and $c' > 0$, then we have the following planar graph for R.

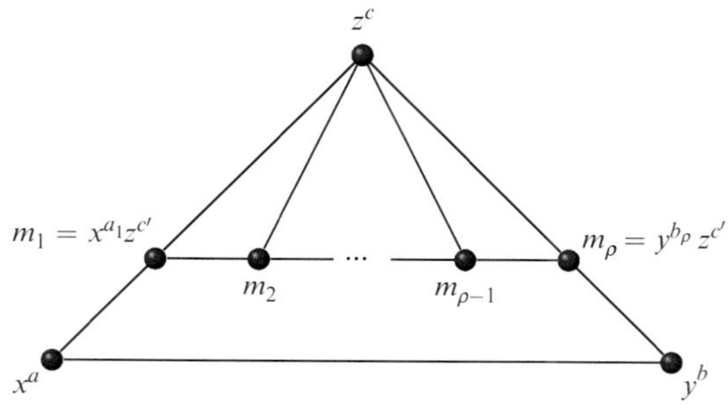

(13)

It seems that this class should generalize further. Specifically, we believe that as long as we only add generators to the ideal so that these would be vertices above the edges from $x^{a_1}z^{c'}$ to $y^{b_p}z^{c'}$, then R would still be in **B**. This is in fact conjectured in [12, 5.6] and will be restated in Sect. 6. The unique part of this ideal is the lower face, as this gives us the entries x^a and y^b in same cycle. For monomial ideals this is the only way in which we can get exactly two different pure power entries from I in f_3.

5 Lower Bass Numbers and Parameters

The Koszul algebra classes are uniquely described by the Bass series of the ring. Because of this the first Bass number, μ_R^1 plays an important role in the methods used in [12]. Here we would like to know what μ_R^1 can be with respect to the number of generators of the ideal and the zeroth Bass number. We should also note that since all of our rings are depth zero, the zeroth Bass number, $\mu_R^0 = m$. We know from the description given in [12, 2.9 and 3.8] that the number r from Definition 3.1 is the number of rows in f_3 that are dependent mod I, see also (9). This provides us with some properties for μ_R^1 and the parameters of A for codepth 3 rings, which we will call a Corollary since they follow immediately from [12, 2.9 and 3.8].

Corollary 5.1. *Let I be an n-generated monomial ideal such that $R = S/I$ has codepth = 3, then the following hold.*

1. $\mu_R^1 = m + n - 1 - r$.
2. *If R is Golod, then* $\mu_R^1 = m + n - 1$.
3. *If I is generic, then* $n - 1 \geq p \geq q$ *and* $n - 2 \geq q$.
4. *If I is generic and* $p > 0$, *then* $p > q$.

We would first like to explore what values for μ_R^1 we can get in terms of the minimal number of generators of I and μ_R^0. We will also see what this says about the parameters p, q, and r of A. Before we begin we would like to note that all of the examples and graphs in this section represent rings that are in Koszul class $\mathbf{H}(p,q)$, and in this class $q = r$, by Theorem 3.3. If I is n-generated and generic, then we have the following possible planar graphs for R.

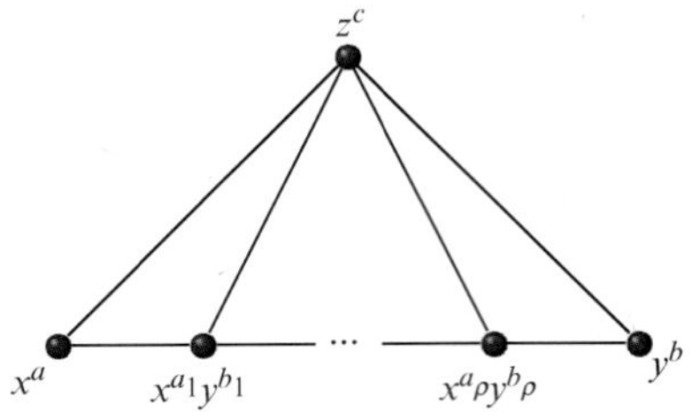

$$I = \langle x^a, y^b, z^c, x^{a_1}y^{b_1}, \ldots, x^{a_\rho}y^{b_\rho}\rangle$$
$$p = n - 1$$
$$q = n - 2 \tag{14}$$
$$\mu_R^1 = m + 1$$

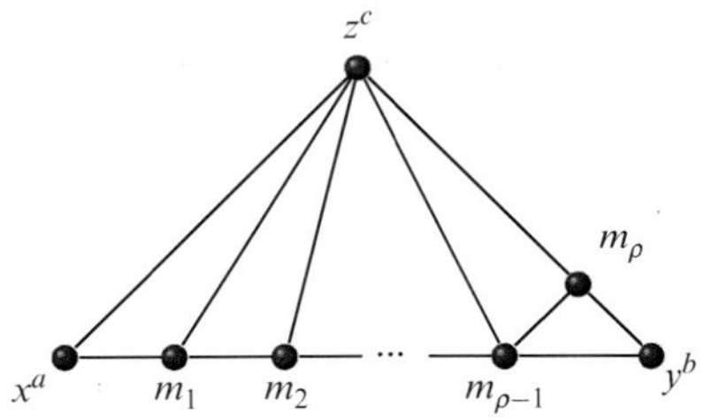

$$I = \langle x^a, y^b, z^c, \{x^{a_i}y^{b_i}\}_{i=1}^{\rho-1}, y^{b_\rho}z^{c_\rho}\rangle$$
$$p = n - 3$$
$$q = n - 4 \tag{15}$$
$$\mu_R^1 = m + 3$$

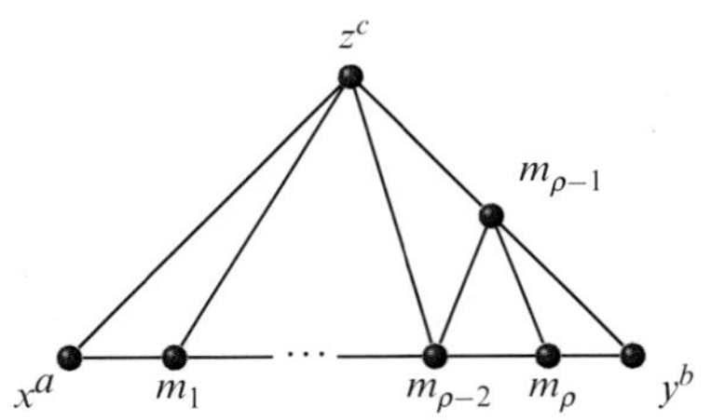

$$I=\langle x^a, y^b, z^c, \{x^{a_i}y^{b_i}\}_{i=1}^{\rho-2}, y^{b_{\rho-1}}z^{c_{\rho-1}}, x^{a_\rho}y^{b_\rho}\rangle$$
$$p = n - 4$$
$$q = n - 5 \tag{16}$$
$$\mu_R^1 = m + 4$$

We note to get the planar graph in (16) we need to enforce the condition that $b_{\rho-2} < b_{\rho-1} < b_\rho$. We can continue this procedure by alternating vertices between the lower outer edges and the right outer edges. We do this by insisting that $b_{i-1} < b_i$ for all $2 \leq i \leq \rho$. Eventually we get the following graph.

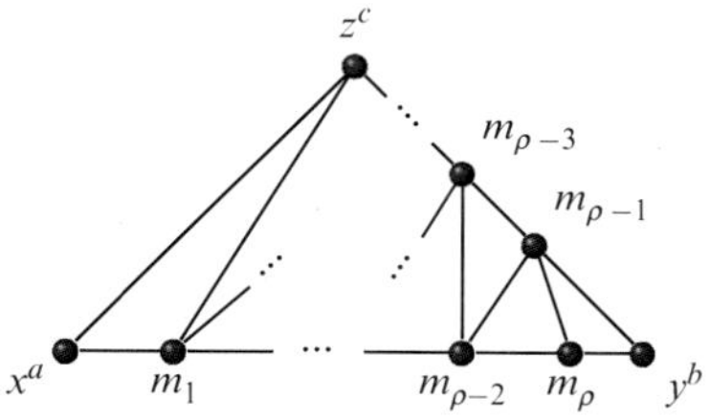

$$
\begin{aligned}
I &= \langle x^a, y^b, z^c, m_1, \ldots, m_p \rangle \\
p &= 2 \\
q &= 1 \\
\mu_R^1 &= m + n - 2
\end{aligned}
\qquad (17)
$$

At this point the next step is clear, we would now add a vertex of the form $x^{a'}z^{c'}$ on the outer left edge. However, we have a choice here. If we choose the vertex close enough to x^a, then we will get $p = 1$ and $q = 0$, (18). On the other hand if we choose the vertex far enough away from x^a, we will satisfy the conditions put forth in Theorem 4.2 and we get $p = q = 0$, (19). The following two graphs illustrate this, though we will omit the generators of I as we have already described them.

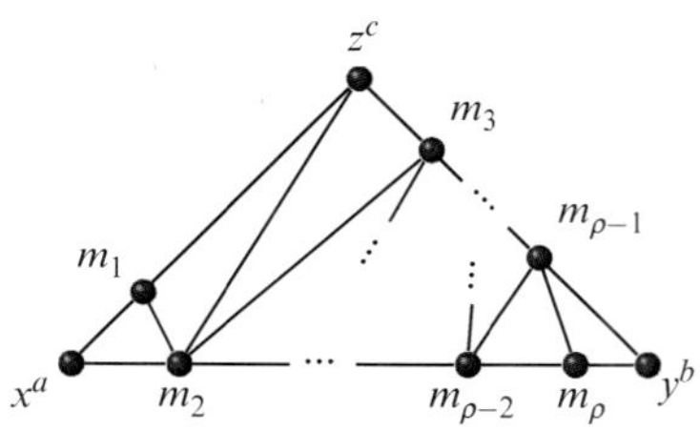

(18)

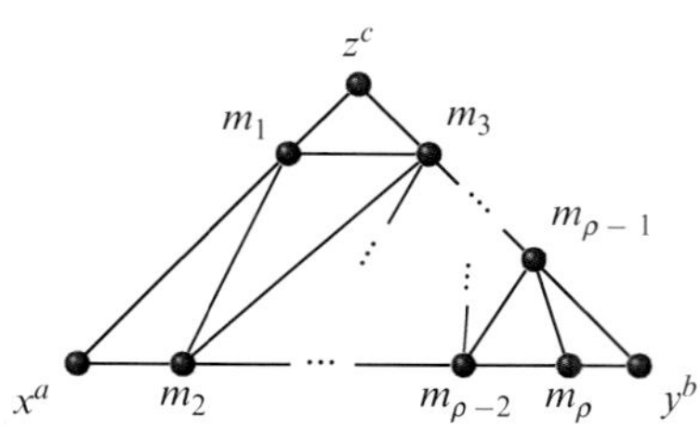

(19)

These graphs seem to suggest that there may be no n-generated generic monomial ideal such that $p = n-2, q = n-3$, and $\mu_R^1 = m+2$. This is not completely true, and we offer the following theorem to describe what happens.

Theorem 5.2. *Let I be an n-generated generic monomial ideal with $n \geq 5$, then the following hold for p, q, and μ_R^1.*

1. *p can be any integer such that $0 \leq p \leq n-1$.*
2. *q can be any integer such that $0 \leq q \leq n-2$ except $q = n-3$.*
3. *μ_R^1 can be any integer such that $m+1 \leq \mu_R^1 \leq m+n-1$ except $\mu_R^1 = m+2$.*

Proof. We have seen that p can indeed be any of the given values listed except for $p = n - 2$. To illustrate this we consider the ideal $I = \langle x^a, y^b, z^c, m_1, \ldots, m_\rho \rangle$, where there is one generator $m_i = x^{a_i} y^{b_i} z^{c_i}$, $a_i, b_i, c_i > 0$, and for every other generator m_j, $j \neq i$ and $m_j = x^{a_j} y^{b_j}$. We will also assume that $a_j > a_{j+1}$ and $b_j < b_{j+1}$ for every $1 \leq j \leq \rho - 1$. This gives the following planar graph for R.

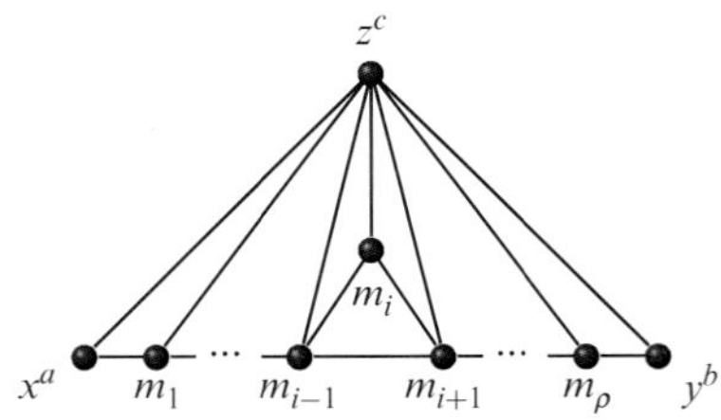

(20)

We notice that there are exactly $\rho + 1 = n - 2$ edges with vertices of the form z^c and $x^{a'} y^{b'}$, thus $p = n-2$. We should also note that there are only $\rho - 1 = n-4$ faces with vertices z^c, $x^{a_j} y^{b_j}$ and $x^{a_{j+1}} y^{b_{j+1}}$, thus we have $q = n - 4$ and $\mu_R^1 = m + 3$. We obtained this graph by choosing a vertex from the graph in (14) and moving it to the interior of the graph.

We must now verify the exceptions in 2 and 3. Since the number of generators is fixed we only need to show that we cannot construct a planar graph with n vertices such that $q = n - 3$. We begin by considering the planar graph where $p = n - 1$ and $q = n - 2$ from (14). If we wanted to drop the number for q by 1, then we must move one of the generators $m_1, \ldots, m_\rho$ since n is fixed. Choose the generator m_i. Note to move this generator in such a way that we only remove one face with vertices z^c, $x^{a_j} y^{b_j}$ and $x^{a_{j+1}} y^{b_{j+1}}$ we have only one option which gives the following graph.

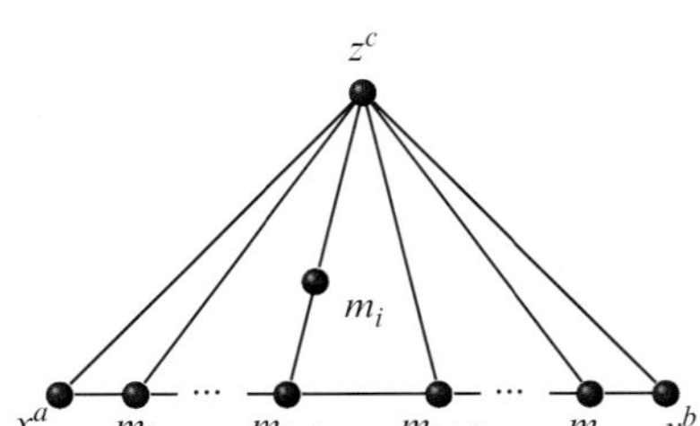

(21)

However, I is generic so there is no possible choice of generators for I which will satisfy this graph. Since this was the only possible option we must have that $q \neq n - 3$ which implies that $\mu_R^1 \neq m + 2$ when I is generic. □

In Theorem 5.2 we restrict to $n \geq 5$ because if $n = 4$ we only have two possibilities. We discuss these possibilities following Question 6.5 in Sect. 6. Since I was generic in Theorem 5.2 we may want to know if $\mu_R^1 = m + 2$ when I is not generic. We do find a positive answer to this question which we will explore in Example 6.6. However, this is the only example we know of where $\mu_R^1 = m + 2$.

We would now like to offer a concrete example satisfying the condition of $p = n-2$.

Example 5.3. Let $I = \langle x^6, y^6, z^6, x^5y, x^4y^2, x^3y^3z, x^2y^4, xy^5\rangle$. Then we get the following planar graph for R.

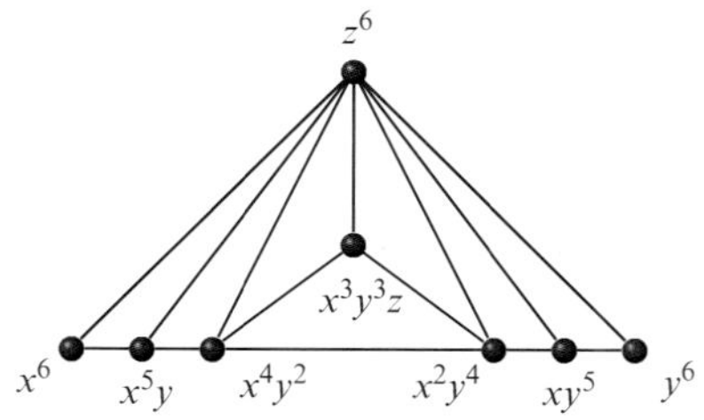

(22)

One can easily check in Macaulay 2 that the minimal free resolution has exactly six columns in f_2 which are contained in I, and exactly four rows in f_3 each of which has a single nonzero entry of z^6. Both these conditions imply that $p = 6$, $q = 4$ and $\mu^1_R = 10$.

In all of the examples thus far the values for p and q were fairly close to each other. This does not need to be true in general. To see this we consider the following general example.

Example 5.4. Let $I = \langle x^a, y^b, z^c, m_1, \ldots, m_\rho\rangle$, where ρ is odd, $m_i = x^{a_i}y^{b_i}z^{c'}$ if i is odd, and $m_i = x^{a_i}y^{b_i}$ if i is even. We also assume that $a_i > a_{i+1}$ and $b_i < b_{i+1}$ for all $1 \le i \le \rho - 1$. With these conditions fixed we get the following planar graph for R.

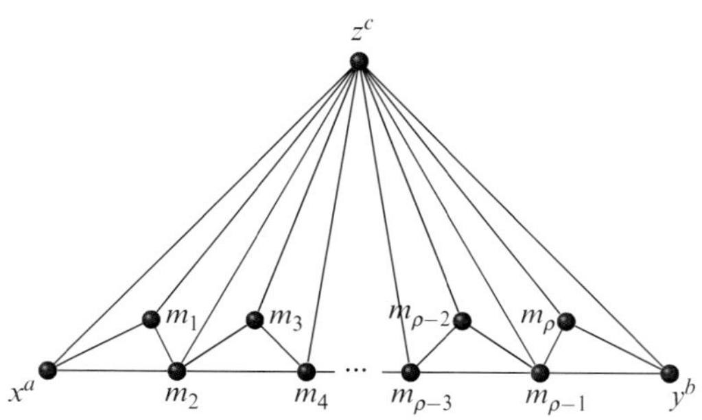

(23)

Our first observation is that I is generic. We also notice that except for the outer left and right edges we get exactly $\frac{\rho-1}{2}$ edges with vertices z^c and $x^{a_i}y^{b_i}$. Since both outer left and right edges also satisfy condition 1 of Corollary 3.2 we have that $p = \frac{\rho-1}{2} + 2$ and $q = 0$. Note that p is indeed an integer here since ρ is odd.

Example 5.4 shows us that p and q need not be close to each other. We observe that p depends only on ρ in the previous example, and thus the number of generators of I, $n = \rho+3$. It should be noted that the number ρ can be as large as we like, given that a and b are large enough. This demonstrates how diverse the Koszul algebra class $\mathbf{H}(p, q)$ is, as all of the previous examples in this section and the constructions used for Theorem 5.2 fall into this class.

6 Questions

In this section we will summarize some of the discussions in this paper and formulate some questions for which we would like answers. As we stated in Sect. 1 we still have many questions about monomial ideals and many of these could be answered if we had more information about the minimal free resolution of these ideals. We know how to find the number p for these ideals generally, but what about q and r? This is not as straightforward. For Artinian monomial ideals we described r as the number of rows in f_3 which are dependent mod I [12, 3.8]. But we only know how to give a complete description for a row of f_3 being dependent mod I when I is generic and in some special cases. This leads us to ask the following questions.

Question 6.1. Let I be an $\mathfrak{m}$-primary, monomial ideal of S.

1. Can we obtain nonzero entries in f_3 from I which are not pure powers of the variables?
2. Can we say that a row in f_3 is dependent mod I if and only if each nonzero entry in this row is also in I?

Since we know of no examples satisfying 1 and no examples where our statement in 2 fails, we will restate this as a conjecture.

Conjecture 6.2. Let I be an $\mathfrak{m}$-primary Artinian monomial ideal minimally generated by n monomials, then the following hold.

1. If f_3 contains nonzero entries from I, each of these entries must be pure power generators of I.
2. $\mu_R^1 = m + n - 1 - r$, where r is the number of rows in f_3 contained in I.

This is essentially the same conjecture that is stated in [12, 5.5]. Since r is the number of rows in f_3 which are dependent mod I, 2 in Question 6.1 is equivalent to 2 in Conjecture 6.2.

While some effort has been put forth to provide an answer to these questions, the work quickly becomes a combinatorial mess. If we could prove Conjecture 6.2 it would allow us to completely classify the Koszul algebra for trivariate monomial ideals. Assuming that Conjecture 6.2 holds we can provide some motivation for Conjecture 6.3. For the Koszul class $\mathbf{T}$ we give the same description of I as in Example 4.1, but we remove the assumption that I is generic. Upon inspection the planar graph for R would be the same as in (10), except all of the faces would no longer need to be triangles. However, if Conjecture 6.2 holds, the planar graph for R would still give $p = 3$ and $r = 0$ by Corollary 3.2, which implies R is in $\mathbf{T}$ by Painter [12, 4.2].

For rings in class **B** we use the template set in Example 4.3. Namely, we need to keep the lower face from (13) intact so that we get exactly two different pure power generators of I as entries in a single column of f_3. If Conjecture 6.2 holds, then these two pure power entries would be the only nonzero entries we could have in f_3 from I. Thus for the Koszul class **B** it seems reasonable that we should be able to add vertices to the planar graph in (13) as long as they fall above the lower face of this graph.

Another important consequence of Conjecture 6.2 would be that $p \geq q$ for all trivariate monomial ideals. This would imply that our ideals will never admit rings in the Koszul class $\mathbf{G}(r)$, since $p < q$ in this class. We also see that for R to be in **H**(0,0) we simply remove the generic assumption from [12, 5.2]. This is due to the fact that if we assume there are no $\sigma_{ij} \in S_2$ that are also contained in I, which implies $p = 0$, then $q = 0$ also since $p \geq q$. Any other rings not satisfying the conditions for **T**, **B**, or **H**(0,0) would have to fall in class $\mathbf{H}(p,q)$ with $p+q \geq 1$ by process of elimination. The following conjecture is a restatement of the conjecture provided in [12, 5.6], which if proven would provide a complete classification of the Koszul algebra structure for trivariate monomial ideals.

Conjecture 6.3. Let I be an $\mathfrak{m}$-primary Artinian monomial ideal minimally generated by n monomials, then we have the following Koszul algebra classifications for R.

1. If $I = \langle x^a, y^b, z^c, x^{a_1}y^{b_1}z^{c_1}, \ldots, x^{a_\rho}y^{b_\rho}z^{c_\rho}\rangle$ with $a,b,c,a_i,b_i,c_i > 0$ for all $1 \leq i \leq \rho$, then R is in **T**.
2. $R = S/I$ is in **B**, if $I = \langle x^a, y^b, z^c, m_1, \ldots, m_j\rangle$ has a subset of the minimal generators $\{m_1, \ldots, m_j\}$ of the form

$$I' = \{x^{a_1}z^{c'}, x^{a_2}y^{b_2}z^{c'}, \ldots, x^{a_{\rho-1}}y^{b_{\rho-1}}z^{c'}, y^{b_\rho}z^{c'}\},$$

   where $\rho \geq 2$, $c' > 0$, and any other minimal generator $m_i \in \{m_1, \ldots, m_j\} - I'$ must be such that $c_i > c'$.
3. If $I = \langle x^a, y^b, z^c, x^{\alpha_1}y^{\beta_1}, x^{\alpha_2}z^{\gamma_1}, y^{\beta_2}z^{\gamma_2}, x^{a_1}y^{b_1}z^{c_1}, \ldots, x^{a_\rho}y^{b_\rho}z^{c_\rho}\rangle$, $a,b,c,\alpha_i,\beta_i,\gamma_i > 0$, then R is in $\mathbf{H}(0,0)$ and is Golod if and only if there is no $\sigma_{ij} \in S_2$ with nonzero entries from I.
4. Otherwise R is in $\mathbf{H}(p,q)$ with $p+q \geq 1$.

We also propose an alternate method to proving Conjecture 6.3 by directly showing that $p \geq q$ for monomial ideals. Again this would rule out the Koszul class $\mathbf{G}(r)$ since in this class $p = 0$ and $q = 1$ with $q \neq r$. In general it is not true that $p \geq q$ for rings in $\mathbf{H}(p,q)$, see [2, 3.4.3]. However, we know of no example where $p < q$ for monomial ideals. Moreover, we know that if I is generic, then $p \geq q$, and if $p > 0$, then $p > q$, see Corollary 5.1 and [12, 5.1]. We propose that this holds for general monomial ideals with a slight adjustment, that $p = q = 1$ if R is in **B**. Since we know exactly what these parameters are in **B**, we will exclude this class from the statement.

Conjecture 6.4. Let I be an $\mathfrak{m}$-primary Artinian monomial ideal of S, with $R = S/I$, such that R is not in Koszul class **B**. Then $p \geq q$ and $p > q$, if $p > 1$.

We are also interested in which general monomial ideals admit Golod rings. We know in general that any such ideals must have minimal generators of the form $x^{\alpha_1}y^{\beta_1}$, $x^{\alpha_2}z^{\gamma_1}$, and $y^{\beta_2}z^{\gamma_2}$, but it is unclear what conditions we would need to impose to get all of these rings even in the generic case.

Question 6.5. Can we describe all $\mathfrak{m}$-primary monomial ideals which admit Golod rings? i.e. R is in **H**(0,0).

Finally, we revisit the question posed about μ_R^1 following Theorem 5.2, namely can we find and ideal so that $\mu_R^1 = m + 2$? The short answer to this question is yes, in a preliminary example. Before we give this example, we recall that we are considering $\mathfrak{m}$-primary monomial ideals in three variables, and that R is not Gorenstein. This implies that I is minimally generated by $n \geq 4$ monomials. If $n = 4$, then I must be generic and we have only two possibilities; $m = 2$, which implies $\mu_R^1 = 3 = m + 1$ and $m = 3$, which implies $\mu_R^1 = 6 = m + 3$. Both cases agree with Theorem 5.2; however, if $n = 5$, then I need not be generic. Consider the following example, which is a base case example for rings in the Koszul class **B**.

Example 6.6. Let $I = \langle x^3, y^3, z^3, x^2z, y^2z\rangle$, then I is not generic and the planar graph for R is.

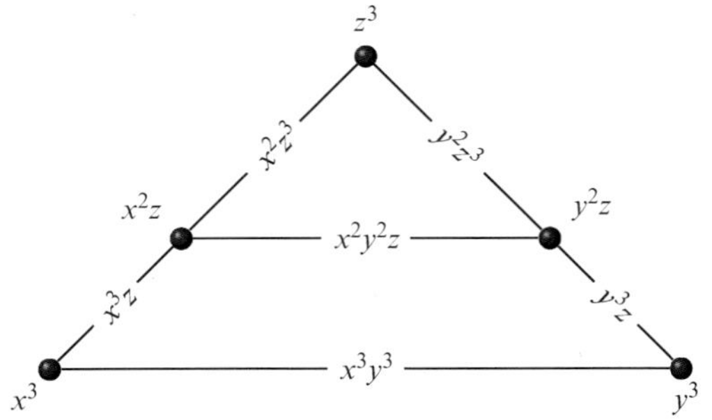

(24)

From this graph it is clear that $p = 1$, $r = 2$, and we can compute $q = 1$, so R is in **B**. Also we see that the graph has five vertices, six edges, and two faces, thus $\mu_R^1 = 4 = m + 2$. This is only case we know of where $\mu_R^1 = m + 2$. We also know from Theorem 5.2 that $\mu_R^1 \neq m + 2$ if I is generic. So we ask the following question.

Question 6.7. If I is not generic and is generated by $n > 5$ monomials, can $\mu_R^1 = m + 2$?

One could possibly use an exhaustive approach to answer this question when $n = 6$. However, we again feel that if we had more information regarding Conjecture 6.2 we would be able to explore this question in more generality.

Acknowledgments I would like to thank David Jorgensen for his guidance and insight on this topic over the past few years. In addition I would like to thank Lars Winther Christensen for his insight on the Koszul classes. Many thanks to Sean Sather-Wagstaff and Susan Cooper for the opportunity to contribute to the Springer volume.

References

1. Avramov, L.L.: Homological asymptotics of modules over local rings. In: Mathematical Sciences Research Institute Publications, vol. 15, pp. 308–327. Springer, Berkeley (1987)
2. Avramov, L.L.: A cohomological study of local rings of embedding codepth 3. J. Pure Appl. Algebra **216**(11), 2489–2506 (2012)
3. Avramov, L.L., Kustin, A.R., Miller, M.: Poincaré series of modules over local rings of small embedding codepth or small linking number. J. Algebra **118**, 162–204 (1988)
4. Bayer, D., Peeva, I., Sturmfels, B.: Monomial resolutions. Math. Res. Lett. **5**, 31–46 (1998)
5. Brown, A.E.: A structure theorem for a class of grade three perfect ideals. J. Algebra **105**, 308–327 (1987)
6. Bruns, W., Herzog, J.: Cohen-Macaulay Rings. Cambridge Studies in Advanced Mathematics, vol. 39. Cambridge University Press, Cambridge (1993)
7. Christensen, L.W., Veliche, O.: Local rings of embedding codepth 3. Examples, arXiv:1209.4256
8. Eisenbud, D.: Introduction to Commutative Algebra with a View Towards Algebraic Geometry. Springer, New York (1995)
9. Miller, E.: Planar graphs as minimal resolutions of trivariate monomial ideals. Doc. Math. **7**, 43–90 (2002)
10. Miller, E., Sturmfels, B.: Combinatorial Commutative Algebra. Graduate Texts in Mathematics. Springer, New York (2000)
11. Painter, J.: On the behavior of minimal free resolutions of trivariate generic monomial ideals. Commun. Algebra. arXiv:1303.0735
12. Painter, J.: On the Koszul algebra for trivariate monomial ideals. arXiv:1303.0083
13. Weyman, J.: On the structure of free resolutions of length 3. J. Algebra **126**, 1–33 (1989)

Non-Gorenstein Isolated Singularities of Graded Countable Cohen–Macaulay Type

Branden Stone

Dedicated to Tony Geramita for his role in shaping mathematics

Abstract In this article we show a partial answer to a question of Huneke and Leuschke (Proc. Am. Math. Soc. 131(10):3003–3007, 2003): Let R be a standard graded Cohen–Macaulay ring of graded countable Cohen–Macaulay representation type, and assume that R has an isolated singularity. Is R then necessarily of graded finite Cohen–Macaulay representation type? In particular, this question has an affirmative answer for standard graded non-Gorenstein rings as well as for standard graded Gorenstein rings of minimal multiplicity. Along the way, we obtain a partial classification of graded Cohen–Macaulay rings of graded countable Cohen–Macaulay type.

1 Motivation and Introduction

For motivational purposes, we first consider local Cohen–Macaulay rings. Throughout, a local Cohen–Macaulay ring is said to have *finite Cohen–Macaulay type* (respectively, *countable Cohen–Macaulay type*) if it has only finitely (respectively, countably) many isomorphism classes of indecomposable maximal Cohen–Macaulay modules.

In [2], Auslander showed that local rings of finite Cohen–Macaulay type must have an isolated singularity. It was later shown in [4] that this was true for standard graded rings as well. Concerning countable Cohen–Macaulay type, it was shown in the excellent case by Huneke and Leuschke that a local ring of countable

B. Stone (✉)
Bard College, P.O. Box 5000, Annandale-on-Hudson, NY, USA
e-mail: bstone@bard.edu

S.M. Cooper and S. Sather-Wagstaff (eds.), *Connections Between Algebra, Combinatorics, and Geometry*, Springer Proceedings in Mathematics & Statistics 76,
DOI 10.1007/978-1-4939-0626-0_9,

Cohen–Macaulay type does not necessarily have an isolated singularity, but the singular locus has dimension at most one [12] (the graded version of this is found in [19]). Given this result, it is natural to try and find a local Cohen–Macaulay ring of countable (and infinite) Cohen–Macaulay type with an isolated singularity. Examining the known examples of countable type rings with an isolated singularity, C. Huneke and G. Leuschke found they were actually finite type! This led to the following question:

Question 1.1 ([12]). Let R be a complete local Cohen–Macaulay ring of countable Cohen–Macaulay representation type, and assume that R has an isolated singularity. Is R then necessarily of finite Cohen–Macaulay representation type?

In this paper, we are mainly concerned with the graded version of this question (see Question 1.2). In Corollary 6.1, we are able to show a positive answer for standard graded non-Gorenstein rings as well as for graded Gorenstein rings of minimal multiplicity. The main strategy is to classify the rings of graded countable Cohen–Macaulay type and compare this list with a classification of graded finite Cohen–Macaulay type in [9].

The sections of this paper break the problem up into cases based on the dimension of the ring, with Sect. 5 discussing the issues surrounding Gorenstein rings. The results are summarized in Sect. 6.

1.1 Preliminaries and Known Results

We say that a ring R is *standard graded* if, as an abelian group, it has a decomposition $R = \bigoplus_{i \geqslant 0} R_i$ such that $R_i R_j \subseteq R_{i+j}$ for all $i, j \geqslant 0$, $R = R_0[R_1]$, and R_0 is a field. Further, we will always assume that a standard graded ring is Noetherian. Unless otherwise stated, we will denote $(R, \mathfrak{m}, \Bbbk)$ by the standard graded ring with $\mathfrak{m}$ being the irrelevant maximal ideal, that is $\mathfrak{m} = \sum_{i=1}^{\infty} R_i$, and $\Bbbk := R_0 = R/\mathfrak{m}$ being an uncountable algebraically closed field of characteristic zero. We let $e(R)$ be the multiplicity of the maximal ideal $\mathfrak{m}$. When R is a d-dimensional Cohen–Macaulay ring, if $e(R) = \dim_{\Bbbk}(\mathfrak{m}/\mathfrak{m}^2) - \dim R + 1$ then R is said to have *minimal multiplicity*. In the case when $\Bbbk$ is infinite, the following are equivalent:

(i) R has minimal multiplicity;
(ii) there exists a regular sequence $x_1, \ldots, x_d$ such that $\mathfrak{m}^2 = (x_1, \ldots, x_d)\mathfrak{m}$;
(iii) the h-vector of R is of the form $(1, n)$.

Here we define the *h-vector* of R to be the sequence $(\dim_{\Bbbk}(R/(x_1, \ldots, x_d))_i)_{i \geqslant 0}$ where $x_1, \ldots, x_d$ is a homogeneous regular sequence of degree one.

A standard graded Cohen–Macaulay ring $(R, \mathfrak{m}, \Bbbk)$ has *graded finite Cohen–Macaulay type* (respectively, *graded countable Cohen–Macaulay type*) if it has only finitely (respectively, countably) many indecomposable, maximal Cohen–Macaulay modules up to a shift in degree.

The graded version of Question 1.1 is as follows.

Question 1.2. Let R be a standard graded Cohen–Macaulay ring of graded countable Cohen–Macaulay representation type, and assume that R has an isolated singularity. Is R then necessarily of graded finite Cohen–Macaulay representation type?

Remark 1.3. It is known that if the completion of a standard graded ring R with respect to the maximal ideal has finite (respectively countable) type, then R must have graded finite (respectively countable) type (see [4, Propositions 8 and 9] or [20, Corollary 2.5]). Hence any affirmative answer to Question 1.1 can be passed to a positive answer of Question 1.2.

With this in mind, Question 1.2 has a positive answer if the ring is a hypersurface. In particular, it was shown in [6, 14] that if R is a complete hypersurface containing an algebraically closed field $\Bbbk$ (of characteristic different from 2), then R is of finite Cohen–Macaulay type if and only if R is the local ring of a simple hypersurface singularity in the sense of [1]. For example, if we let $\Bbbk = \mathbb{C}$, then R is one of the complete ADE singularities over $\mathbb{C}$. That is, R is isomorphic to $\Bbbk[\![x, y, z_2, \ldots, z_r]\!]/(f)$, where f is one of the following polynomials:

$$(A_n) : x^{n+1} + y^2 + z_2^2 + \cdots + z_r^2,\ n \geqslant 1;$$

$$(D_n) : x^{n-1} + xy^2 + z_2^2 + \cdots + z_r^2,\ n \geqslant 4;$$

$$(E_6) : x^4 + y^3 + z_2^2 + \cdots + z_r^2;$$

$$(E_7) : x^3y + y^3 + z_2^2 + \cdots + z_r^2;$$

$$(E_8) : x^5 + y^3 + z_2^2 + \cdots + z_r^2.$$

It was further shown in [6] that a complete hypersurface singularity over an algebraically closed uncountable field $\Bbbk$ has (infinite) countable Cohen–Macaulay type if and only if it is isomorphic to one of the following:

$$(A_\infty) : \Bbbk[\![x, y, z_2, \ldots, z_r]\!]/(y^2 + z_2^2 + \cdots + z_r^2); \tag{1}$$

$$(D_\infty) : \Bbbk[\![x, y, z_2, \ldots, z_r]\!]/(xy^2 + z_2^2 + \cdots + z_r^2). \tag{2}$$

As both (A_∞) and (D_∞) are non-isolated singularities, if a hypersurface has an isolated singularity and is countable type, then it must have finite type.

In 2011, Karr and Wiegand [13, Theorem 1.4] showed the one-dimensional case as well. This was under the assumption that the integral closure of the ring R is finitely generated as an R-module. We recover these results in Sect. 3.

2 Zero Dimensional Rings

It is well known that a zero dimensional equicharacteristic local ring R is a hypersurface if and only if R is of finite Cohen–Macaulay type [11, Satz 1.5]. We show the graded countable analog to this statement in Proposition 2.1.

Proposition 2.1. *Let* $(R, \mathfrak{m}, \Bbbk)$ *be a zero-dimensional standard graded Cohen–Macaulay ring. Further assume that* $\Bbbk$ *is an uncountable field. Then the following are equivalent:*

(1) R is of graded finite Cohen–Macaulay type;
(2) R is of graded countable Cohen–Macaulay type;
(3) R is a hypersurface ring.

Proof. The implication (1) implies (2) is straightforward. To show (2) implies (3), assume that R is not a hypersurface. Thus there must be two linear forms $a, b \in \mathfrak{m} \setminus \mathfrak{m}^2$ that are basis elements of $\mathfrak{m}/\mathfrak{m}^2$. By [20, Lemma 4.1], there are uncountably many distinct homogeneous ideals $\{I_\alpha\}_{\alpha \in \Bbbk}$ in R. In this context, we have that $I_\alpha = (a + \alpha b)R$. Consider the graded indecomposable maximal Cohen–Macaulay modules $\{R/I_\alpha\}_{\alpha \in \Bbbk}$. As each of these modules have different annihilators, we know they are not isomorphic. A contradiction as we assumed that R was of graded countable type.

To prove that (3) implies (1), we consider the $\mathfrak{m}$-adic completion of R and then apply [11, Satz 1.5] to see that the completion is of finite Cohen–Macaulay type. By Remark 1.3, we know that R also has graded finite Cohen–Macaulay type.

We thus have a complete classification of graded countable Cohen–Macaulay type for zero dimensional standard graded rings. Further, Proposition 2.1 gives a positive answer to Question 1.2 for zero dimensional standard graded rings with uncountable residue field.

3 One-Dimensional Rings

In the one-dimensional case, Question 1.2 has a positive answer as shown by Karr and Wiegand [13, Theorem 1.4]. In this section, we examine the Drozd–Roĭter conditions and give a classification of one-dimensional reduced standard graded rings of graded countable Cohen–Macaulay type. Thus retrieving the result of R. Karr and R. Wiegand.

3.1 Finite Type and the Drozd–Roĭter Conditions

As detailed by Cimen et al. [7], if $(R, \mathfrak{m}, \Bbbk)$ is a one-dimensional, Noetherian, reduced, local Cohen–Macaulay ring such that the integral closure of R, say S, is finitely generated as an R-module, then we know precisely when R has finite Cohen–Macaulay type. This happens when the following conditions occur:

DR1 S is generated by three elements as an R-module;
DR2 the intersection of the maximal R-submodules of S/R is a cyclic R-module.

These conditions are called the *Drozd-Roĭter conditions*. It was further shown in [7, Proposition 1.12] that the Drozd-Roĭter conditions are equivalent to the following:

$$\dim_{\Bbbk}(S/\mathfrak{m}S) \leqslant 3; \tag{dr1}$$

$$\dim_{\Bbbk}\left(\frac{R+\mathfrak{m}S}{R+\mathfrak{m}^2 S}\right) \leqslant 1. \tag{dr2}$$

In order to have a different grasp of what it means to satisfy the Drozd–Roĭter conditions, we state another set of equivalent conditions. This result is stated in the next proposition where λ represents length as an R-module, and $\overline{*}$ is the integral closure of ideals.

Proposition 3.1. *Let $(R, \mathfrak{m}, \Bbbk)$ be a one-dimensional, local, Noetherian, reduced, Cohen–Macaulay ring with finite integral closure and uncountable residue field $\Bbbk$. Let x be a minimal reduction of the maximal ideal $\mathfrak{m}$. The Drozd–Roĭter conditions are equivalent to the following:*

$$e(R) \leqslant 3; \tag{3}$$

$$\lambda(\overline{\mathfrak{m}^2}/x\mathfrak{m}) \leqslant 1. \tag{4}$$

Proof. To show that (3) holds, we will show that $e(R) = \dim_{\Bbbk}(S/\mathfrak{m}S)$ where S is the integral closure of R. As x is a reduction of $\mathfrak{m}$, we know that xS is also a reduction of $\mathfrak{m}S$. But this holds if and only if $\mathfrak{m}S \subseteq \overline{xS}$. As principal ideals are integrally closed in S, we have that

$$xS \subseteq \mathfrak{m}S \subseteq \overline{xS} = xS,$$

and hence $xS = \mathfrak{m}S$. By assumption, S is finitely generated as an R-module. Therefore we have that the map $S \to S$ defined by multiplication by x is an injection. Hence, we have the following commutative diagram in Fig. 1.

Consider the exact sequence in the right side of Fig. 1,

$$0 \longrightarrow K \longrightarrow R/xR \longrightarrow S/xS \longrightarrow C \longrightarrow 0. \tag{5}$$

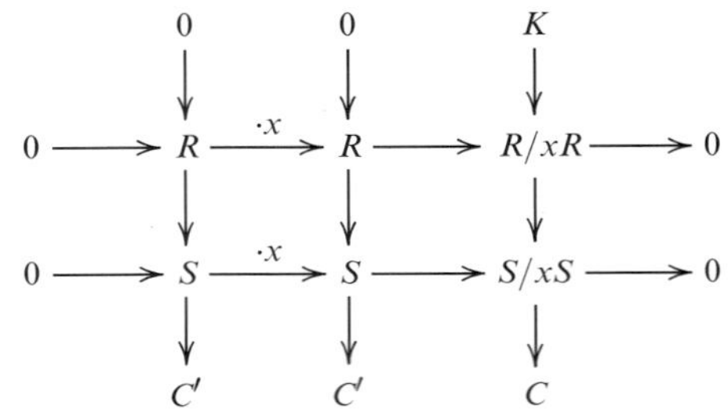

Fig. 1 A commutative diagram where K, C', and C are the respective kernel and cokernels

Notice by the Snake Lemma applied to Fig. 1 that $\lambda(K) = \lambda(C)$; here λ represents length as R-modules. As C and K have the same length, (5) shows that R/xR and S/xS also have the same length. Since x is a minimal reduction of the maximal ideal, we know that $e(R) = \lambda(R/xR)$. Therefore,

$$e(R) = \lambda(R/xR) = \lambda(S/xS) = \lambda(S/\mathfrak{m}S) = \dim_{\Bbbk}(S/\mathfrak{m}S).$$

In order to show (4), first notice that

$$\frac{R + \mathfrak{m}S}{R + \mathfrak{m}^2 S} \simeq \frac{\mathfrak{m}S}{\mathfrak{m}^2 S + (R \cap \mathfrak{m}S)} \simeq \frac{\mathfrak{m}S}{\mathfrak{m}^2 S + \mathfrak{m}} \simeq \frac{xS}{x^2 S + \mathfrak{m}}. \tag{6}$$

For simplicity, we define B as follows,

$$B := \dim_{\Bbbk}\left(\frac{R + \mathfrak{m}S}{R + \mathfrak{m}^2 S}\right) = \lambda\left(\frac{xS}{x^2 S + \mathfrak{m}}\right).$$

We now consider the short exact sequence

$$0 \longrightarrow \frac{\mathfrak{m}^2 S + \mathfrak{m}}{\mathfrak{m}^2 S} \longrightarrow \frac{\mathfrak{m}S}{\mathfrak{m}^2 S} \longrightarrow \frac{\mathfrak{m}S}{\mathfrak{m}^2 S + \mathfrak{m}} \longrightarrow 0.$$

Rewriting the two terms on the left gives us

$$\frac{\mathfrak{m}^2 S + \mathfrak{m}}{\mathfrak{m}^2 S} \simeq \frac{\mathfrak{m}}{\mathfrak{m}^2 S \cap \mathfrak{m}} \simeq \mathfrak{m}/\overline{\mathfrak{m}^2}; \tag{7}$$

$$\frac{\mathfrak{m}S}{\mathfrak{m}^2 S} \simeq \frac{S}{\mathfrak{m}S} = \frac{S}{xS}. \tag{8}$$

Combining (7) and (8) with the above short exact sequence yields

$$\lambda\left(\frac{S}{xS}\right) = \lambda\left(\mathfrak{m}/\overline{\mathfrak{m}^2}\right) + B. \tag{9}$$

On the other hand, consider the following short exact sequence

$$0 \longrightarrow \frac{\overline{\mathfrak{m}^2} + xR}{xR} \longrightarrow \frac{R}{xR} \longrightarrow \frac{R}{\overline{\mathfrak{m}^2} + xR} \longrightarrow 0, \tag{10}$$

along with the isomorphisms

$$\frac{\overline{\mathfrak{m}^2} + xR}{xR} \simeq \frac{\overline{\mathfrak{m}^2}}{xR \cap \overline{\mathfrak{m}^2}} = \frac{\overline{\mathfrak{m}^2}}{x\mathfrak{m}}. \tag{11}$$

Note that the equality in (11) can be justified as follows. Since $xS = \mathfrak{m}S$, we know that $\overline{\mathfrak{m}^2} = \mathfrak{m}^2 S \cap R = x^2 S \cap R$. If $y \in xR \cap \overline{\mathfrak{m}^2}$, then $y = xr \in \overline{\mathfrak{m}^2} = x^2 S \cap R$. This forces $r \in xS \cap R = \mathfrak{m}$. Hence $y \in x\mathfrak{m}$. Equality follows as $x\mathfrak{m} \subseteq xR \cap \overline{\mathfrak{m}}$.

Computing length in (10) gives us

$$\lambda\left(\frac{R}{xR}\right) = \lambda\left(\frac{\overline{\mathfrak{m}^2}}{x\mathfrak{m}}\right) + \lambda\left(\frac{R}{\overline{\mathfrak{m}^2} + xR}\right).$$

We can repeat the above steps with the following short exact sequence and isomorphisms:

$$0 \longrightarrow \frac{\overline{\mathfrak{m}^2} + xR}{\overline{\mathfrak{m}^2}} \longrightarrow \frac{R}{\overline{\mathfrak{m}^2}} \longrightarrow \frac{R}{\overline{\mathfrak{m}^2} + xR} \longrightarrow 0;$$

$$\frac{\overline{\mathfrak{m}^2} + xR}{\overline{\mathfrak{m}^2}} \simeq \frac{xR}{\overline{\mathfrak{m}^2} \cap xR} \simeq \frac{xR}{\mathfrak{m}xR} \simeq \frac{R}{\mathfrak{m}}.$$

Once again, if we compute the length, we have that

$$\lambda\left(\frac{R}{\overline{\mathfrak{m}^2}}\right) = 1 + \lambda\left(\frac{R}{\overline{\mathfrak{m}^2} + xR}\right) \tag{12}$$

Combining (9) and (12) with the fact that $\lambda\left(\frac{R}{xR}\right) = \lambda\left(\frac{S}{xS}\right)$ and $\lambda\left(\frac{\mathfrak{m}}{\overline{\mathfrak{m}^2}}\right) = \lambda\left(\frac{R}{\overline{\mathfrak{m}^2}}\right) - 1$, we have

$$\lambda\left(\frac{\overline{\mathfrak{m}^2}}{x\mathfrak{m}}\right) + \lambda\left(\frac{R}{\overline{\mathfrak{m}^2} + xR}\right) = \lambda\left(\frac{\mathfrak{m}}{\overline{\mathfrak{m}^2}}\right) + B$$

$$= \lambda\left(\frac{R}{\overline{\mathfrak{m}^2}}\right) - 1 + B$$

$$= 1 + \lambda\left(\frac{R}{\overline{\mathfrak{m}^2 + xR}}\right) - 1 + B.$$

Simplifying we see that $\lambda\left(\frac{\overline{\mathfrak{m}^2}}{x\mathfrak{m}}\right) = B$. We now have

$$e(R) = \dim_{\Bbbk}(S/\mathfrak{m}S);$$

$$\lambda\left(\frac{\overline{\mathfrak{m}^2}}{x\mathfrak{m}}\right) = \dim_{\Bbbk}\left(\frac{R + \mathfrak{m}S}{R + \mathfrak{m}^2 S}\right).$$

Applying (dr1) and (dr2) yields the desired result.

Given Proposition 3.1, we can construct a couple of examples.

Example 3.2. Consider the ring $R = \Bbbk[\![t^3, t^7]\!]$. This is a one-dimensional domain with $e(R) = 3$. The element t^3 is a minimal reduction of the maximal ideal $(t^3, t^7)R$. If we compute the length, we see that $\lambda\left(\overline{\mathfrak{m}^2}/t^3\mathfrak{m}\right) = 2$. Hence, by Proposition 3.1, we have that R is not of finite type.

Example 3.3. Let $R = \Bbbk[\![x, y]\!]/(x^3y - xy^3)$. This ring is one-dimensional and reduced. If we compute the multiplicity, we find that $e(R) = 4$. Thus, we immediately have from Proposition 3.1 that R is not of finite type. Computing the length nonetheless, we find that $\lambda\left(\overline{\mathfrak{m}^2}/(x + 2y)\mathfrak{m}\right) = 1$.

3.2 Graded Countable Type

In [20, Theorem 5.3], one-dimensional standard graded rings of graded countable type were shown to either be of minimal multiplicity or a hypersurface. Turning to the case of minimal multiplicity, we find the following.

Theorem 3.4. *Let $(R, \mathfrak{m}, \Bbbk)$ be a standard graded one-dimensional Cohen–Macaulay ring with uncountable residue field $\Bbbk$. If the h-vector of R is $(1, n)$ with $n \geqslant 3$, then R is not of graded countable Cohen–Macaulay type.*

Proof. Let x be a minimal homogeneous reduction of the maximal ideal $\mathfrak{m}$, and let $x, u, v, w, c_4, \cdots, c_n \in \mathfrak{m}$ be elements of a minimal k-basis of $\mathfrak{m}/\mathfrak{m}^2$. By assumption $n \geqslant 3$, so we are guaranteed at least four elements in the basis of $\mathfrak{m}/\mathfrak{m}^2$. Without losing any generality, we assume that w is the fourth basis element. Assume that there is a graded isomorphism

$$I_\alpha = (x, u + \alpha v) \simeq (x, u + \beta v) = I_\beta$$

where α, β are elements of $\mathbb{k}$. As $\dim(R) = 1$, these ideals are graded indecomposable maximal Cohen–Macaulay modules. Since this isomorphism is graded of degree 0, we have that

$$\begin{aligned} x &\mapsto \delta_1 x + \delta_2(u + \beta v) \\ u + \alpha v &\mapsto \delta_3 x + \delta_4(u + \beta v) \end{aligned}$$

where $\det(\delta_i)$ is a unit and δ_i are elements of $\mathbb{k}$. We have that

$$0 = \delta_3 x^2 + \delta_4 x(u + \beta v) - \delta_1 x(u + \alpha v) - \delta_2(u + \alpha v)(u + \beta v). \tag{13}$$

Notice that $(u + \alpha v)(u + \beta v)$ is an element of $\mathfrak{m}^2$. As R is of minimal multiplicity, we have that $x\mathfrak{m} = \mathfrak{m}^2$. Hence we can view elements of $\mathfrak{m}^2$ as elements of $x\mathfrak{m}$. In particular we view u^2, uv, v^2 in the following way

$$\begin{pmatrix} u^2 \\ uv \\ v^2 \end{pmatrix} = \begin{pmatrix} x(a_{10}x + a_{11}u + a_{12}v + a_{13}w + a_{15}c_4 + \cdots + a_{1n}c_n) \\ x(a_{20}x + a_{21}u + a_{22}v + a_{23}w + a_{25}c_4 + \cdots + a_{2n}c_n) \\ x(a_{30}x + a_{31}u + a_{32}v + a_{33}w + a_{35}c_4 + \cdots + a_{3n}c_n) \end{pmatrix} = x \cdot A \cdot \begin{pmatrix} x \\ u \\ v \\ w \\ c_4 \\ \vdots \\ c_n \end{pmatrix}$$

where the matrix $A = (a_{ij})$, $1 \leqslant i \leqslant 3$, $0 \leqslant j \leqslant n$. Since u^2, uv, v^2 are homogeneous elements, the grading forces the entries of A to be elements of $\mathbb{k}$. Further, if we let

$$\Phi = \begin{pmatrix} 1 & \alpha + \beta & \alpha\beta \end{pmatrix} \text{ and } \mathbf{b} = \begin{pmatrix} x & u & v & w & c_4 & \cdots & c_n \end{pmatrix}^t,$$

then we can use matrix notation to write

$$(u + \alpha v)(u + \beta v) = u^2 + (\alpha + \beta)uv + \alpha\beta v^2 = x \cdot \Phi \cdot A \cdot \mathbf{b}.$$

We can now cancel the x in Eq. (13) and rewrite it as

$$0 = \begin{pmatrix} \delta_3 - \Phi A_0 \delta_2 \\ \delta_4 - \delta_1 - \Phi A_1 \delta_2 \\ \beta\delta_4 - \alpha\delta_1 - \Phi A_2 \delta_2 \\ -\Phi A_3 \delta_2 \\ -\Phi A_4 \delta_2 \\ \vdots \\ -\Phi A_n \delta_2 \end{pmatrix}^t \cdot \mathbf{b} \tag{14}$$

where A_i are the columns of the matrix A. All of the elements in the coefficient matrix of (14) are elements of k and hence equal zero as $x, u, v, w, c_4, \ldots, c_n$ form a $\Bbbk$-basis.

At this point we focus on δ_2. If $\delta_2 \neq 0$, then the fact that $\Phi A_3 \delta_2 = 0$ implies that

$$a_{13} + (\alpha + \beta)a_{23} + \alpha\beta a_{33} = 0 \tag{15}$$

in the field $\Bbbk$. As the a_{ij} are independent of our choice of α and β, Eq. (15) shows that every α, β such that $I_\alpha \simeq I_\beta$ is a solution to

$$f(X, Y) = a_{13} + (X + Y)a_{23} + XYa_{33} \in \Bbbk[X, Y].$$

This forces $f(X, Y)$ to be identically zero, a contradiction. Hence there are uncountably many I_α that are not isomorphic.

If we let $\delta_2 = 0$, then Eq. (15) becomes the relation

$$\delta_3 x + (\delta_4 - \delta_1)u + (\beta\delta_4 - \alpha\delta_1)v = 0. \tag{16}$$

As x, u, v are independent over $\Bbbk$, we have that the coefficients are zero. In particular $\delta_4 - \delta_1 = 0$. Since $\delta_1\delta_4 - \delta_2\delta_3 \neq 0$, we know that $\delta_1 = \delta_4 \neq 0$. Thus the fact that $\beta\delta_4 - \alpha\delta_1 = 0$ implies that $\alpha = \beta$. Hence there are uncountably many non-isomorphic ideals I_α.

Given the above results, we are now ready to characterize one-dimensional standard graded rings of graded countable Cohen–Macaulay type.

Corollary 3.5. *Let* $(R, \mathfrak{m}, \Bbbk)$ *be a one-dimensional standard graded Cohen–Macaulay ring with uncountable residue field* $\Bbbk$*. If* R *is of graded countable Cohen–Macaulay type, then* R *is either of minimal multiplicity with* h*-vector* $(1, 2)$*, or isomorphic to one of the following hypersurfaces:*

(1) $\Bbbk[x]$*;*
(2) $\Bbbk[x, y]/(xy)$*;*
(3) $\Bbbk[x, y]/(xy(x + y))$*;*
(4) $\Bbbk[x, y]/(xy^2)$*;*
(5) $\Bbbk[x, y]/(y^2)$.

Proof. A direct application of [20, Theorem 5.3] and Theorem 3.4 show that R is either a hypersurface ring, or has minimal multiplicity with h-vector $(1, 2)$.

Concerning the hypersurfaces, items (1)–(3) have graded finite Cohen–Macaulay type as can be seen from [6] or [9]. The hypersurfaces (4) and (5) are not graded finite Cohen–Macaulay type, but their completions are the one dimensional (A_∞) and (D_∞) hypersurface singularities shown in (1) and (2) on page 301. It was shown by Buchweitz et al. in [6] that these are the only hypersurfaces that are countable but not finite Cohen–Macaulay type. Hence by Remark 1.3, the rings (4) and (5) are of graded countable Cohen–Macaulay type.

Corollary 3.6. *Let $(R, \mathfrak{m}, \Bbbk)$ be a one-dimensional standard graded Cohen–Macaulay ring with uncountable residue field $\Bbbk$. If R is of graded countable Cohen–Macaulay type then, $e(R) \leqslant 3$.*

Proof. As shown in [20, Corollary 4.5], the possible h-vectors are (1), $(1, n)$, or $(1, n, 1)$ for some integer n. Combining this with Corollary 3.5, we know that the possible h-vectors of R are the following:

$$(1),\ (1,1),\ (1,2),\ (1,1,1).$$

Hence we have that $e(R) \leqslant 3$.

An obvious improvement to Corollary 3.5 would be to classify the rings whose h-vector is $(1, 2)$. Currently, there are two known examples of one-dimensional standard graded Cohen–Macaulay rings of graded countable Cohen–Macaulay type having h-vector $(1, 2)$:

$$\Bbbk[x,y,z]/(xy, yz, z^2); \tag{17}$$

$$\Bbbk[x,y,z]/\det_2\begin{pmatrix} x & y & z \\ y & z & x \end{pmatrix}. \tag{18}$$

The completion of (17) is the local endomorphism ring of $(x, y)R$ where $R = \Bbbk[\![x, y]\!]/(xy^2)$, the one-dimensional (D_∞) hypersurface. This endomorphism ring is known to be of countable Cohen–Macaulay type (see [15, Theorem 1.5] or [16, Example 14.23]) and thus by Remark 1.3, (17) is of graded countable Cohen–Macaulay type. It is worth noting that (17) is not reduced.

As for (18), this is the only one-dimensional ring of finite type and h-vector $(1, 2)$ found in the list of graded finite Cohen–Macaulay type rings [9]. Given this result, if we assume our ring to have an isolated singularity, then we have a positive answer to Question 1.2.

Corollary 3.7. *Let $(R, \mathfrak{m}, \Bbbk)$ be a one-dimensional, reduced, standard graded Cohen–Macaulay ring with uncountable residue field $\Bbbk$. Further assume that R has a finite integral closure. If R is of graded countable Cohen–Macaulay type, then R is of graded finite type and isomorphic to one of the following:*

(1) $\Bbbk[x]$;
(2) $\Bbbk[x, y]/(xy)$;
(3) $\Bbbk[x, y]/(xy(x + y))$;
(4) $\Bbbk[x,y,z]/\det_2\begin{pmatrix} x & y & z \\ y & z & x \end{pmatrix}$.

Proof. By Remark 1.3, we can pass to the completion. Since local rings of minimal multiplicity with h-vector $(1, 2)$ have $\lambda\left(\overline{\mathfrak{m}^2/x\mathfrak{m}}\right) \leqslant 1$, we can apply Proposition 3.1 and Corollary 3.5 to obtain the desired result.

4 Non-Gorenstein Rings of Dimension At Least Two

We are unable to classify standard graded rings of graded countable type with dimension larger than one. However, if we assume an isolated singularity, we have a positive answer to Question 1.2 in the non-Gorenstein case.

4.1 *Non-Gorenstein Rings of Dimension Two*

As shown in [9], two dimensional standard graded Cohen–Macaulay rings with an isolated singularity must be a domain and have minimal multiplicity. In [8], it can be seen that standard graded Cohen–Macaulay domains of minimal multiplicity have the following classification:

(i) quadratic hypersurfaces (not necessarily nondegenerate);
(ii) cones over the Veronese embedding $\mathbb{P}^2 \hookrightarrow \mathbb{P}^5$. That is, rings isomorphic to

$$\Bbbk[x_0, x_1, \ldots, x_n] \big/ \det_2 \begin{pmatrix} x_0 & x_1 & x_2 \\ x_1 & x_3 & x_4 \\ x_2 & x_4 & x_5 \end{pmatrix},$$

where for $n \geqslant 5$;
(iii) rational normal scrolls.

The ring defined in (i) is Gorenstein and is discussed in Sect. 5. The ring in (ii) is of graded finite (hence countable) Cohen–Macaulay type when $n = 5$ as can be seen in [9] or [21, Example 17.6.1]. For $n > 6$, the dimension of the singular locus is larger than 1, hence they are not of graded countable type. However when $n = 6$, it is unclear whether the ring is graded countable type or not. Notice that this case does not concern Question 1.2 as the ring does not have an isolated singularity, but does play a role in a general classification. Further, the dimension of each ring in (ii) is at least three. As such, it is left to consider the rational normal scrolls defined as follows.

Definition 4.1. Let $0 \leqslant a_0 \leqslant a_1 \leqslant \cdots \leqslant a_k$ be given integers and let $\{x^i_j \mid 0 \leqslant j \leqslant a_i, 0 \leqslant i \leqslant k\}$ be a set of variables over a field $\Bbbk$. Then, take the ideal I of the polynomial ring $S = \Bbbk[x^i_j \mid 0 \leqslant j \leqslant a_i, 0 \leqslant i \leqslant k]$ generated by all the 2×2-minors of the matrix:

$$\left(\begin{array}{cccc|c|cccc} x^0_0 & x^0_1 & \cdots & x^0_{a_0-1} & \cdots & x^k_0 & x^k_1 & \cdots & x^k_{a_k-1} \\ x^0_1 & x^0_2 & \cdots & x^0_{a_0} & \cdots & x^k_1 & x^k_2 & \cdots & x^k_{a_k} \end{array}\right).$$

Define the graded ring R to be the quotient S/I with $\deg(x^i_j) = 1$ for all i, j, and call R the *scroll* of type $(a_0, a_1, \ldots, a_k)$.

Notice that the polynomial ring $\Bbbk[x_0, x_1, \ldots, x_N]$ is a rational normal scroll of type $(0, \cdots, 0, 1)$, where $N = \sum_{i=0}^{k} a_i + k + 1$. Further, a scroll is of type $(0, \cdots, 0, 1)$ if and only if it is regular (for basic properties of scrolls, see [8, 17]). In general, it is known that scrolls are integrally closed Cohen–Macaulay domain of dimension $k + 2$.

Proposition 4.2. *Let $(R, \mathfrak{m}, \Bbbk)$ be a standard graded Cohen–Macaulay ring that is not Gorenstein and* $\dim(R) = 2$*. Further assume that R has an isolated singularity and that $\Bbbk$ is an uncountable field. If R is of graded countable Cohen–Macaulay type, then R is of graded finite type and is isomorphic to*

$$\Bbbk[x_0, x_1, \ldots, x_n] \Big/ \det_2 \begin{pmatrix} x_0 & \ldots & x_{n-1} \\ x_1 & \ldots & x_n \end{pmatrix},$$

where $n \geqslant 2$.

Proof. It was shown in [9] that two dimensional, non-Gorenstein rings with an isolated singularity have minimal multiplicity. By the above discussion, we know that R must be isomorphic to a two dimensional scroll. The only non-Gorenstein scrolls of dimension two are of type (m) where $m \geqslant 2$. These are the rings listed in the statement of Proposition 4.2. It is known that these rings are of graded finite Cohen–Macaulay type (see [9] or [3, Theorem 2.3]).

Proposition 4.2 gives a partial (positive) answer to Question 1.2. Ideally though, we would like to remove the isolated singularity condition from the hypothesis of Proposition 4.2. Doing so would show that graded countable Cohen–Macaulay type implies graded finite Cohen–Macaulay type for non-Gorenstein two dimensional rings.

4.2 Non-Gorenstein Rings of Dimension At Least Three

Let $(R, \mathfrak{m}, \Bbbk)$ be a standard graded Cohen–Macaulay ring of graded countable Cohen–Macaulay type, that is not Gorenstein and $\dim R \geqslant 3$. By [20, Proposition 4.6], we know that R must be a domain and have minimal multiplicity. Thus we only have to consider the three classes of rings listed in Sect. 4.1 in order to obtain the following.

Proposition 4.3. *Let $(R, \mathfrak{m}, \Bbbk)$ be a standard graded Cohen–Macaulay ring with uncountable residue field $\Bbbk$ and an isolated singularity. Further assume that R is not Gorenstein and* $\dim R \geqslant 3$*. If R is of graded countable Cohen–Macaulay type, then R is of graded finite type and is isomorphic to one of the following rings:*

(1) $\Bbbk[x_1, \ldots, x_5]/\det_2 \begin{pmatrix} x_1 & x_2 & x_4 \\ x_2 & x_3 & x_5 \end{pmatrix}$;
(2) $\Bbbk[x_1, \ldots, x_6]/\det_2(\mathit{sym}\ 3 \times 3)$.

Proof. Consider the classes of rings listed in Sect. 4.1. We can ignore the rings in (i) as they are Gorenstein. The rings in (ii) only have an isolated singularity when $n = 5$. This ring is known to be of graded finite Cohen–Macaulay type [9]. As such, we list it above.

Concerning the scrolls in (iii), Auslander and Reiten classify the scrolls of finite type in [5]. In particular, they show that if $k \geqslant 1$ and R is not of type $(1, 1)$ or $(1, 2)$, then R has $|\Bbbk|$ many indecomposable graded Cohen–Macaulay modules, up to shifts. As it is, the graded scrolls of type $(1, 1)$ and $(1, 2)$ are the only graded scrolls of dimension at least 3 that have graded countable Cohen–Macaulay type. Notice that a scroll of type $(1, 1)$ is Gorenstein and hence we are left with the scroll of type $(1, 2)$.

Proposition 4.3 gives another case where Question 1.2 has a positive answer. Concerning the general classification, in order to remove the isolated singularity condition from the hypothesis, one only needs to determine the graded Cohen–Macaulay type of

$$\Bbbk[x_0, x_1, \ldots, x_6] \Big/ \det{}_2 \begin{pmatrix} x_0 & x_1 & x_2 \\ x_1 & x_3 & x_4 \\ x_2 & x_4 & x_5 \end{pmatrix}. \tag{19}$$

Knowing this information would give a complete classification for non-Gorenstein rings of dimension at least three.

5 Gorenstein Rings

Not much is known about Gorenstein rings of graded countable Cohen–Macaulay type. However, it is well known that standard graded Gorenstein rings of graded finite Cohen–Macaulay type are hypersurfaces [11, Satz 1.2]. This fact is heavily exploited in the classification of standard graded Cohen–Macaulay rings of graded finite Cohen–Macaulay type [9]. The countable analog of this fact is still unknown.

Conjecture 5.1. A Gorenstein ring of countable Cohen–Macaulay type is a hypersurface.

Using the concept of super-stretched, Conjecture 5.1 was shown to be true for standard graded rings of dimension at most one [20], but the conjecture remains open for higher dimensions. By the results of [6, 14] on hypersurfaces, a positive result to Conjecture 5.1 would completely classify Gorenstein rings of graded countable Cohen–Macaulay type.

If we restrict to rings of minimal multiplicity, then we are only dealing with hypersurfaces, as h-vectors of Gorenstein rings are symmetric. As such, by [6, 14] we have a complete characterization. In particular, consider a standard graded Gorenstein ring $(R, \mathfrak{m}, \Bbbk)$ of minimal multiplicity, i.e. the h-vector is $(1, 1)$. Hence if R is of graded countable Cohen–Macaulay type, then R is isomorphic to one of the following hypersurfaces:

$$(A_1) : \Bbbk[x_1, \ldots, x_n]/(x_1^2 + \cdots + x_n^2),\ n \geqslant 1;$$

$$(A_\infty) : \Bbbk[x_1, \ldots, x_n]/(x_2^2 + \cdots + x_n^2),\ n \geqslant 1.$$

Remark 5.2. If we further assume that a standard graded Gorenstein ring of minimal multiplicity has an isolated singularity, then graded countable Cohen–Macaulay type implies graded finite Cohen–Macaulay type. To see this, notice that only (A_1) has a zero dimensional singular locus.

It is worth noting that standard graded rings of dimension at least 2 and with graded finite Cohen–Macaulay type have minimal multiplicity. This can be seen by examining the list of graded rings of graded finite Cohen–Macaulay type in [9]. With these results, we ask the following question.

Question 5.3. If $(R, \mathfrak{m}, \Bbbk)$ is a standard graded Gorenstein ring of graded countable Cohen–Macaulay type and $\dim(R) \geqslant 2$, is R necessarily of minimal multiplicity?

A positive answer to Question 5.3 would ultimately force the ring to be a hypersurface and affirm Conjecture 5.1. Further it would show that Question 1.2 has an affirmative answer for Gorenstein rings of graded countable Cohen–Macaulay type.

Remark 5.4. Since Conjecture 5.1 is still open, the natural place to look are rings of complete intersection. According to [20], we know that standard graded complete intersection with graded countable Cohen–Macaulay type are either a hypersurface or defined by two quadrics. With this fact, one could use tools from representation theory [18] and the analysis of the h-vector, to show that the ring is indeed a hypersurface no matter what the dimension is. As these techniques are outside the scope of the paper, we leave it here as a remark.

6 Summary of Results

In Corollary 6.1 we summarize the previous results in relation to Question 1.2. As noted earlier, proving Conjecture 5.1 would give a complete answer to the question. In the rest of the section, we consider what can be shown when the ring in question does not necessarily have an isolated singularity. In particular, we state a partial classification of standard graded rings of graded countable Cohen–Macaulay type.

Corollary 6.1. *Let $(R, \mathfrak{m}, \Bbbk)$ be a standard graded ring of graded countable Cohen–Macaulay type. Further assume that R has an isolated singularity and that $\Bbbk$ is an uncountable algebraically closed field of characteristic zero. Then R is of graded finite type if one of the following holds:*

(1) $\dim(R) \leqslant 1$*;*
(2) *R is not Gorenstein;*
(3) *R is of minimal multiplicity.*

Proof. If the dimension of R is at most 1, then we can directly apply Proposition 2.1 and Corollary 3.7. For non-Gorenstein rings of dimension at least 2, we can apply Propositions 4.2 and 4.3. If R is Gorenstein of minimal multiplicity and $\dim(R) \geqslant 2$, then we have graded finite Cohen–Macaulay type from Remark 5.2 and the discussion preceding it.

6.1 Partial Classification in the General Case

Throughout this paper, we have obtained a partial classification of standard graded rings of graded countable Cohen–Macaulay type. For uniformity's sake, we state the partial classification in the same format as the classification of graded finite Cohen–Macaulay type found in [9]. As such, notice that rings listed in the arbitrary dimension section are omitted from the other cases. In particular, the A_∞ hypersurface is only listed in the arbitrary dimension section. For more details on rings of countable Cohen–Macaulay type, see [16].

Arbitrary Dimension

The following rings are shown to be graded countable type in [6, 9, 14]. This list is not known to be complete.

$$\mathbb{k}[x_1, \ldots, x_n],\ n \geqslant 1;$$

$$\mathbb{k}[x_1, \ldots, x_n]/(x_1^2 + \cdots + x_n^2),\ n \geqslant 1;$$

$$\mathbb{k}[x_1, \ldots, x_n]/(x_2^2 + \cdots + x_n^2),\ n \geqslant 1.$$

Dimension Zero

As seen in Proposition 2.1, graded countable type is the same as graded finite type. By [9], we have the following complete list.

$$\mathbb{k}[x]/(x^m),\ m \geqslant 1.$$

Dimension One

We do not have a complete list of dimension one graded countable type rings. However, by Corollary 3.7 and the discussion before it, we know these rings must have h-vector $(1, 2)$ or be isomorphic to

$$\Bbbk[x, y]/(xy);$$
$$\Bbbk[x, y]/(xy(x + y));$$
$$\Bbbk[x, y]/(xy^2);$$
$$\Bbbk[x, y, z]/\det_2\begin{pmatrix} x & y & z \\ y & z & x \end{pmatrix};$$
$$\Bbbk[x, y, z]/(xy, yz, z^2).$$

Dimension Two

According to Proposition 4.2, the non-Gorenstein graded countable type rings with an isolated singularity are known. Besides the ring given in the Arbitrary Dimension section, we are not aware of any other graded countable type ring with a non-isolated singularity. Thus we state what was already recorded in [9].

$$\Bbbk[x_1, \ldots, x_{n+1}]\Big/\det_2\begin{pmatrix} x_1 & \ldots & x_n \\ x_2 & \ldots & x_{n+1} \end{pmatrix}, \quad n \geqslant 2.$$

Dimension Three

If we restrict the proof of Proposition 4.3 to three dimensional rings, we can omit the isolated singularity condition. As such, we know the rings below are the only non-Gorenstein graded countable type rings of dimension three. As the Gorenstein case is open, it is not known whether this is a complete list or not.

$$\Bbbk[x_1, \ldots, x_5]\Big/\det_2\begin{pmatrix} x_1 & x_2 & x_4 \\ x_2 & x_3 & x_5 \end{pmatrix};$$
$$\Bbbk[x_1, \ldots, x_6]/\det_2(\text{sym } 3 \times 3).$$

Remark 6.2. According to the remarks after Proposition 4.3, for non-Gorenstein rings of dimension at least four, the only possible ring is (19).

Acknowledgments The author would like to thank the anonymous referee for valuable feedback and suggestions regarding the content of this paper. For instance, the referee pointed out the existence of the ring (17) as well as some discrepancies in the initial version of Sect. 4. Further, the author is especially thankful to Craig Huneke for several useful conversations. Additionally, results in this note were inspired by many Macaulay2 [10] computations.

The author was partially funded by the NSF grant, Kansas Partnership for Graduate Fellows in K-12 Education (DGE-0742523).

References

1. Arnol'd, V.I.: Critical points of smooth functions. In: Proceedings of the International Congress of Mathematicians (Vancouver, BC, 1974), vol. 1, pp. 19–39. Canadian Mathematical Congress, Montreal, QC (1975)
2. Auslander, M.: Isolated singularities and existence of almost split sequences. In: Representation Theory, II (Ottawa, ON, 1984). Lecture Notes in Mathematics, vol. 1178, pp. 194–242. Springer, Berlin (1986)
3. Auslander, M., Reiten, I.: Almost split sequences for **Z**-graded rings. In: Singularities, Representation of Algebras, and Vector Bundles (Lambrecht, 1985). Lecture Notes in Mathematics, vol. 1273, pp. 232–243. Springer, Berlin (1987)
4. Auslander, M., Reiten, I.: Cohen-Macaulay modules for graded Cohen-Macaulay rings and their completions. In: Commutative Algebra (Berkeley, CA, 1987). Mathematical Sciences Research Institute Publications, vol. 15, pp. 21–31. Springer, New York (1989)
5. Auslander, M., Reiten, I.: The Cohen-Macaulay type of Cohen-Macaulay rings. Adv. Math. **73**(1), 1–23 (1989)
6. Buchweitz, R.-O., Greuel, G.-M., Schreyer, F.-O.: Cohen-Macaulay modules on hypersurface singularities. II. Invent. Math. **88**(1), 165–182 (1987)
7. Cimen, N., Wiegand, R., Wiegand, S.: One-dimensional rings of finite representation type. In: Abelian Groups and Modules (Padova, 1994). Mathematics and Its Applications, vol. 343, pp. 95–121. Kluwer Academic, Dordrecht (1995)
8. Eisenbud, D., Harris, J.: On varieties of minimal degree (a centennial account). In: Algebraic Geometry, Bowdoin, 1985 (Brunswick, Maine, 1985). Proceedings of Symposia in Pure Mathematics, vol. 46, pp. 3–13. American Mathematical Society, Providence (1987)
9. Eisenbud, D., Herzog, J.: The classification of homogeneous Cohen-Macaulay rings of finite representation type. Math. Ann. **280**(2), 347–352 (1988)
10. Grayson, D.R., Stillman, M.E.: Macaulay2, a software system for research in algebraic geometry. http://www.math.uiuc.edu/Macaulay2/ (2013)
11. Herzog, J.: Ringe mit nur endlich vielen Isomorphieklassen von maximalen, unzerlegbaren Cohen-Macaulay-Moduln. Math. Ann. **233**(1), 21–34 (1978)
12. Huneke, C., Leuschke, G.J.: Local rings of countable Cohen-Macaulay type. Proc. Am. Math. Soc. **131**(10), 3003–3007 (2003)
13. Karr, R., Wiegand, R.: Direct-sum behavior of modules over one-dimensional rings. In: Commutative Algebra—Noetherian and Non-Noetherian Perspectives, pp. 251–275. Springer, New York (2011)
14. Knörrer, H.: Cohen-Macaulay modules on hypersurface singularities. I. Invent. Math. **88**(1), 153–164 (1987)
15. Leuschke, G.J., Wiegand, R.: Local rings of bounded Cohen-Macaulay type. Algebra Represent. Theory **8**(2), 225–238 (2005)
16. Leuschke, G.J., Wiegand, R.: Cohen-Macaulay Representations. Mathematical Surveys and Monographs, vol. 181. American Mathematical Society, Providence (2012)
17. Miró-Roig, R.M.: The representation type of rational normal scrolls. Rend. Circ. Mat. Palermo **62**(1), 153–164 (2013)
18. Stevenson, G.: Subcategories of singularity categories via tensor actions. arXiv.org, math.AC, May 2012. arXiv:1105.4698

19. Stone, B.: Super-stretched and graded countable Cohen-Macaulay type. Ph.D. thesis, University of Kansas (2012)
20. Stone, B.: Super-stretched and graded countable Cohen-Macaulay type. arXiv.org, math.AC, January 2013. arXiv:1301.3593
21. Yoshino, Y.: Cohen-Macaulay Modules Over Cohen-Macaulay Rings. London Mathematical Society Lecture Note Series, vol. 146. Cambridge University Press, Cambridge (1990)

Printed by Printforce, the Netherlands